A Geometrical Introduction t

TENSOR CALCULUS

A Geometrical Introduction to

TENSOR CALCULUS

Jeroen Tromp

Princeton University Press
Princeton and Oxford

Published by Princeton University Press
41 William Street, Princeton, New Jersey 08540
99 Banbury Road, Oxford OX2 6JX

press.princeton.edu

Library of Congress Cataloging-in-Publication Data

Names: Tromp, Jeroen, author.
Title: A geometrical introduction to tensor calculus / Jeroen Tromp.
Description: Princeton : Princeton University Press, [2025] |
Includes bibliographical references and index.
Identifiers: LCCN 2024023689 (print) | LCCN 2024023690 (ebook) |
ISBN 9780691267982 (pbk) | ISBN 9780691267975 | ISBN 9780691269221 (ebook)
Subjects: LCSH: Calculus of tensors. | Fluid mechanics. | Geophysics. |
BISAC: SCIENCE / Mechanics / Fluids | MATHEMATICS / Calculus
Classification: LCC QC20.7.C28 T76 2024 (print) | LCC QC20.7.C28 (ebook) |
DDC 515/.63—dc23/eng20240814
LC record available at https://lccn.loc.gov/2024023689
LC ebook record available at https://lccn.loc.gov/2024023690

British Library Cataloging-in-Publication Data is available

Editorial: Ingrid Gnerlich and Whitney Rauenhorst
Production Editorial: Karen Carter
Text Design: Wanda España
Jacket/Cover Design: Wanda España
Production: Jacqueline Poirier
Publicity: William Pagdatoon

Cover Image: Beverly Pepper, *Thetis Circle*, 2013, COR-TEN steel, stone base in grey pietra serena. © National Academy of Design, New York / Bridgeman Images

This book has been composed in Minion Pro and Kohinoor Devanagari

Printed in the United States of America

10 9 8 7 6 5 4 3 2 1

To my creative muse, Catrinel, *cu dragoste infinită*

CONTENTS

List of Examples xiii
Preface xvii

1 **Introduction** 1

2 **Linear Spaces and Transformations** 3

2.1 Properties of Linear Spaces 3
2.2 Vector Spaces 4
2.3 Linear Transformations 4

3 **Differentiable Manifolds** 7

3.1 Charts and Coordinates 7
3.2 Definition 10
3.3 Local Coordinate Changes 11
3.4 Functions on Manifolds 13
3.5 Orientable Manifolds 13

4 **Vectors and One-Forms** 15

4.1 Vectors 15
4.1.1 Vectors as Tangents to Curves 16
4.1.2 Bases and Coordinates 17

4.1.3 Vector Field 19
4.1.4 Transformations 20
4.2 One-Forms 23
4.2.1 Duality 23
4.2.2 Bases 24
4.2.3 Transformations 26
4.3 Alternative Perspective 28
4.4 Lie Bracket 29

5 Tensors 33
5.1 Definition 33
5.2 Operations on Tensors 35
5.2.1 Addition 35
5.2.2 Tensor Product 35
5.2.3 Contraction 36
5.2.4 Transpose of (2,0) and (0,2) Tensors 38
5.2.5 Transpose of a (1,1) Tensor 40
5.3 Transformations 41
5.3.1 Tetrad Formalism 43
5.3.2 Pseudotensors 46
5.4 Kronecker or Identity Tensor 46
5.5 Logarithms and Exponentials of (1,1) Tensors 46
5.6 Tensor Densities and Capacities 47
5.6.1 Pseudotensor Densities and Capacities 48
5.7 Levi-Civita Density and Capacity 48
5.8 Determinant of Rank-2 Tensors 50
5.9 Inverse of Rank-2 Tensors 50
5.10 Metric Tensor 51
5.10.1 Formulation 51
5.10.2 Geometrical Meaning 55
5.10.3 Norm of Vectors and One-Forms 56
5.10.4 Metric in Tetrads 56
5.11 Adjoint of a (1,1) Tensor 57

5.12 Tensor Densities and Capacities Revisited 58
5.13 Levi-Civita Pseudotensor 58
5.14 Kronecker Determinants 60
5.15 Rotations 63
5.15.1 Euler Angles 64
5.15.2 Rodrigues's Formula 66

6 Maps between Manifolds 71

6.1 Maps 71
6.2 Maps between Manifolds of Different Dimensions 72
6.2.1 Pullback 72
6.2.2 Pushforward 74
6.3 Maps between Manifolds of the Same Dimensions 76

7 Differentiation on Manifolds 85

7.1 Covariant Derivative 85
7.1.1 Formulation 85
7.1.2 Transformation of Connection Coefficients 89
7.1.3 Divergence 91
7.1.4 Parallel Transport 91
7.1.5 Torsion and Curvature Tensors 94
7.1.6 Bianchi Identities 97
7.1.7 Torsion-Free Connection 97
7.1.8 Covariant Derivative of the Metric Tensor 98
7.1.9 Mixed Covariant Derivative in Tetrad Basis 101
7.1.10 Spin Connection 102
7.1.11 Contracted Bianchi Identities 102
7.1.12 Covariant Derivative of Tensor Densities and Capacities 107
7.1.13 Nonmetricity 107
7.2 Euler Derivative 112
7.3 Lie Derivative 114
7.3.1 Lie Derivative of Vectors 114
7.3.2 Geometrical Interpretation 117

7.3.3 Autonomous Lie Derivative 118
7.3.4 Lie Derivative of One-Forms 118
7.3.5 Lie Derivative of (p, q) Tensors 119
7.3.6 Lie Derivative of Functions 120
7.3.7 Lie Derivative of Metric Tensors 120
7.3.8 Lie Derivative of Levi-Civita Tensor 123

8 Differential Forms 127

8.1 Definition 127
8.2 Operations on Forms 130
8.2.1 Addition 130
8.2.2 Exterior Product 130
8.2.3 Interior Product 133
8.3 *k*-Vectors 133
8.4 Hodge Dual 134
8.5 Volumes 135
8.5.1 Properties 136
8.6 Surfaces 137
8.7 Exterior Derivative 139
8.7.1 Coordinate-Free Definition 140
8.7.2 Exact Forms 142
8.7.3 Commutativity with Pullback and Pushforward 144
8.8 Lie Derivative of a Form 148
8.9 Vector- and Tensor-Valued Forms 150
8.9.1 Transformations of Tensor-Valued Forms 151
8.9.2 Operations on Tensor-Valued Forms 159
8.9.3 Connection One-Forms 161
8.9.4 Torsion Two-Forms 162
8.9.5 Exterior Covariant Derivative 164
8.9.6 Covariant Lie Derivative 167
8.9.7 Curvature Two-Forms 169
8.9.8 Commutator of Covariant Lie and Exterior Covariant Derivatives 171

8.9.9 Bianchi Identities Revisited ... 172
8.9.10 Nonmetricity Revisited ... 173
8.10 Integration of Forms ... 174
8.10.1 Line Integrals ... 174
8.10.2 Surface Integrals ... 175
8.10.3 Volume Integrals ... 177
8.11 Stokes's Theorem ... 178
8.11.1 Fundamental Theorem of Calculus ... 178
8.11.2 Green's Theorem ... 179
8.11.3 Gauss's Theorem ... 180
8.11.4 Stokes's Theorem ... 180
8.11.5 Variational Principles ... 180
8.11.6 Noether's Theorem ... 181
8.11.7 Applications ... 183

Glossary 211
Bibliography 219
Author Index 223
General Index 225

LIST OF EXAMPLES

3.1 Charts for the Circle 8
3.2 Charts in Continuum Mechanics 9
3.3 Charts in General Relativity 10
3.4 Coordinate Changes in Continuum Mechanics 12
4.1 Material Velocity in Continuum Mechanic 18
4.2 Material Velocity Transformations 21
4.3 One-Forms in Continuum Mechanics 27
4.4 Bases in General Relativity 29
4.5 Anholonomicity 31
4.6 Compatibility 32
5.1 Curvature Tensor 34
5.2 Curvature Tensor in Components 36
5.3 Ricci Tensor 37
5.4 Hooke's Law 38
5.5 Moment Tensor 39
5.6 Transformations of the Curvature Tensor 41
5.7 Transformations of the Stress Tensor 42
5.8 Tetrads in General Relativity 44
5.9 Tetrads in Continuum Mechanics 45
5.10 Cross Product 49
5.11 Metric Tensor in Continuum Mechanics 53
5.12 Metric Tensor in Relativity 54
5.13 Volume Form in Continuum Mechanics 59
5.14 Surface One-Form in Continuum Mechanics 59
5.15 Cross Product Revisited 60
5.16 Cramer's Rule 62
5.17 Elastic Tensor 67
6.1 Metric on the Sphere 73

6.2 Referential Manifold ... 77
6.3 Cauchy–Green Tensor ... 81
6.4 Definitions of Strain ... 81
6.5 Definitions of Stress ... 82
6.6 Two-Point Tensors in Continuum Mechanics ... 83
7.1 Material Velocity Gradient ... 90
7.2 Deformation Rate and Vorticity ... 90
7.3 Geodesics in General Relativity ... 94
7.4 Ricci Identity for Material Velocity ... 95
7.5 Polar Coordinates ... 100
7.6 Lagrangian Connection Coefficients ... 100
7.7 Field Equations ... 104
7.8 Cosmological Constant and Dark Energy ... 105
7.9 Dynamic Equations ... 105
7.10 Inertial Frames ... 106
7.11 Metric-Affine Gravitation Theory ... 108
7.12 Nonmetricity in Continuum Mechanics ... 108
7.13 Maxwell's Equations ... 108
7.14 Material Derivative ... 112
7.15 Corotational Material Derivative ... 112
7.16 Definition of the Lie Derivative in Continuum Mechanics ... 119
7.17 Lie Derivative of the Metric in Continuum Mechanics ... 121
7.18 Distinguishing Rank-2 Tensors ... 123
7.19 Lie Derivative of the Volume Form ... 124
7.20 Lie Derivative of the Jacobian of the Motion ... 124
8.1 Properties of the Four-Dimensional Volume Form ... 137
8.2 Grad, Div, and All That ... 141
8.3 Classical Vector Calculus ... 142
8.4 Form-versions of Maxwell's Equations ... 143
8.5 Defects ... 145
8.6 Stress as a Vector-Valued Two-Form ... 151
8.7 Strain as a One-Form-Valued Two-Vector ... 156
8.8 Form Version of Hooke's Law ... 159
8.9 Eulerian and Lagrangian Exterior Covariant Derivatives ... 166
8.10 Mixed Exterior Covariant Derivative ... 166
8.11 Spin Connection One-Forms ... 167
8.12 Eulerian and Lagrangian Covariant Lie Derivatives ... 168
8.13 Bianchi Identities in General Relativity ... 173
8.14 Nonmetricity in General Relativity ... 173
8.15 Line Integral ... 175

8.16 Surface Integral 176
8.17 Volume Integral 177
8.18 Reynolds Transport Theorem 183
8.19 Conservation of Mass 186
8.20 Conservation of Linear Momentum 186
8.21 Conservation of Angular Momentum 188
8.22 First Law of Thermodynamics 190
8.23 First Law of Thermodynamics with Spin 192
8.24 Second Law of Thermodynamics 194
8.25 Caloric Equations of State and Gibbs Relations 195
8.26 Action in Continuum Mechanics 201
8.27 Action in Continuum Mechanics with Spin 202
8.28 Burgers and Frank Vectors 204
8.29 Action in General Relativity 206

PREFACE

I have a background in geophysics, where the theory of continuum mechanics plays a central role. Initially, I learned continuum mechanics using classical tensor calculus, which presents tensors in a coordinate-dependent manner, relying heavily on the concepts of basis vectors and components to represent them. In this approach, tensors are introduced as multidimensional arrays of numbers, and their properties and transformations are analyzed using index notation. The focus is on manipulating these numerical components, following specific transformation rules under changes in coordinate systems. This method involves detailed calculations and computations that make it ideal for solving practical problems in various scientific and engineering disciplines, especially geophysics.

Later in my career, I discovered the geometrical description of tensors, which emphasizes a coordinate-independent approach, focusing on the intrinsic geometric properties of tensors and their behavior under transformations. This approach is prevalent in the field of differential geometry and fundamental to the geometric structures of manifolds. In this methodology, tensors are multilinear maps without reference to any coordinate system. They are geometric objects that are independent of any specific choice of coordinates. This viewpoint emphasizes the geometric significance of tensors. This allows for a more abstract and elegant understanding of tensors and their relations, often leading to deeper insights into the underlying geometric structures and the related physics.

The purpose of this book is to help readers to understand and use tensors geometrically. It can serve as the basis for a course on tensor calculus for physicists, applied mathematicians, and engineers or as a complementary reference when studying specific areas of physics that rely on tensor calculus, such as fluid and continuum mechanics, electrodynamics, and general relativity. More than eighty examples illustrate the geometrical use of tensors. These examples are highlighted in the text in the form of "boxes" and center on two leitmotivs. The first takes the reader through all aspects of geometrical continuum mechanics, and the second educates the reader in the field and dynamic equations of Einstein, Einstein–Cartan, and metric-affine theories of general relativity. Thus, the book can form the basis of courses in both these subjects. The content is from a physicist's perspective, with the expectation that it will be valuable to scientists, applied mathematicians, and engineers with similar training.

Acknowledgments

First and foremost, I thank my wife, Catrinel, whose wit, erudition, creativity, and grace inspire me and fill my life with immense happiness. I am one lucky man.

I am grateful for the help and support of my colleagues throughout the creation of this book. I would like to thank Niall Coffey, Will Eaton, Stephen Griffies, Kevin Liu, Frederik Simons, Josh Rines, and Yuri Tamama for carefully reviewing parts of this material. It has been a pleasure working with the wonderful staff at Princeton University Press, particularly Ingrid Gnerlich, who has guided this book to publication, and Gráinne O'Shea, who did a splendid job of copy editing. I especially appreciate the level of detail and care given to the design and layout of the book.

On the cover of the book, there is an image of "Thetis Circle," a 4-meter Cor-Ten steel artwork created by Beverly Pepper (1922–2020) and displayed on the Princeton University campus. This artwork perfectly captures the concepts of *torsion* and *curvature*. The gap between the two endpoints of the artwork represents torsion, whereas the distance between the heads of two people standing awkwardly perpendicular to its outer surface at the two endpoints represents curvature. In the description of material defects, the gap represents a *dislocation*, whereas the distance between the two heads represents a *disclination*.

I wrote parts of this book back in 2008 and 2009, while Albert Tarantola and Michael Slawinski were on sabbatical leave at Princeton University. We explored plans for a book on continuum mechanics, which petered out after Albert's untimely death on December 6, 2009. However, I am grateful for their constructive comments and suggestions during that period.

Parts of the book were written when the author was on sabbatical at Universiteit Utrecht (2022 & 2023), Institut de Physique du Globe de Paris (2022), Laboratoire GéoAzur (2022 & 2023), Istituto Nazionale di Geofisica e Vulcanologia Roma (2023), and ETH Zürich (2023). The hospitality afforded at these European institutions is greatly appreciated.

Chapter 1

INTRODUCTION

You have to spend some energy and effort to see the beauty of math.
—MARYAM MIRZAKHANI

Unfortunately for Calvin, every job requires *some* math. Monday, September 17, 1990, Calvin and Hobbes. © Watterson. Reprinted with permission of ANDREWS MCMEEL SYNDICATION. All rights reserved.

In the late nineteenth century, Woldemar Voigt introduced the word "tensor" in the context of elasticity theory, a name that reflects its mechanical origins (Maugin, 2013). At that time, the mathematical world was striving toward generalizations, and physical interest in the theory of elasticity was driven by the now outdated notion of ether. The appearance of tensors in the mathematical language came from these developments.

Tensors have become a fundamental tool for science and engineering. They are widely used to describe physical quantities and their relationships in multidimensional spaces. Tensors are essential for studying fluid dynamics, continuum mechanics, electromagnetism, and general relativity. They enable precise descriptions of physical properties that have multiple dimensions and magnitudes.

In recent years, tensors have become increasingly important for data analytics and machine learning due to their ability to represent and analyze complex data and relationships in multidimensional spaces. Tensors serve as data structures in deep learning and machine learning algorithms, allowing for the representation and processing of data in neural networks and enabling the modeling of complex relationships and patterns in data.

The geometrical description of tensors used in this book provides an elegant and powerful tool for comprehending their behavior, relationships, and applications in various fields. It aids in linking abstract mathematical concepts to physical phenomena, enabling researchers, scientists, and engineers to understand and utilize tensors in their respective domains.

This book provides the mathematical background for geometrical descriptions of tensors and their calculus. Its approach is inspired by various sources, including Schutz (1980), Carroll (2004), and Misner et al. (2017). For precise mathematical definitions of the concepts, the reader is directed toward more mathematical textbooks, such as Bishop & Goldberg (1980), Dubrovin et al. (1985), Lovelock & Rund (1989), Marsden & Hughes (1983), Martin (2002), and Stone & Goldbart (2009).

The main objective is to expand $\mathbb{R}^n$, *Euclidean space*, into a *manifold*, a curved space that behaves like $\mathbb{R}^n$ locally. The sphere and torus are examples of manifolds. Manifolds are invoked for several reasons. Deformed bodies can adopt forms that cannot be described within $\mathbb{R}^n$. Certain mathematical spaces—such as *Hamiltonian mechanics' phase space*, which is spanned by the components of coordinates and momenta, or *spacetime* in the theory of *general relativity*—are not $\mathbb{R}^n$.

The book is organized as follows. Chapter 2 discusses *linear spaces* and *coordinate transformations*, while subsequent chapters explore *tensors*, which can be viewed as *linear operators* acting on *vectors* and their *duals*. Chapter 3 is devoted to manifolds and their coordinate descriptions. Chapter 4 extends the concept of vectors in $\mathbb{R}^n$ to manifolds with curvilinear coordinates, introducing the dual entities of vectors: one-forms. The focus is on coordinate systems, including their *bases* and *components* in terms of these bases. Locally, any differentiable manifold of n dimensions behaves like $\mathbb{R}^n$, so attention is paid to the neighborhoods of points of the manifold. In chapter 5, tensors are introduced and viewed as *multilinear maps* from vectors and one-forms to numbers, $\mathbb{R}$. Chapter 6 introduces maps of tensors between manifolds, and differentiation on manifolds is discussed in chapter 7. Chapter 8 introduces exterior *differential forms* and the *exterior derivative*, which acts on k-forms to produce $(k+1)$-forms. The gradient is an example of the exterior derivative, and other familiar operations, such as the divergence and curl in $\mathbb{R}^3$, can be viewed as particular cases of exterior differentiation. The *fundamental theorem of exterior calculus* generalizes vector-calculus integral theorems, such as Gauss's and Stokes's theorems. The book contains over eighty examples from continuum mechanics, sourced from the book by Tromp (2025), and general relativity, which are used to illustrate the theory.

Chapter 2

LINEAR SPACES AND TRANSFORMATIONS

The only way to learn mathematics is to do mathematics.
—PAUL HALMOS

Tensor calculus is a language used to describe physical phenomena in fluid dynamics, continuum mechanics, electromagnetism, and general relativity. In this brief chapter, we will discuss the concept of *linear spaces*, which are critical to the foundation of tensor theory. Linear spaces and their linear algebras are important because they can represent a wide range of physical objects abstractly. This enables us to study them using a unified framework. This level of abstraction allows us to analyze the same structures of linear spaces and transformations regardless of the specific physical object under consideration.

2.1 Properties of Linear Spaces

A linear space is a set of elements, $\mathsf{u}, \mathsf{v}, \mathsf{w}, \ldots,$[1] that exhibits the following properties.

- There is an internal operation, called *addition* and denoted by $+$, such that for any two elements u and v of the linear space, $\mathsf{u} + \mathsf{v}$ is also an element of this space. The addition is *commutative*, which means that, for any two elements,

$$\mathsf{u} + \mathsf{v} = \mathsf{v} + \mathsf{u}, \tag{2.1}$$

  and *associative*, which means that, for any three elements,

$$(\mathsf{u} + \mathsf{v}) + \mathsf{w} = \mathsf{u} + (\mathsf{v} + \mathsf{w}). \tag{2.2}$$

[1] In this chapter, elements of a linear space are denoted by sans serif font.

- There exists a *zero element*, also called a *null element*, and denoted by 0, such that, for any u,

$$0 + \mathrm{u} = \mathrm{u} + 0 = \mathrm{u}. \tag{2.3}$$

- For every element, u, there exists another element, called the *opposite of* u and denoted by $-\mathrm{u}$, such that

$$\mathrm{u} + (-\mathrm{u}) = 0. \tag{2.4}$$

There are additional properties that concern an external operation, namely, the multiplication of elements of the linear space by elements of $\mathbb{R}$, the real numbers, $a, b, c, \ldots$. We require that if u is an element of the linear space, then so is $a\,\mathrm{u}$. Furthermore, this operation has the following properties:

$$0\,\mathrm{u} = 0 \qquad \text{and} \qquad 1\,\mathrm{u} = \mathrm{u}; \tag{2.5}$$

in other words, the multiplication of any element by zero produces the null element and by unity leaves this element unchanged. The latter property requires that $a = 1$ be a dimensionless quantity—an important consideration in physical applications.

Finally, considering real numbers a and b, we have the *distributive* properties

$$a\,(b\,\mathrm{u}) = (a\,b)\,\mathrm{u}; \quad a\,\mathrm{u} + b\,\mathrm{u} = (a + b)\,\mathrm{u}; \quad a\,\mathrm{u} + a\,\mathrm{v} = a\,(\mathrm{u} + \mathrm{v}). \tag{2.6}$$

2.2 Vector Spaces

A *vector space* is an illustration of a linear space. An example of a vector space is the space representing the addition of forces acting on a particle. Elements of vector spaces are *vectors*, $\mathbf{u}, \mathbf{v}, \mathbf{w}, \ldots$.[2] When discussing vectors and their extensions in chapters 4 and 5, we use the fact that they form linear spaces.

A quintessential vector space is $\mathbb{R}^n$, which is the set of all *n-tuples* of real numbers, $(u^1, \ldots, u^n)$, with $u^i \in \mathbb{R}$. As required by the first point in section 2.1,

$$(u^1, \ldots, u^n) + (w^1, \ldots, w^n) = (u^1 + w^1, \ldots, u^n + w^n), \tag{2.7}$$

in other words, adding two n-tuples results in another n-tuple that belongs to the same space. Similarly,

$$a(u^1, \ldots, u^n) = (au^1, \ldots, au^n), \tag{2.8}$$

multiplying an n-tuple by a scalar, we get another n-tuple belonging to the same space.

2.3 Linear Transformations

Transformation, $T: U \to W$, is a function from U to W, where U and W are linear spaces. For T to be linear, we require that, for any two elements of U,

$$T(a\,\mathrm{u}_1 + b\,\mathrm{u}_2) = a\,T(\mathrm{u}_1) + b\,T(\mathrm{u}_2). \tag{2.9}$$

[2] Throughout the book, we use the **bold** font to denote vectors.

An $m \times n$ matrix, A, is an example of a linear transformation between U and W, which—for this example—we let be $\mathbb{R}^n$ and $\mathbb{R}^m$, respectively. Concisely, we write $A\mathbf{u} = \mathbf{w}$, where $\mathbf{u}$ and $\mathbf{w}$ are vectors, such that $\mathbf{u} \in \mathbb{R}^n$ and $\mathbf{w} \in \mathbb{R}^m$. Also, as required,

$$A(a\,\mathbf{u}_1 + b\,\mathbf{u}_2) = a\,A\mathbf{u}_1 + b\,A\mathbf{u}_2. \tag{2.10}$$

In concluding this chapter, it is worth mentioning the concept of *dimension* in vector spaces. For instance, if a vector $\mathbf{u}$ belongs to an n-dimensional space, we can express it as a linear combination of n other vectors belonging to the same space, using a unique set of scalars $\{a_1, \ldots, a_n\} \in \mathbb{R}^n$: $\mathbf{u} = a_1\mathbf{u}_1 + \ldots + a_n\mathbf{u}_n$. The vectors $\mathbf{u}_1, \ldots, \mathbf{u}_n$ form a basis of the space, and the number of vectors required, in this case n, is the dimension of the space.

Chapter 3

DIFFERENTIABLE MANIFOLDS

Above all, don't fear difficult moments. The best comes from them.
—RITA LEVI-MONTALCINI

Vectors are essential in physics and engineering and form linear spaces, discussed in chapter 2. Throughout this book, we refer to *manifolds* as a general term for spaces and use other terms such as *charts* and *atlases* in this context. This chapter formalizes the concept of manifolds and their associated entities, which generalizes the properties of n-dimensional Euclidean space, $\mathbb{R}^n$. While we provide only a brief introduction to this subject, readers can refer to books such as Spivak (1965), Do Carmo (1976), Munkres (1990), Do Carmo (1994), Martin (2002), and Torres del Castillo (2012) for more detailed discussions.

3.1 Charts and Coordinates

Let us explore functions and their differentiation and integration in spaces beyond $\mathbb{R}^n$, including spheres, tori, and other complex topologies. Curved *spacetime*, for instance, is a famous example of a manifold used to describe physical phenomena in general relativity. On an n-dimensional manifold, we assume that functions behave *locally* as they do in $\mathbb{R}^n$. By defining a *local coordinate system* or *chart* on a portion of the manifold, we can assign n quantities to every point $\mathcal{S}$ in the manifold: the coordinates of the point in the chart. These quantities can be pure real numbers, such as angles in radians, or they can have physical dimensions, such as length. Generally, a global coordinate system cannot be defined on a manifold. For example, certain points may have pathological coordinate values, such as the two poles of a sphere described in spherical coordinates. However, we can always cover the manifold with a patchwork of local charts or an *atlas*, in each of which local coordinates can be defined.

A chart on an open set,[1] N, of an n-dimensional manifold is a collection of n functions,

$$x^i = x^i(\mathcal{S}), \qquad i = 1, \ldots, n, \tag{3.1}$$

assigning to every point $\mathcal{S} \in N$ its coordinate values $x = \{x^i\}$. These functions are assumed to be differentiable and to have an inverse,

$$\mathcal{S} = \mathcal{S}(x), \tag{3.2}$$

so that, given coordinate values x, a unique point $\mathcal{S}$ is identified.[2] The uniqueness requirement leads us to exclude from the definition of a manifold objects that exhibit points without a well-defined notion of a tangent space, introduced in section 4.1. Consequently, we focus our attention on *differentiable manifolds*.

Example 3.1 Charts for the Circle

To gain insight into charts, consider a circle in $\mathbb{R}^2$. Figure 3.1 illustrates how it may be covered by two overlapping charts. One can also consider a set of coordinate systems $\alpha \in (0, 2\pi)$ and $\beta \in (-\pi, \pi)$. The charts cover the circle, and—on their overlap—$\alpha \mapsto \beta$ is a *diffeomorphism*,[a] since $\beta = \alpha$ for $\alpha \in (0, \pi)$, and $\beta = \alpha - 2\pi$ for $\alpha \in (\pi, 2\pi)$. Thus, these two charts form an atlas.

[a] Diffeomorphism: an invertible function that maps one differentiable manifold to another, such that both the function and its inverse are smooth.

Combining charts, we obtain an atlas of the manifold: a system of overlapping and smoothly related local coordinate systems. Each chart must be of the same dimension as the other charts—and of the same dimension as the manifold itself—and the map that relates two given charts over the regions of their overlap must be *one-to-one*.[3] We are careful to distinguish the *map* $\Phi : N \to M$, which relates points $\mathcal{S} \in N$ and $\mathcal{S}' = \Phi(\mathcal{S}) \in M$, from a *function* $\mathrm{f} : N \mapsto \mathbb{R}$, which returns a value, $\mathrm{f}(\mathcal{S})$, at a given location $\mathcal{S} \in N$, by using Greek or Roman letters for a function.[4] The dimension of the manifold corresponds to the dimension of the charts. Consequently, an important property of a manifold is that it allows us to discuss properties of spaces with no reference to an ambient space or, in other words, without embedding them in a higher-dimensional space. For instance, a two-dimensional sphere can be considered on its own, not as an object residing in a three-dimensional space.

[1] Open set: any point therein can be moved by a small amount in any direction and remain within the set. Strictly speaking, this requires that some notion of "distance" be defined.

[2] Expression (3.2) is a shorthand notation, popular among physicists, for $\mathcal{S} = \Psi(x)$, where Ψ is the inverse of function ψ, which is defined by $x^i = \psi^i(\mathcal{S})$.

[3] One-to-one map $\Phi : N \to M$, also called an *injective function*, associates any given point $\mathcal{S}$ in its domain N with a distinct point $\mathcal{S}' = \Phi(\mathcal{S})$ in its codomain M.

[4] In chapter 4 we identify functions with $(0, 0)$ tensors.

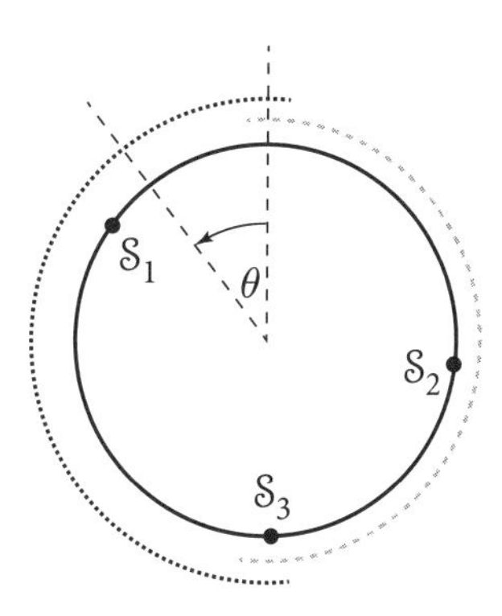

Figure 3.1: The short-dashed chart maps point $\mathcal{S}_1$ to the local coordinate system $\theta \in (-\epsilon, \pi + \epsilon) \subset \mathbb{R}^1$, whereas the long-dashed chart maps point $\mathcal{S}_2$ to the local coordinate system $\theta \in (\pi - \epsilon, 2\pi + \epsilon) \subset \mathbb{R}^1$, where θ is measured counterclockwise from the top, and ϵ is a small positive number that produces two overlapping charts. Point $\mathcal{S}_3$ is captured by both charts. Together, the charts cover the entire circle, thereby defining an atlas capturing any point $\mathcal{S}$ in the circle.

Many other sets of charts may be employed—a property that leads us to the definition of a manifold, discussed in section 3.2. However, we assume that differentiable manifolds of interest can be covered by a countable number of charts, which is a crucial axiom in studying the topology of manifolds.

Example 3.2 Charts in Continuum Mechanics

In continuum mechanics, all physical quantities—for example, mass density, material velocity, and stress—are defined as unique tensors with respect to an *inertial* or *Galilean reference frame*, independent of any coordinate system. Throughout the examples from continuum mechanics in this book, we consider two primary classes of coordinate systems within a *spatial manifold* $\mathcal{S}$. The first class comprises inertial *spatial* or *Eulerian* coordinates, denoted by the set $\{r^i\}$, which remain unaffected by the continuum's motion. The second class encompasses *comoving* or *Lagrangian* coordinates (see, e.g., Sedov, 1966; Weile et al., 2013), denoted by the set $\{X^I\}$,[a] which are "attached" to an element of the continuum, thereby providing a "particle label." Thus, a spatial point $\mathcal{S}$ may be represented in two complementary charts, namely, an Eulerian chart

$$r^i = r^i(\mathcal{S}), \qquad i = 1, 2, 3, \tag{3.3}$$

and a Lagrangian chart

$$X^I = X^I(\mathcal{S}), \qquad I = 1, 2, 3, \tag{3.4}$$

as illustrated in figure 3.2.

[a] It is conventional in continuum mechanics to identify Lagrangian coordinates by the symbol X and uppercase Roman superscripts, X^I.

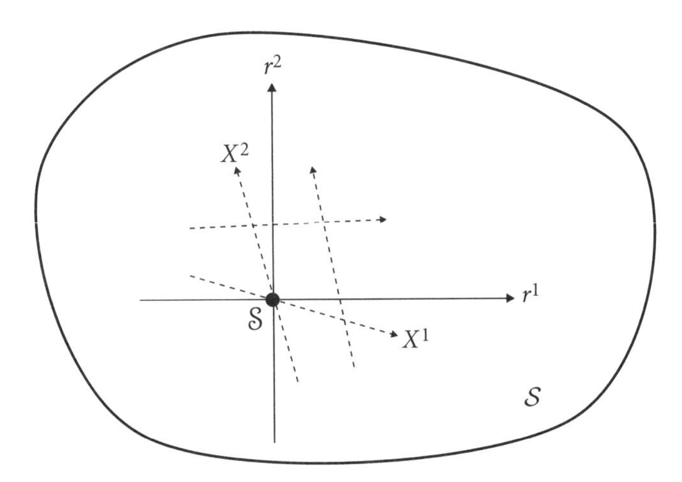

Figure 3.2: In continuum mechanics, a spatial point $\mathcal{S}$ in the spatial manifold $\mathscr{S}$ may be identified with either a set of Eulerian coordinates $\{r^i\}$ or a set of comoving or convected Lagrangian coordinates $\{X^I\}$. The Eulerian coordinates are chosen to be Cartesian in this example, but this is not required.

Example 3.3 Charts in General Relativity

In general relativity, an *event* $\mathcal{S}$ in a four-dimensional *spacetime manifold* may be captured by a set of coordinates

$$\{x^\mu\} \equiv \{x^0, x^1, x^1, x^3\} = \{ct, x^i\}. \tag{3.5}$$

In this expression, and in the rest of the book, lowercase Greek indices identify four-dimensional coordinates $\{x^\mu\}$, lowercase Roman indices identify three-dimensional spatial coordinates $\{x^i\}$, $x^0 = ct$, time is denoted by t, and c is the speed of light.

Thus, an event $\mathcal{S}$ may be represented in the spacetime chart

$$x^\mu = x^\mu(\mathcal{S}), \qquad \mu = 0, 1, 2, 3. \tag{3.6}$$

3.2 Definition

To formalize the concept of a manifold, let us consider a map Φ from an n-dimensional open set N to an m-dimensional open set M, such that

$$\mathcal{S}' = \Phi(\mathcal{S}). \tag{3.7}$$

As illustrated in figure 3.3, this map is a rule that associates a point $\mathcal{S}(x) \in N$ with a point $\mathcal{S}'(x') \in M$. In coordinates,

$$x^{i'} = \Phi^{i'}(x), \tag{3.8}$$

where $\{x^{i'}\} \in \mathbb{R}^m$.

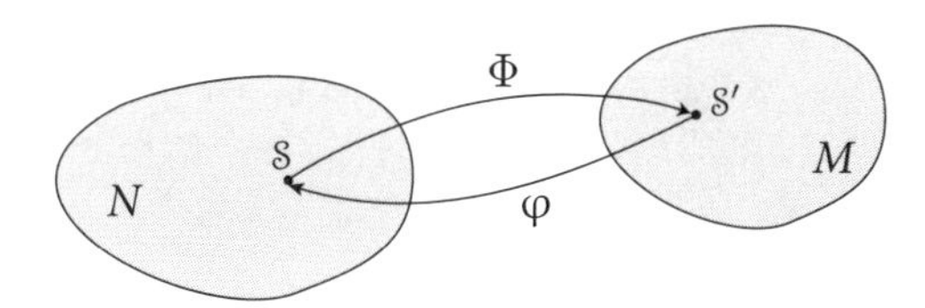

Figure 3.3: Map Φ taking a point $\mathcal{S}$ in open set N to a point $\mathcal{S}' = \Phi(\mathcal{S})$ in open set M. If N and M are diffeomorphic, then the map Φ is bijective and has inverse $\Phi^{-1} = \varphi$.

We say that Φ is a C^p map if its pth partial derivatives exist and are continuous. Thus, a C^0 map is continuous but not differentiable, and a C^∞ map is continuous and infinitely differentiable; we refer to the latter as *smooth*.

The sets N and M are diffeomorphic if map (3.8) is a smooth *bijective*[5] map that possesses a smooth inverse,

$$\mathcal{S} = \Phi^{-1}(\mathcal{S}') = \varphi(\mathcal{S}'), \tag{3.9}$$

or in coordinates

$$x^i = \varphi^i(x'), \tag{3.10}$$

where $\{x^i\} \in \mathbb{R}^n$.

If two sets of points are diffeomorphic they have the same dimension; furthermore, they constitute the same manifold.

Mapping is crucial in understanding differentiable manifolds, as a single chart may not cover the entire manifold unless it is $\mathbb{R}^n$ or its open subset. For instance, the inadequacy of a single chart is illustrated by the circle in figure 3.1 and the Mercator projection of Earth's surface, which leaves out the north and south poles. As articulated by Do Carmo (1994), the distinction between a regular surface and a differentiable manifold lies in the emphasis on the change of coordinates through mapping.

In summary, a C^p n-dimensional manifold is a set of points, N, such that in some open neighborhood of each point $\mathcal{S} \in N$ there is a one-to-one C^p map onto an open subset of $\mathbb{R}^n$.

We see that, locally, a manifold resembles $\mathbb{R}^n$. Also, we limit our interest to manifolds for which any two distinct points have neighborhoods that do not intersect; these are so-called *Hausdorff manifolds*.

3.3 Local Coordinate Changes

The location of a point $\mathcal{S}$ on a manifold is independent of its coordinates. Since there is an infinity of local charts that capture this point, let us examine transformations among them. Let $x = \{x^i\}$ and $x' = \{x^{i'}\}$ be two overlapping coordinate systems related via local diffeomorphic transformations

$$x^i = \varphi^i(x') \qquad \text{and} \qquad x^{i'} = \Phi^{i'}(x). \tag{3.11}$$

[5] Bijection: an invertible map $\Phi : N \to M$ is bijective if every element $\mathcal{S}'$ in its codomain M is mapped to a corresponding element $\mathcal{S}$ in its domain N by the inverse map, $\Phi^{-1} = \varphi : M \to N$.

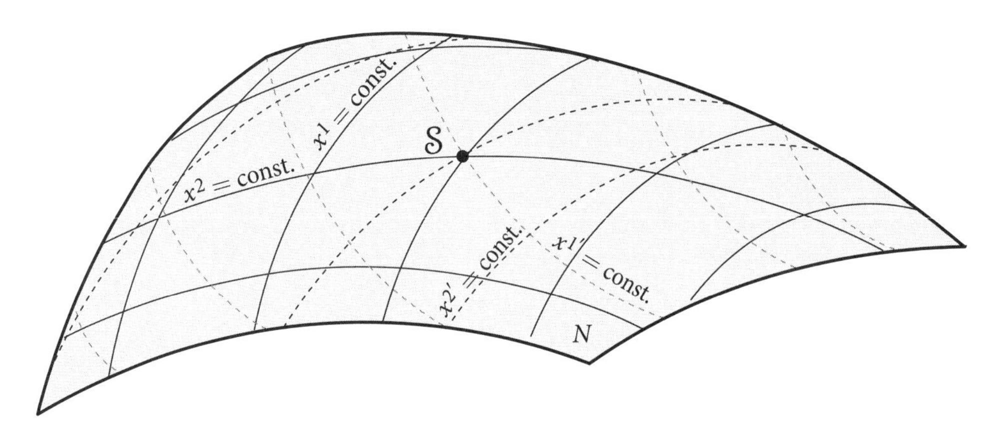

Figure 3.4: Two overlapping coordinate systems or charts are used to identify a spatial point $\mathcal{S}$ in the manifold N: $\{x^1, x^2\}$ (solid lines) and $\{x^{1'}, x^{2'}\}$ (dashed lines).

Figure 3.4 illustrates the representation of a single *spatial*[6] point $\mathcal{S} = \mathcal{S}(x) = \mathcal{S}(x')$ by two overlapping coordinate systems, x and x'.

Example 3.4 Coordinate Changes in Continuum Mechanics

In continuum mechanics, the Eulerian and Lagrangian coordinates introduced in example 3.2 are related via the *motion* of the continuum, which is captured by the transformation

$$\begin{aligned} r^i &= \varphi^i(X, T), \\ t &= T. \end{aligned} \tag{3.12}$$

This time-dependent transformation, φ^i, takes us from the Lagrangian coordinates $\{X^I\}$, which are "attached" to an element of the continuum, to the inertial Eulerian coordinates $\{r^i\}$ of point $\mathcal{S}$ in the spatial manifold S, at a given time $t = T$, as illustrated in figure 3.2. Effectively, the motion tracks particle trajectories. We assume that the motion is invertible (no tearing of the continuum), such that

$$\begin{aligned} X^I &= \Phi^I(r, t), \\ T &= t. \end{aligned} \tag{3.13}$$

The inverse motion, Φ^I, takes us from the inertial Eulerian coordinates $\{r^i\}$, which identify a location $\mathcal{S}$ in the spatial manifold S, to the Lagrangian coordinates $\{X^I\}$ of the element of the continuum that happens to be occupying that spatial location at time $t = T$.

[6] Emphasizing spatial, as opposed to *temporal*.

3.4 Functions on Manifolds

A domain of a function on a manifold, N, might be the entire manifold; hence, we can write $\mathrm{f}(\mathcal{S})$,[7] which is the value of this function at a particular point $\mathcal{S}$; it is independent of the coordinate system in which it is expressed. In other words, $\mathrm{f}(\mathcal{S}(x)) = f(x)$, and $\mathrm{f}(\mathcal{S}(x')) = f'(x')$, where—in general—$f$ and f' are different functions that result in the same value for each point of their overlap on N. As stated in section 3.3, a sufficient condition to express a function in two coordinate systems is their being related by a bijective map.

3.5 Orientable Manifolds

In scientific applications, *orientability* of manifolds is a crucial requirement. In the case of a surface, it implies that the normal may change continuously but never vanishes. For any orientable manifold, there are two possible orientations, depending on the chosen direction of the normal or tangent vector, as is the case for a curve, discussed in section 4.1.1. Additionally, any manifold with a boundary induces the orientation of that boundary, as shown in Stokes's theorem (section 8.11). To achieve orientability, we must be able to smoothly vary the orientation of each tangent space (section 4.1). This means we can move the basis smoothly on the manifold. The concept of charts provides a way to quantify the concept of orientability.

The *Jacobian* of a bijective mapping between overlapping charts $x^i(\mathcal{S}) \in N$ and $x^{i'}(\mathcal{S}) \in N$, that is, two charts containing the same point $\mathcal{S}$ related by the mapping $x^i = \varphi^i(x')$, is defined as the determinant of the matrix of partial derivatives:[8]

$$\lambda \equiv \det\left(\frac{\partial \varphi^i}{\partial x^{i'}}\right). \tag{3.14}$$

A manifold is orientable if there exists an atlas such that for each pair of overlapping charts, the corresponding coordinate systems are *consistently oriented*: Jacobian (3.14) is positive, which means that the coordinate systems are related by rotations without reflections; the systems remain right-handed.

In view of figure 3.1 and example 3.1, since $\partial\beta/\partial\alpha = 1$, a circle is an orientable manifold. Famous nonorientable manifolds are the Möbius strip and the Klein bottle.

[7]Here we use the Roman font to distinguish a function defined at a point $\mathcal{S}$ of N, $\mathrm{f}(\mathcal{S})$—a $(0,0)$ tensor, as discussed in chapter 5—from a function of specific coordinates x, which we denote by $f(x)$.

[8]Throughout this book we use the relation operator $\equiv$ to define a symbol. We do not use $:=$ or $\stackrel{\text{def}}{=}$, used commonly by mathematicians.

Chapter 4

VECTORS AND ONE-FORMS

It is not enough to have a good mind. The main thing is to use it well.
—RENÉ DESCARTES

In this chapter, we delve into the geometry of manifolds and introduce vectors and one-forms as fundamental entities. Unlike coordinates, these geometric objects are independent of any specific coordinate system and are associated with spatial locations, such as a point $\mathcal{S}$ in a manifold N. In this book, we consistently represent scalar functions with Greek or Roman letters and use **bold font** for vectors, one-forms, and tensors. Therefore, the temporal and spatial variations of these entities are described using *only* the notation $\mathbf{u}(\mathcal{S}, t)$, which denotes the vector $\mathbf{u}$ as a function of spatial point $\mathcal{S}$ and time t. However, we can express vectors in terms of their components relative to the associated basis vectors, given the introduction of a coordinate system. These components depend explicitly on the coordinates chosen.

4.1 Vectors

Many areas of science and engineering demand a distinctive definition of *vectors* that differs from some introductory explanations. A vector represents an element of a linear space linked to a point $\mathcal{S}$ on a manifold. Unlike an arrow extending from one point to another, a vector is envisioned as an arrow connected to a particular point of the manifold. On a *differentiable manifold* N, there is a *tangent space* $T_{\mathcal{S}}$, depicted in figure 4.1, which is itself a linear vector space. It means that for any two vectors $\mathbf{u}$ and $\mathbf{w}$ in $T_{\mathcal{S}}$ and any two real numbers a and b, the following property holds:

$$(a+b)(\mathbf{u}+\mathbf{w}) = a\,\mathbf{u} + b\,\mathbf{u} + a\,\mathbf{w} + b\,\mathbf{w}, \tag{4.1}$$

which is a concise way of writing the essence of the properties of linear spaces, discussed in chapter 2.

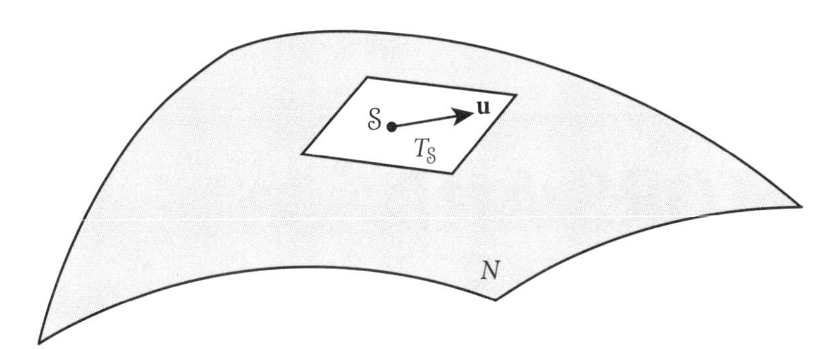

Figure 4.1: The space tangent to manifold N at point $\mathcal{S}$ is $T_\mathcal{S}$, and **u** illustrates a vector in the tangent space.

The collection of tangent spaces at all points on the manifold is the *tangent bundle*, denoted by TN. A tangent bundle is a *product space* of manifolds N and $T_\mathcal{S}$. Since N and $T_\mathcal{S}$ are manifolds, so is their product (see, e.g., Schutz, 1980, Chapter 2.11). The dimension of the tangent space $T_\mathcal{S}$ of vectors is the same as that of the manifold. The tangent bundle TN consists of all points and vectors. In $\mathbb{R}^3$, the tangent bundle is thus a six-dimensional manifold, and, in general, TN is a $2\,n$-dimensional manifold.

Classical mechanics serves as an illustration of manifolds and their corresponding tangent spaces. Consider a point $\mathcal{S}$ in physical space that can be represented in a local coordinate system of a three-dimensional manifold N as (x^1, x^2, x^3). The velocity vector at this point $(\dot{x}^1, \dot{x}^2, \dot{x}^3)$ belongs to the tangent space of $\mathcal{S}$, denoted as $T_\mathcal{S}$. The framework of Lagrangian mechanics operates in the tangent bundle TN, which comprises all the tangent spaces at each point $\mathcal{S} \in N$. Tangent spaces are geometrical entities that are associated with manifolds, independent of any functions that connect them. In classical mechanics, these spaces are connected via the Lagrangian $L(x, \dot{x})$.

4.1.1 Vectors as Tangents to Curves

Following Schutz (1980), a *curve* is a mapping from $\mathbb{R}$ to the manifold prescribed by a continuous mapping from point $\lambda \in \mathbb{R}$ to point $\mathcal{S}$ on N, which we denote as $\mathcal{S} = \mathcal{S}(\lambda)$.[1] As the value of the parameter λ changes, the mapping traces a curve on the manifold.

Consider a curve $\mathcal{S}(\lambda)$, parameterized in terms of a generating parameter λ, on a manifold N. At every point $\mathcal{S}$ on N along this curve, we identify the directional derivative $\mathrm{d}/\mathrm{d}\lambda$ with a vector **u**:

$$\mathbf{u} \equiv \frac{\mathrm{d}}{\mathrm{d}\lambda}. \tag{4.2}$$

Hence, a vector at a spatial point $\mathcal{S}$ may be viewed as a directional derivative operator along a path on the manifold through that point; in other words, it can be viewed as an arrow tangent to a curve, not an arrow stretching between two points, as illustrated in figure 4.2.[2]

Differentiability of the manifold ensures that we can draw such an arrow at each location along the curve.

[1] To avoid using too many parentheses, we abuse the notation: $\mathcal{S}(\lambda) \equiv \mathcal{S}(x(\lambda))$.
[2] See Munkres (1990) for a more rigorous discussion and definition.

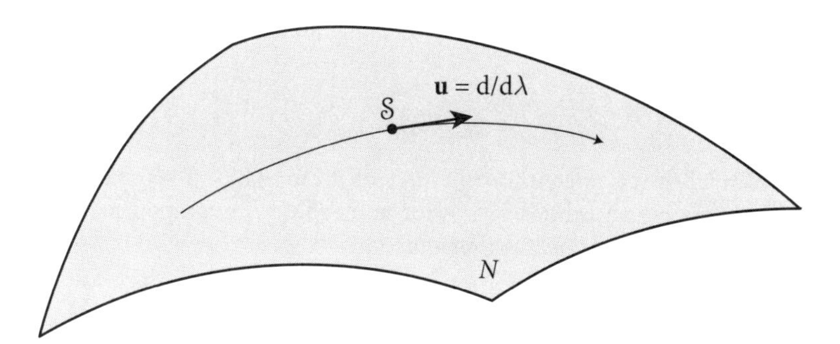

Figure 4.2: A vector $\mathbf{u}$ may be regarded as a directional derivative along a curve $\mathcal{S} = \mathcal{S}(\lambda)$ on the manifold N. This vector is an element of the tangent space $T_{\mathcal{S}}$, as illustrated in figure 4.1. Abstractly, we make the identification $\mathbf{u} \equiv \mathrm{d}/\mathrm{d}\lambda$.

4.1.2 Bases and Coordinates

Vectors exist independently of their representation, but for the sake of computation or learning, it is useful to express them as components relative to *basis vectors.* A *basis*, $\{\mathbf{e}_i\}$, is a set of vectors that spans the vector space, and any vector in that space can be written as a linear combination of the basis vectors, which must be *linearly independent*. For example, in $\mathbb{R}^3$, three vectors $\mathbf{u}$, $\mathbf{v}$, and $\mathbf{w}$ are linearly independent if and only if there are no nonzero constants a, b, c, such that $a\,\mathbf{u} + b\,\mathbf{v} + c\,\mathbf{w} = \mathbf{0}$, where $\mathbf{0}$ denotes the zero vector. In an n-dimensional manifold, $\{\mathbf{e}_i\}$ form a basis if there is no set of nonzero constants $\{c^i\}$ such that $\sum_{i=1}^{n} c^i\,\mathbf{e}_i = \mathbf{0}$. Although there are infinitely many bases for a given vector space, they all have the same number of elements, which is the *dimension* of the vector space. Any vector $\mathbf{u}$ can be expressed as a linear combination of basis vectors:

$$\mathbf{u} = \sum_i u^i\,\mathbf{e}_i \equiv u^i\,\mathbf{e}_i. \tag{4.3}$$

Here we introduced the *summation convention*, also known as the *Einstein summation convention*, which enables us to eliminate the need for writing the summation symbol by implying summation over the repeated upper and lower index i. Although we refer to u^i as components of $\mathbf{u}$, strictly speaking, they are coefficients of basis vectors. We define the space tangent to a point on a manifold as the space of directional derivative operators through that point, which, as shown in equation (4.13), is a linear space with the same dimension as the manifold.[3]

To determine the basis of this space, we use a chart $x^i = x^i(\mathcal{S})$ introduced in chapter 3. We consider a differentiable function f in coordinates x, that is, $f(x)$, and the value of this function at each point along a curve identified by specifying $x(\lambda) = \{x^i(\lambda)\}$ is determined by

$$g(\lambda) = f(x(\lambda)). \tag{4.4}$$

Differentiating along this curve, using the chain rule, we obtain the directional derivative of f:

$$\frac{\mathrm{d}g}{\mathrm{d}\lambda} = \frac{\mathrm{d}x^i}{\mathrm{d}\lambda}\frac{\partial f}{\partial x^i}. \tag{4.5}$$

[3]Interested readers might refer to Carroll (2004), pp. 63–64.

Since f is an arbitrary function and g is its subset along the curve, we may concisely write

$$\frac{\mathrm{d}}{\mathrm{d}\lambda} = \frac{\mathrm{d}x^i}{\mathrm{d}\lambda}\,\partial_i, \tag{4.6}$$

where $\partial_i \equiv \partial/\partial x^i$. We recognize the definition of vector (4.2). Thus, comparing equations (4.3) and (4.6), we regard $\mathrm{d}/\mathrm{d}\lambda$ as vector $\mathbf{u}$, $\mathrm{d}x^i/\mathrm{d}\lambda$ as its components u^i, and ∂_i as the corresponding basis $\mathbf{e}_i$, as illustrated in figure 4.3. By stating

$$\mathbf{e}_i \equiv \partial_i \tag{4.7}$$

we identify partial differentiation ∂_i in the direction defined by coordinate x^i with basis vector $\mathbf{e}_i$. Basis vectors are not necessarily unit vectors, but they can be.

Example 4.1 Material Velocity in Continuum Mechanics

The temporal derivative of the motion (3.12), that is, the partial derivative of $\varphi^i(X^I, T)$ with respect to time T, holding the Lagrangian coordinates X^I fixed, defines the Eulerian components of the *material velocity*:

$$v^i \equiv \partial_T \varphi^i. \tag{4.8}$$

The material velocity is a vector:

$$\mathbf{v} = v^i\,\mathbf{e}_i, \tag{4.9}$$

where the Eulerian basis vectors $\mathbf{e}_i$ are defined in terms of the Eulerian coordinates $\{r^i\}$, illustrated in figure 3.2, as

$$\mathbf{e}_i \equiv \partial_i, \tag{4.10}$$

where we have used the shorthand notation $\partial_i = \partial_{r^i}$.

To show the linearity of directional derivatives, let us consider a second curve through the same point; this curve is parameterized by μ, namely $x^i(\mu)$. Following an approach analogous to the one for obtaining equations (4.5) and (4.6), we see that the directional derivative along this curve is

$$\frac{\mathrm{d}}{\mathrm{d}\mu} = \frac{\mathrm{d}x^i}{\mathrm{d}\mu}\,\partial_i, \tag{4.11}$$

which we denote by $\mathbf{v}$. Let us add the two vectors multiplied by real numbers a and b to find

$$a\,\mathbf{u} + b\,\mathbf{v} = a\,\frac{\mathrm{d}}{\mathrm{d}\lambda} + b\,\frac{\mathrm{d}}{\mathrm{d}\mu} = \left(a\,\frac{\mathrm{d}x^i}{\mathrm{d}\lambda} + b\,\frac{\mathrm{d}x^i}{\mathrm{d}\mu}\right)\partial_i, \tag{4.12}$$

which is a vector $\mathbf{w}$ whose components, $w^i = a\,\mathrm{d}x^i/\mathrm{d}\lambda + b\,\mathrm{d}x^i/\mathrm{d}\mu$, correspond to a third curve passing through the point. This curve could be parameterized by τ, and thus we may write

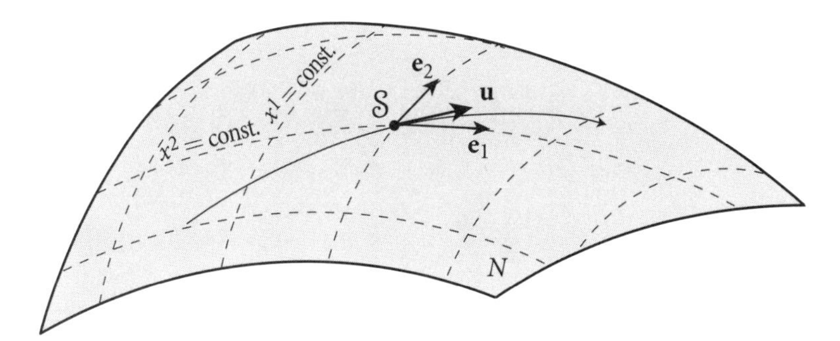

Figure 4.3: A vector $\mathbf{u}$ is identified with the directional derivative $\mathrm{d}/\mathrm{d}\lambda$ along a curve $\mathcal{S} = \mathcal{S}(\lambda)$ on the manifold N. With the introduction of a local coordinate system $x = \{x^i\}$ on N in the neighborhood of $\mathcal{S}$, we may consider a function $f(x)$, such that $g(\lambda) = f(x(\lambda))$. Then $\mathrm{d}g/\mathrm{d}\lambda = (\mathrm{d}x^i/\mathrm{d}\lambda)\,\partial_i f$. Since the function f is generic, we may write $\mathrm{d}/\mathrm{d}\lambda = (\mathrm{d}x^i/\mathrm{d}\lambda)\,\partial_i$, expressing the directional derivative in terms of partial derivatives. We make the identification $\mathbf{u} = \mathrm{d}/\mathrm{d}\lambda$, and thus we recognize the components of the vector as $u^i = \mathrm{d}x^i/\mathrm{d}\lambda$ and the basis vectors as $\mathbf{e}_i = \partial_i$, such that $\mathbf{u} = u^i\,\mathbf{e}_i$.

$$a\,\frac{\mathrm{d}}{\mathrm{d}\lambda} + b\,\frac{\mathrm{d}}{\mathrm{d}\mu} = \frac{\mathrm{d}}{\mathrm{d}\tau}. \tag{4.13}$$

This relation shows the linearity of directional derivatives and thus demonstrates that they form a linear space. Note that vectors can be added together only at the same point of the manifold, that is, they have to be members of the same tangent space $T_{\mathcal{S}}$, as illustrated in figure 4.4; addition of vectors located at two distinct points is not a defined operation.

Notation (4.7) emphasizes that vectors exist in the tangent space and are tangent to curves contained in the manifold. We maintain this image throughout the book; however, for notational convenience, we denote vector bases by $\mathbf{e}_i$ rather than by ∂_i.

4.1.3 Vector Field

A *vector field* is a rule that assigns a vector to every spatial point of a manifold. Thus, a vector field $\mathbf{u}$ on a manifold N is a mapping from the manifold to its tangent bundle TN; in other words $\mathbf{u} : N \to TN$, such that at a given spatial location $\mathcal{S} \in N$ the vector $\mathbf{u}$ is an element of the tangent space $\mathbf{u} \in T_{\mathcal{S}}$.

Given a vector field $\mathbf{u} = u^i\,\mathbf{e}_i$ at all points $\mathcal{S} \in N$, we can find a curve $x(\lambda) = \{x^i(\lambda)\}$ such that

$$\frac{\mathrm{d}x^i}{\mathrm{d}\lambda} = u^i(x), \tag{4.14}$$

as illustrated in figure 4.3. This is a set of ordinary differential equations whose solution is a curve. Since it is obtained by solving first-order differential equations (4.14), it is referred to as the *integral curve*. By construction, this curve is tangent everywhere to the vector field. There is an integral curve associated with each spatial point $\mathcal{S}$, which may be obtained by integration of the system (4.14), starting at point $\mathcal{S}$ (see, e.g., Schutz, 1980, Chapter 2.12). Curves $x^i(\lambda) = \text{const.}$ define the coordinate lines, and a manifold-filling set of integral curves is called a *congruence*.

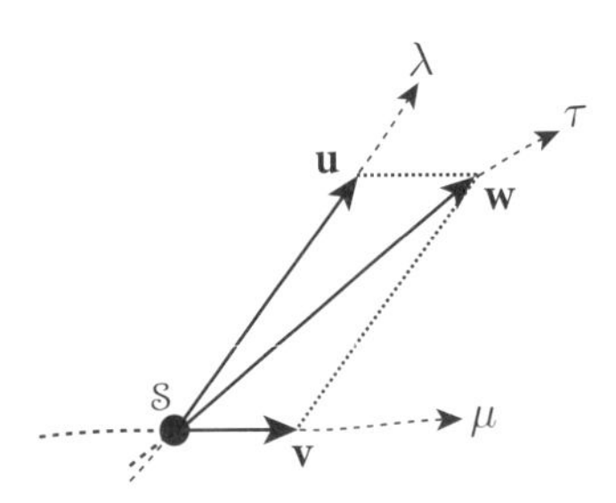

Figure 4.4: The vector $\mathbf{u}$ corresponds to $\mathrm{d}/\mathrm{d}\lambda$ evaluated at point $\mathcal{S}$ denoted by the black dot. Similarly, the vector $\mathbf{v}$ corresponds to $\mathrm{d}/\mathrm{d}\mu$. Differentiation along the curve parameterized by τ corresponds to the vector $\mathbf{w}$, which is the sum $\mathbf{w}=\mathbf{u}+\mathbf{v}$ and is equivalent to writing $\mathrm{d}/\mathrm{d}\tau=\mathrm{d}/\mathrm{d}\lambda+\mathrm{d}/\mathrm{d}\mu$. The addition operation is defined only at a single point in $T_{\mathcal{S}}$, as illustrated in figure 4.1.

4.1.4 Transformations

Let $x=\{x^i\}$ and $x'=\{x^{i'}\}$ be two overlapping systems of coordinates related via the local diffeomorphic transformations (3.11). Figure 4.5 illustrates the representation of a spatial point $\mathcal{S}$ in a manifold N in two such coordinate systems, and a vector $\mathbf{u}$ may be expressed in terms of basis vectors associated with either coordinate system.

The basis vectors associated with coordinates x^i are $\partial_i\equiv\mathbf{e}_i$, and the basis vectors associated with coordinates $x^{i'}$ are $\partial_{i'}\equiv\mathbf{e}_{i'}$. The two sets of basis vectors are related via

$$\partial_{i'}=\lambda^i{}_{i'}\,\partial_i \qquad \text{and} \qquad \partial_i=\Lambda^{i'}{}_i\,\partial_{i'}, \tag{4.15}$$

which we may write as

$$\mathbf{e}_{i'}=\mathbf{e}_i\,\lambda^i{}_{i'} \qquad \text{and} \qquad \mathbf{e}_i=\mathbf{e}_{i'}\,\Lambda^{i'}{}_i, \tag{4.16}$$

where we have defined the transformation "matrices,"

$$\lambda^i{}_{i'}\equiv\partial_{i'}\varphi^i \qquad \text{and} \qquad \Lambda^{i'}{}_i\equiv\partial_i\Phi^{i'}. \tag{4.17}$$

The order of symbols in equation (4.16) emphasizes the repeated subscript and superscript: the placement of the matrix elements $\lambda^i{}_{i'}$ and $\Lambda^{i'}{}_i$ is such that repeated upper and lower indices are adjacent to one another. Matrices $\lambda^i{}_{i'}$ and $\Lambda^{i'}{}_i$ are inverses of one another, in the sense that

$$\lambda^i{}_{i'}\,\Lambda^{i'}{}_j=\delta^i{}_j \qquad \text{and} \qquad \Lambda^{i'}{}_i\,\lambda^i{}_{j'}=\delta^{i'}{}_{j'}. \tag{4.18}$$

Here, we encounter the *Kronecker delta symbol*, whose property is

$$\delta^i{}_j=\begin{cases}1 & i=j\\ 0 & i\neq j\end{cases}. \tag{4.19}$$

The Kronecker delta symbol corresponds to the identity tensor, as described in section 5.4. Since a vector is a well-defined geometrical object and there is an infinity of orientations of

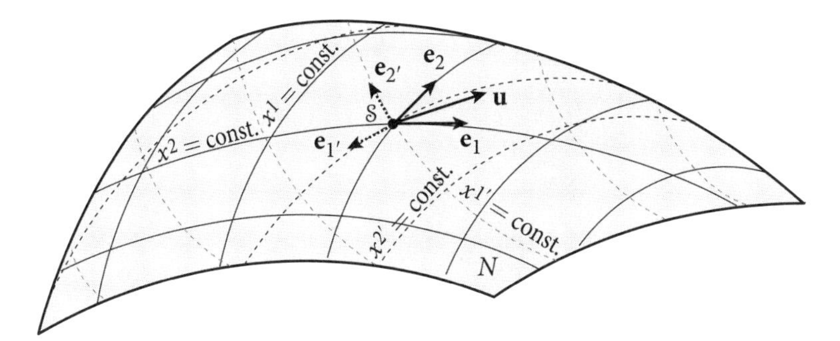

Figure 4.5: Two overlapping coordinate systems or charts are used to identify a spatial point $\mathcal{S}$ in the manifold N: $\{x^1, x^2\}$ and $\{x^{1'}, x^{2'}\}$. The basis vectors associated with the two coordinate systems are $\{\mathbf{e}_1, \mathbf{e}_2\}$ and $\{\mathbf{e}_{1'}, \mathbf{e}_{2'}\}$, respectively. Just as a point $\mathcal{S}$ may be identified in either coordinate system based on relationships $\mathcal{S}(x^1, x^2)$ and $\mathcal{S}(x^{1'}, x^{2'})$, respectively, a vector $\mathbf{u}$ may be expressed in terms of the basis vectors associated with either coordinate system: $\mathbf{u} = u^i\,\mathbf{e}_i = u^{i'}\,\mathbf{e}_{i'}$.

coordinate systems, it follows that a given vector can be expressed in a multitude of systems. Herein, we focus on the process of changing components. We may express the vector $\mathbf{u}$ at spatial point $\mathcal{S}$ in terms of either coordinate system as

$$\mathbf{u}(\mathcal{S}) = u^i(x)\,\mathbf{e}_i(x) = u^{i'}(x')\,\mathbf{e}_{i'}(x'). \tag{4.20}$$

Note how the vector itself is a function of the spatial point $\mathcal{S}$, whereas the components and basis vectors depend explicitly on the chosen coordinate system, that is, x or x'. Basis vectors and components of vectors with unprimed indices, that is, $\mathbf{e}_i$ and u^i, depend implicitly on x, and basis vectors and components of vectors with primed indices, that is, $\mathbf{e}_{i'}$ and $u^{i'}$, depend implicitly on x'. With this understanding, we generally write less cluttered expressions such as

$$\mathbf{u} = u^i\,\mathbf{e}_i = u^{i'}\,\mathbf{e}_{i'}. \tag{4.21}$$

Since the basis vectors transform according to rules (4.16), it follows from equation (4.21) that the components of the vector $\mathbf{u}$ transform according to

$$u^{i'} = \Lambda^{i'}{}_i\,u^i, \qquad u^i = \lambda^i{}_{i'}\,u^{i'}. \tag{4.22}$$

Examining equations (4.16) and (4.22), we see that bases transform with the inverses of matrices used in the transformation of components, and vice versa.

Example 4.2 Material Velocity Transformations

In example 4.1, we encountered the material velocity, $\mathbf{v}$, of a continuum. The corresponding vector field, $\mathbf{v}(\mathcal{S}, t)$, defines the *flow of matter*, which may be expressed in either Eulerian or Lagrangian coordinates as

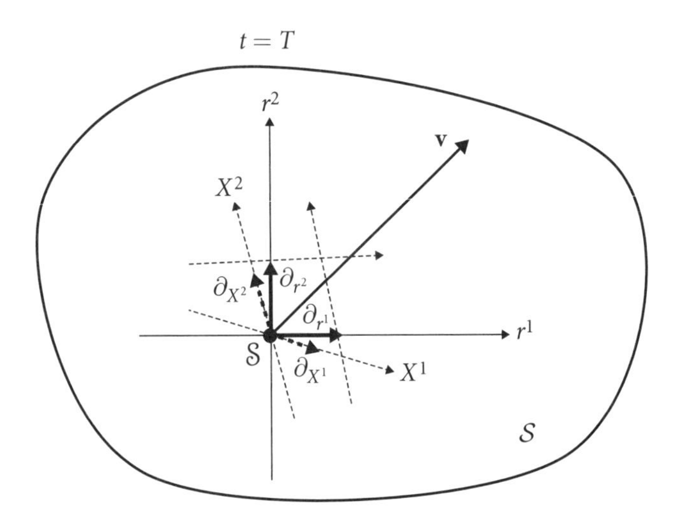

Figure 4.6: In continuum mechanics, a spatial point $\mathcal{S}$ in the spatial manifold $\mathcal{S}$ at time $t = T$ may be identified with either a set of Eulerian coordinates $\{r^i\}$ or a set of Lagrangian coordinates $\{X^I\}$, as previously illustrated in figure 3.2. These coordinates induce a set of Eulerian basis vectors $\{\partial_{r^i}\}$ (shown in solid arrows) or Lagrangian basis vectors $\{\partial_{X^I}\}$ (shown in dashed arrows) in the tangent space at $\mathcal{S}$. The material velocity $\mathbf{v}$ (shown by the thin arrow) is a geometrical object that lives in the tangent space at $\mathcal{S}$ and may be expressed in either set of basis vectors, as stated mathematically in equation (4.23). The Eulerian coordinates are chosen to be Cartesian in this example, but this is not required.

$$\mathbf{v} = v^i\,\mathbf{e}_i = v^I\,\mathbf{e}_I, \tag{4.23}$$

as illustrated in figure 4.6. Analogous to definition (4.10) of the Eulerian basis vectors $\mathbf{e}_i$, Lagrangian basis vectors $\mathbf{e}_I$ are defined in terms of the Lagrangian coordinates $\{X^I\}$ as

$$\mathbf{e}_I \equiv \partial_I, \tag{4.24}$$

where we have used the shorthand notation $\partial_I = \partial_{X^I}$.

Eulerian and Lagrangian basis vectors are related via the transformations

$$\mathbf{e}_I = F^i{}_I\,\mathbf{e}_i \qquad \text{and} \qquad \mathbf{e}_i = (F^{-1})^I{}_i\,\mathbf{e}_I, \tag{4.25}$$

where we have defined the *deformation gradient* in terms of the motion (3.12) as

$$F^i{}_I \equiv \partial_I \varphi^i, \tag{4.26}$$

with an inverse defined in terms of the inverse motion (3.13) as

$$(F^{-1})^I{}_i \equiv \partial_i \Phi^I. \tag{4.27}$$

Since the basis vectors transform according to rules (4.25), the components of the material velocity $\mathbf{v}$ transform according to

$$v^I = (F^{-1})^I{}_i\, v^i, \qquad v^i = F^i{}_I\, v^I. \tag{4.28}$$

These transformations are a specific example of the general vector transformation laws (4.22).

4.2 One-Forms

A *one-form* $\boldsymbol{\omega}$ at point $\mathcal{S}$ is a linear map from vectors in $T_{\mathcal{S}}$ to the real numbers $\mathbb{R}$. In other words, a one-form is a linear real-valued function of vectors. Thus, given two vectors **u** and **w** and two real numbers a and b,

$$\boldsymbol{\omega}\,(a\,\mathbf{u} + b\,\mathbf{w}) = a\,\boldsymbol{\omega}\,(\mathbf{u}) + b\,\boldsymbol{\omega}\,(\mathbf{w}), \tag{4.29}$$

where $\boldsymbol{\omega}\,(\mathbf{u})$ and $\boldsymbol{\omega}\,(\mathbf{w})$ are real numbers.[4] One-forms define a linear space, such that for two one-forms $\boldsymbol{\omega}$ and $\boldsymbol{\eta}$ we may write

$$(a\,\boldsymbol{\omega} + b\,\boldsymbol{\eta})(\mathbf{u}) = a\,\boldsymbol{\omega}(\mathbf{u}) + b\,\boldsymbol{\eta}(\mathbf{u}), \tag{4.30}$$

which is similar to equation (4.1) for vectors. One-forms are also referred to as *dual* vectors, for the reason illustrated by equation (4.31), below, and as differential one-forms.

4.2.1 Duality

The *dual* to the tangent space $T_{\mathcal{S}}$ of vectors at a spatial point $\mathcal{S}$ on a manifold N is its *cotangent space*, also known as a *dual space* $T^*_{\mathcal{S}}$ of one-forms, and the collection of these dual spaces at all points is the *cotangent bundle*, which we denote by T^*N. A *one-form field* $\boldsymbol{\omega}$ on a manifold N is a mapping from the manifold to its cotangent bundle T^*N, that is, $\boldsymbol{\omega} : N \to T^*N$, such that at $\mathcal{S} \in N$, the one-form $\boldsymbol{\omega}$ is an element of the cotangent space, $\boldsymbol{\omega} \in T^*_{\mathcal{S}}$.

Cotangent spaces are geometric entities associated with manifolds, independent of functions that relate them to manifolds and tangent spaces. In other words, cotangent spaces are distinct geometrical objects. For example, the Lagrangian $L = L(x^i, p_i)$ in classical mechanics depends on generalized coordinates x^i, which define a point $\mathcal{S}$ in a manifold N, and generalized momenta $p_i \equiv \partial L/\partial x^i$, which define one-forms that live in the cotangent space $T^*_{\mathcal{S}}$, at that point.

Following Misner et al. (2017), we view a one-form $\boldsymbol{\omega}$, as a machine with a single "slot" that takes a vector **u**, to produce a real-valued number $\boldsymbol{\omega}(\mathbf{u})$. By analogy, one can regard a vector **u**, as a linear real-valued function of one-forms $\boldsymbol{\omega}$: a linear machine with a single "slot" that accepts one-forms to produce a real-valued number $\mathbf{u}(\boldsymbol{\omega})$. Since the same number is produced in both cases, we write

$$\mathbf{u}(\boldsymbol{\omega}) = \boldsymbol{\omega}(\mathbf{u}) \equiv \langle \boldsymbol{\omega}, \mathbf{u} \rangle, \tag{4.31}$$

[4] In this book, we use Greek letters in bold font to denote one-forms.

which shows that vectors and one-forms are *duals* of one another. The product $\langle \boldsymbol{\omega}, \mathbf{u} \rangle$ in equation (4.31) is called the *duality product*. It is linear in both slots in the sense that—for two vectors, $\mathbf{u}$ and $\mathbf{w}$, two one-forms, $\boldsymbol{\omega}$ and $\boldsymbol{\eta}$, and four real-valued numbers, a, b, c, and d—we have

$$\langle a\,\boldsymbol{\omega} + b\,\boldsymbol{\eta}, c\,\mathbf{u} + d\,\mathbf{w} \rangle = a\,c\,\langle \boldsymbol{\omega}, \mathbf{u} \rangle + a\,d\,\langle \boldsymbol{\omega}, \mathbf{w} \rangle + b\,c\,\langle \boldsymbol{\eta}, \mathbf{u} \rangle + b\,d\,\langle \boldsymbol{\eta}, \mathbf{w} \rangle. \tag{4.32}$$

Note that the first slot of the duality product accepts only one-forms, whereas the second slot accepts only vectors.

4.2.2 Bases

Let us define a one-form basis $\mathbf{e}^i$, by the *duality condition*,

$$\langle \mathbf{e}^i, \mathbf{e}_j \rangle = \delta^i{}_j, \tag{4.33}$$

where $\mathbf{e}_j$ is a vector basis.

In a manner analogous to equation (4.3), we write a one-form $\boldsymbol{\omega}$ in terms of its components ω_i and basis $\mathbf{e}^i$ as

$$\boldsymbol{\omega}(\mathcal{S}) = \omega_i(x)\,\mathbf{e}^i(x), \tag{4.34}$$

where again we emphasize the dependence of the one-form $\boldsymbol{\omega}$ on spatial points $\mathcal{S}$, independent of any coordinate system. With the implicit understanding that components and bases with unprimed indices are functions of coordinates x, we write the last equation in the less cluttered form

$$\boldsymbol{\omega} = \omega_i\,\mathbf{e}^i. \tag{4.35}$$

Given a vector $\mathbf{u} = u^j\,\mathbf{e}_j$, and a one-form in equation (4.35), we write their duality product, given in formula (4.31), as

$$\langle \boldsymbol{\omega}, \mathbf{u} \rangle = \langle \omega_i\,\mathbf{e}^i, u^j\,\mathbf{e}_j \rangle = \omega_i\,u^j \langle \mathbf{e}^i, \mathbf{e}_j \rangle = \omega_i\,u^j\,\delta^i{}_j = \omega_i\,u^i. \tag{4.36}$$

The components u^i of vector $\mathbf{u}$ are

$$u^i = \langle \mathbf{e}^i, \mathbf{u} \rangle, \tag{4.37}$$

and the components ω_i of one-form $\boldsymbol{\omega}$ are

$$\omega_i = \langle \boldsymbol{\omega}, \mathbf{e}_i \rangle. \tag{4.38}$$

Thus, there are two standard types of bases: vector and one-form bases. In Cartesian coordinates, the former consists of vectors tangent to the axes, the latter of sheets parallel to the planes. An illustration of vector and one-form bases and their duality condition in $\mathbb{R}^3$ is given in figure 4.7.

In equation (4.7), we stated that partial derivatives along the coordinate axes ∂_i can be identified with a basis vector $\mathbf{e}_i$ of a tangent space. In a manner similar to the approach discussed on page 17, consider the *differential* or *exterior derivative* of the function f in

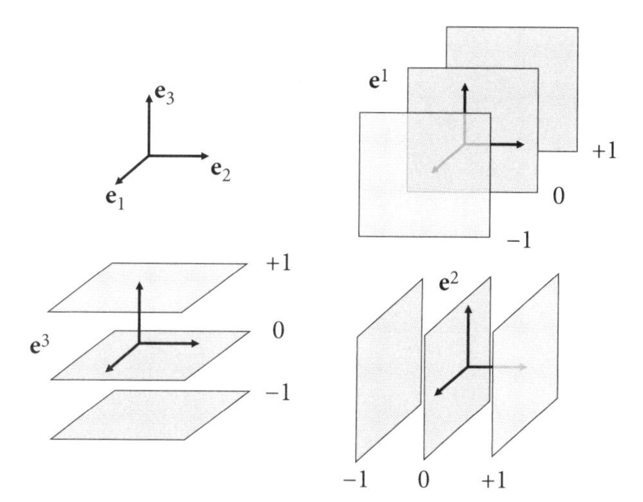

Figure 4.7: In Cartesian coordinates, basis vectors $\mathbf{e}_1$, $\mathbf{e}_2$, and $\mathbf{e}_3$ are denoted by three orthogonal arrows, as illustrated in the top left. Basis one-forms $\mathbf{e}^1$, $\mathbf{e}^2$, and $\mathbf{e}^3$ are denoted by sets of "sheets". Basis vectors and one-forms satisfy the duality condition (4.33), namely, $\langle \mathbf{e}^i, \mathbf{e}_j \rangle = \delta^i{}_j$. For example, in the bottom left the basis vector $\mathbf{e}_3$ "pierces" one level of the basis one-form $\mathbf{e}^3$, since $\langle \mathbf{e}^3, \mathbf{e}_3 \rangle = 1$, whereas it pierces zero levels of the basis forms $\mathbf{e}^1$ and $\mathbf{e}^2$, since $\langle \mathbf{e}^1, \mathbf{e}_3 \rangle = 0$ and $\langle \mathbf{e}^2, \mathbf{e}_3 \rangle = 0$. Similar duality relations are illustrated for the basis one-forms $\mathbf{e}^1$ (top right) and $\mathbf{e}^2$ (bottom right). The vector basis and the one-form basis are drawn at right angles to one another; this need not be the case in a general manifold. Adapted from Albert Tarantola.

coordinates x, that is, $f(x)$:

$$\mathrm{d}f = \partial_i f \mathrm{d}x^i. \tag{4.39}$$

Comparing this expression to equation (4.35), we identify df with a one-form, $\partial_i f$ with its components, and $\mathrm{d}x^i$ with the associated basis. In other words, equation (4.39) states the one-form field df, with respect to the basis $\mathrm{d}x^i$; it is not a first-order approximation to a change in f.

Just as vectors may be visualized as "arrows" emanating from a point, one-forms may be regarded as "sheets" centered on the same point. Using this image, the duality product is illustrated in figure 4.7. One may think of the value of the duality product between one-form ω and vector $\mathbf{u}$ as the number of sheets of ω "pierced" by arrow $\mathbf{u}$. The addition of one-forms and the nature of the duality product are illustrated further in figure 4.8.

Analogous to the identification of basis vector $\mathbf{e}_i$ with the partial derivative in the x^i direction, ∂_i, herein, we appreciate the geometrical insight into the basis of one-forms implied by the identification

$$\mathbf{e}^i \equiv \mathrm{d}x^i; \tag{4.40}$$

notably, we can view $\mathrm{d}x^i$ as a collection of sheets on which coordinate x^i is constant. The action of basis one-form $\mathrm{d}x^i$ on $\mathbf{u}$ returns the components of the vector:

$$\mathrm{d}x^i(\mathbf{u}) = u^i. \tag{4.41}$$

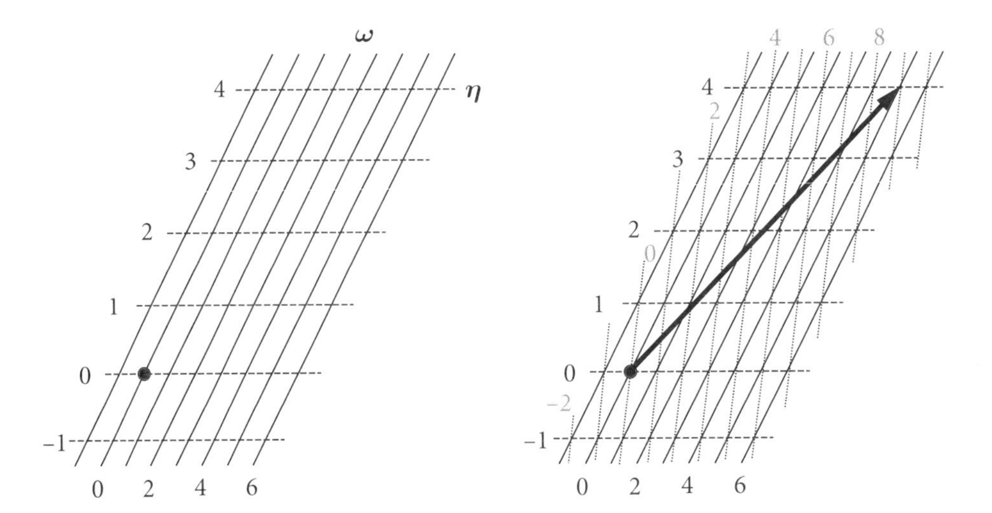

Figure 4.8: A graphical rule in two dimensions for the sum of two one-forms, represented by a linear collection of sheets. (A rule for the graphical sum of vectors, represented by arrows, is illustrated in figure 4.4.) Left: one-form $\boldsymbol{\omega}$ is represented by the solid lines, one-form $\boldsymbol{\eta}$ is represented by the dashed lines. Right: $\boldsymbol{\omega}+\boldsymbol{\eta}$ is represented by the dotted lines. The vector $\mathbf{u}$ is represented by the thick black arrow. Counting the number of levels of a form "pierced" by the vector, one deduces that $\langle\boldsymbol{\omega},\mathbf{u}\rangle=5$, $\langle\boldsymbol{\eta},\mathbf{u}\rangle=4$, and $\langle\boldsymbol{\omega}+\boldsymbol{\eta},\mathbf{u}\rangle=9$, in agreement with a defining property for the sum of one-forms: $\langle\boldsymbol{\omega}+\boldsymbol{\eta},\mathbf{u}\rangle=\langle\boldsymbol{\omega},\mathbf{u}\rangle+\langle\boldsymbol{\eta},\mathbf{u}\rangle$.

Similarly, the action of the basis vector ∂_i on $\boldsymbol{\omega}$ returns the components of the one-form:

$$\partial_i(\boldsymbol{\omega})=\omega_i. \tag{4.42}$$

Since these are alternative ways of writing equations (4.37) and (4.38), the action of basis one-forms on basis vectors and vice versa is

$$\mathrm{d}x^i(\partial_{x^j})=\delta^i{}_j \qquad \text{and} \qquad \partial_{x^i}(\mathrm{d}x^j)=\delta^j{}_i, \tag{4.43}$$

which are fancy ways of writing the duality condition (4.33). For notational convenience, we denote basis one-forms by $\mathbf{e}^i$ rather than by $\mathrm{d}x^i$.

4.2.3 Transformations

The correspondence between one-forms and vectors allows us to write the transformation rules for bases and components. By analogy to rules (4.16) and (4.22), we have

$$\mathbf{e}^{i'}=\Lambda^{i'}{}_i\,\mathbf{e}^i \qquad \text{and} \qquad \mathbf{e}^i=\lambda^i{}_{i'}\,\mathbf{e}^{i'}, \tag{4.44}$$

and

$$\omega_{i'}=\omega_i\,\lambda^i{}_{i'} \qquad \text{and} \qquad \omega_i=\omega_{i'}\,\Lambda^{i'}{}_i, \tag{4.45}$$

respectively. Examining these two transformations, we see that transformation matrices for the components and bases of one-forms are inverses of one another, as is the case for vectors.

In old terminology, one-forms, whose components transform according to formula (4.45), are referred to as *covariant vectors*, *covectors*, and *dual vectors*. We do not use these terms because they contain the word "vector," which is contrary to our mental image of a one-form as a collection of sheets. Also, as articulated by Schutz (1980) on page 62,

> The modern viewpoint emphasizes the fact that neither the vector nor the one-form is in fact changed by a basis transformation: they are coordinate-independent geometrical objects. Therefore, modern terminology has dropped the old names because they overemphasize the coordinate-dependent description of these objects.

Example 4.3 One-Forms in Continuum Mechanics

One-forms naturally arise in continuum mechanics. Consider the differential of a scalar field q, for example, the mass density. In Eulerian coordinates, $\{r^i, t\}$, we express the functional dependence of this field as $q(r^i, t)$, so we have the differential

$$\mathrm{d}q = \partial_i q \, \mathrm{d}r^i. \tag{4.46}$$

Alternatively, in Lagrangian coordinates, $\{X^I, T\}$, we express the functional dependence of this field as $Q(X^I, T)$, and we have

$$\mathrm{d}q = \partial_I Q \, \mathrm{d}X^I. \tag{4.47}$$

Using the chain rule, being mindful of the nature of coordinate "slots," we have the relationship

$$\partial_I Q = F^i{}_I \, \partial_i q, \tag{4.48}$$

where $F^i{}_I$ denotes elements of the deformation gradient (4.26). Because equations (4.46) and (4.47) both represent the *same* differential scalar field dq, the Eulerian and Lagrangian differentials $\mathrm{d}r^i$ and $\mathrm{d}X^I$ must be related via

$$\mathrm{d}X^I = (F^{-1})^I{}_i \, \mathrm{d}r^i, \qquad \mathrm{d}r^i = F^i{}_I \, \mathrm{d}X^I. \tag{4.49}$$

Upon comparing these expressions to the transformation rule for basis vectors (4.25), we note that the rules appear to be "reversed." This motivates us to introduce two new sets of basis elements defined in terms of differentials $\mathrm{d}r^i$ and $\mathrm{d}X^I$, namely,

$$\mathbf{e}^i \equiv \mathrm{d}r^i, \tag{4.50}$$

and

$$\mathbf{e}^I \equiv \mathrm{d}X^I. \tag{4.51}$$

We conclude that the differential dq should be regarded as a new form of tensor called a one-form. Such a $(0, 1)$ tensor, say ω, may be expressed in either Eulerian or

Lagrangian coordinates as

$$\boldsymbol{\omega} = \omega_i\, \mathbf{e}^i = \omega_I\, \mathbf{e}^I, \tag{4.52}$$

and its components transform according to the rules

$$\omega_I = \omega_i\, F^i{}_I, \qquad \omega_i = \omega_I\, (F^{-1})^I{}_i. \tag{4.53}$$

Duality of the Eulerian and Lagrangian basis vectors and one-forms may be expressed as

$$\mathrm{d}r^i(\,\partial_{r^j}\,) = \delta^i{}_j \qquad \text{or} \qquad \partial_{r^i}(\,\mathrm{d}r^j\,) = \delta^j{}_i, \tag{4.54}$$

and

$$\mathrm{d}X^I(\,\partial_{X^J}\,) = \delta^I{}_J \qquad \text{or} \qquad \partial_{X^I}(\,\mathrm{d}X^J\,) = \delta^J{}_I. \tag{4.55}$$

4.3 Alternative Perspective

In this section, we present an alternative perspective on vectors and bases (see also Misner et al., 2017, §2.3). A curve in a manifold, parameterized in terms of a generating parameter λ, assigns a point $\mathcal{S}(\lambda)$, to each value of λ. A tangent vector to this curve is defined by

$$\mathbf{u} \equiv \frac{\mathrm{d}\mathcal{S}}{\mathrm{d}\lambda}. \tag{4.56}$$

To each point $\mathcal{S}$ we assign coordinates $x = \{x^i\}$ according to rule (3.1), with inverse (3.2). At every $\mathcal{S}$, we consider the tangent space $T_{\mathcal{S}}$, which contains vectors whose origin is at the tangency point, as illustrated in figure 4.1. Using the notation of Cartan (see, e.g., Misner et al., 2017), the natural vector basis $\{\mathbf{e}_i\}$ associated with the coordinate system $\{x^i\}$ is defined by

$$\mathbf{e}_i \equiv \frac{\partial \mathcal{S}}{\partial x^i}. \tag{4.57}$$

Regarding point $\mathcal{S}$ as a function of coordinates $\{x^i\}$, as in rule (3.1), the basis vector $\mathbf{e}_i$, points in the direction in which $\mathcal{S}$ moves along the curve if x^i is incremented while holding the remaining coordinates fixed.

Considering a cotangent space $T^*_{\mathcal{S}}$, and using again the notation of Cartan, the natural one-form basis associated with the coordinate system $\{x^i\}$, is defined by

$$\mathbf{e}^i \equiv \frac{\mathrm{d}x^i}{\mathrm{d}\mathcal{S}}. \tag{4.58}$$

Regarding coordinates $\{x^i\}$ as a function of $\mathcal{S}$, as in (3.2), the basis form $\mathbf{e}^i$ identifies level surfaces on which coordinate x^i is constant as the point $\mathcal{S}$ moves along the curve. Taking the differential of a function f naturally leads to the notion of a one-form:

$$\frac{\mathrm{d}f}{\mathrm{d}\mathcal{S}} = \frac{\partial f}{\partial x^i}\,\frac{\mathrm{d}x^i}{\mathrm{d}\mathcal{S}}, \tag{4.59}$$

which is an alternative view of equation (4.39).

Basis vectors and one-forms are *duals* of one another, in the sense that

$$\frac{\mathrm{d}x^i}{\mathrm{d}\mathcal{S}}\frac{\partial \mathcal{S}}{\partial x^j} = \frac{\partial x^i}{\partial x^j} = \delta^i{}_j, \tag{4.60}$$

which is condition (4.33),

$$\langle \mathbf{e}^i, \mathbf{e}_j \rangle \equiv \frac{\mathrm{d}x^i}{\mathrm{d}\mathcal{S}}\frac{\partial \mathcal{S}}{\partial x^j}, \tag{4.61}$$

as expected.

Example 4.4 Bases in General Relativity

A curve in a spacetime manifold, called a *worldline*, assigns an event $\mathcal{S}(\tau)$, to each value of the *proper time* τ. The *four-velocity* is a vector tangent to this curve defined by (e.g., Misner et al., 2017)

$$\mathbf{u} \equiv \frac{\mathrm{d}\mathcal{S}}{\mathrm{d}\tau}. \tag{4.62}$$

The natural vector basis $\{\mathbf{e}_\mu\}$ associated with the spacetime coordinate system $\{x^\mu\}$ is defined by

$$\mathbf{e}_\mu \equiv \frac{\partial \mathcal{S}}{\partial x^\mu}, \tag{4.63}$$

and the natural one-form basis $\{\mathbf{e}^\mu\}$ associated with this spacetime coordinate system is defined by

$$\mathbf{e}^\mu \equiv \frac{\mathrm{d}x^\mu}{\mathrm{d}\mathcal{S}}. \tag{4.64}$$

Basis vectors and one-forms are *duals* in the sense

$$\frac{\mathrm{d}x^\mu}{\mathrm{d}\mathcal{S}}\frac{\partial \mathcal{S}}{\partial x^\nu} = \frac{\partial x^\mu}{\partial x^\nu} = \delta^\mu{}_\nu. \tag{4.65}$$

4.4 Lie Bracket

The duality product (4.31) of a form, $\boldsymbol{\omega}$, and a vector, $\mathbf{u}$, produces a scalar, $\langle \boldsymbol{\omega}, \mathbf{u} \rangle$. Let us introduce an operation involving two vectors, $\mathbf{u}$ and $\mathbf{w}$, that produces a third vector in the same tangent space, the *Lie bracket* $[\mathbf{u}, \mathbf{w}]$.

Consider two vectors, $\mathbf{u}$ and $\mathbf{w}$, along curves defined by parameters, λ and μ, respectively, as illustrated in figure 4.9. The Lie bracket is defined by the *commutator*

$$[\mathbf{u}, \mathbf{w}] \equiv \frac{\mathrm{d}}{\mathrm{d}\mu}\frac{\mathrm{d}}{\mathrm{d}\lambda} - \frac{\mathrm{d}}{\mathrm{d}\lambda}\frac{\mathrm{d}}{\mathrm{d}\mu}. \tag{4.66}$$

If we express the vectors $\mathbf{u}$ and $\mathbf{w}$ in components relative to a basis, it follows that

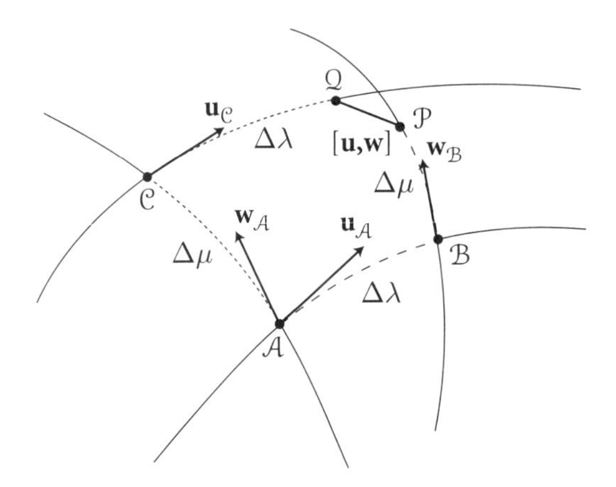

Figure 4.9: The Lie bracket of vectors $\mathbf{u}$, with a congruence generated by a parameter λ, and $\mathbf{w}$, with a congruence generated by a parameter μ, is denoted by the vector $[\mathbf{u}, \mathbf{w}]$. At point $\mathcal{A}$, the vectors are labeled $\mathbf{u}_{\mathcal{A}}$ and $\mathbf{v}_{\mathcal{A}}$. If we travel along the long-dashed integral curve defined by $\mathbf{u}_{\mathcal{A}}$ a step $\Delta\lambda$, we arrive at location $\mathcal{B}$, and if we continue from there along the long-dashed integral curve defined by $\mathbf{v}_{\mathcal{B}}$ a step $\Delta\mu$, we arrive at location $\mathcal{P}$. In contrast, if we travel along the short-dashed integral curve defined by $\mathbf{v}_{\mathcal{A}}$ a step $\Delta\mu$, we arrive at location $\mathcal{C}$, and if we continue from there along the short-dashed integral curve defined by $\mathbf{u}_{\mathcal{C}}$ a step $\Delta\lambda$, we arrive at location $\mathcal{Q}$. In general, points $\mathcal{P}$ and $\mathcal{Q}$ do not coincide, and the quadrilateral is not closed: $[\mathbf{u}, \mathbf{w}] \neq \mathbf{0}$. If $\mathbf{e}_i$ and $\mathbf{e}_j$ are basis vectors of a holonomic coordinate system, $[\mathbf{e}_i, \mathbf{e}_j] = \mathbf{0}$, whereas for an anholonomic system, $[\mathbf{e}_i, \mathbf{e}_j] = \tau_{ij}^k\, \mathbf{e}_k$, where τ_{ij}^k denote the structure coefficients.

$$[\mathbf{u}, \mathbf{w}] = (u^i\, \partial_i w^j - w^i\, \partial_i u^j)\, \partial_j, \tag{4.67}$$

illustrating that the Lie bracket of the vectors $\mathbf{u} = u^i\, \partial_i$ and $\mathbf{w} = w^i\, \partial_i$ produces a third vector $\mathbf{v} = v^j\, \partial_j$ with components $v^j = u^i\, \partial_i w^j - w^i\, \partial_i u^j$. The Lie bracket is antisymmetric:

$$[\mathbf{u}, \mathbf{w}] = -\,[\mathbf{w}, \mathbf{u}]. \tag{4.68}$$

Applied to a function f, the Lie bracket is the difference between taking the derivative of f in the direction of $\mathbf{u}$ followed by the derivative in the direction of $\mathbf{w}$ and the derivative in the direction of $\mathbf{w}$ followed by the derivative in the direction of $\mathbf{u}$; it is generally nonzero due to noncommutativity. The Lie bracket $[\mathbf{u}, \mathbf{w}]$, illustrated in figure 4.9, may be viewed as the difference between the function evaluated at points $\mathcal{P}$ and $\mathcal{Q}$, which results in a function evaluated at $\mathcal{A}$,

$$\mathrm{f}(\mathcal{P}) - \mathrm{f}(\mathcal{Q}) = [\mathbf{u}, \mathbf{w}]\, \mathrm{f}\, \big|_{\mathcal{A}}, \tag{4.69}$$

plus higher-order terms. The accuracy of this description increases as $\mathbf{u}$ and $\mathbf{w}$ shrink.

Thus far in this section, we have tacitly assumed that the Lie bracket of two basis vectors, $\mathbf{e}_i = \partial_i$ and $\mathbf{e}_j = \partial_j$, is zero due to the commutativity of partial derivatives:

$$[\mathbf{e}_i, \mathbf{e}_j] = \mathbf{0}, \tag{4.70}$$

implying that

$$\partial_i \partial_j - \partial_j \partial_i = 0. \tag{4.71}$$

Such a basis is called *holonomic*. Sometimes we consider basis vectors which are *nonholonomic*, or *anholonomic*, in which case

$$[\mathbf{e}_i, \mathbf{e}_j] = \tau_{ij}^{k}\, \mathbf{e}_k, \tag{4.72}$$

or

$$\partial_i \partial_j - \partial_j \partial_i = \tau_{ij}^{k}\, \partial_k. \tag{4.73}$$

The parameters τ_{ij}^{k} are known as *structure coefficients*. In a nonholonomic basis, the Lie bracket of two vectors (4.67) becomes

$$[\mathbf{u}, \mathbf{w}] = (u^i\, \partial_i w^j - w^i\, \partial_i u^j + u^i\, w^k\, \tau_{ik}^{j})\, \partial_j. \tag{4.74}$$

Structure coefficients are antisymmetric in their lower two indices.

Example 4.5 Anholonomicity

Consider the transformation from Cartesian coordinates $\{x, y, z\}$ to spherical coordinates $\{r, \theta, \phi\}$. The associated holonomic basis vectors are $\{\partial_x, \partial_y, \partial_z\}$ and $\{\partial_r, \partial_\theta, \partial_\phi\}$, respectively. We have the relationships

$$x = r\, \sin\theta \cos\phi, \qquad y = r\, \sin\theta\, \sin\phi, \qquad z = r\, \cos\theta, \tag{4.75}$$

with inverse

$$r = \sqrt{x^2 + y^2 + z^2}, \qquad \theta = \arctan\left(\sqrt{x^2 + y^2}/z\right), \\ \phi = \arctan(y/x). \tag{4.76}$$

The spherical basis vectors are related to the Cartesian basis vectors via

$$\mathbf{e}_r \equiv \partial_r = \sin\theta \cos\phi\, \partial_x + \sin\theta \sin\phi\, \partial_y + \cos\theta\, \partial_z, \tag{4.77}$$

$$\mathbf{e}_\theta \equiv \partial_\theta = r\, \cos\theta \cos\phi\, \partial_x + r\, \cos\theta \sin\phi\, \partial_y - r\, \sin\theta\, \partial_z, \tag{4.78}$$

$$\mathbf{e}_\phi \equiv \partial_\phi = -r\, \sin\theta \sin\phi\, \partial_x + r\, \sin\theta \cos\phi\, \partial_y. \tag{4.79}$$

It is important to note that these basis vectors are not all "unit"[a] vectors. Specifically,

$$\mathbf{e}_r = \hat{\mathbf{r}}, \qquad \mathbf{e}_\theta = r\, \hat{\boldsymbol{\theta}}, \qquad \mathbf{e}_\phi = r\, \sin\theta\, \hat{\boldsymbol{\phi}}, \tag{4.80}$$

where $\hat{\mathbf{r}}$, $\hat{\boldsymbol{\theta}}$, and $\hat{\boldsymbol{\phi}}$ denote traditional unit vectors in the directions of increasing r, θ, and ϕ, respectively. The basis vectors $\{\mathbf{e}_r, \mathbf{e}_\theta, \mathbf{e}_\phi\}$ are holonomic, thanks to the commutativity of the partial derivatives ∂_r, ∂_θ, and ∂_ϕ. However, the unit basis vectors

$$\hat{\mathbf{r}} = \partial_r, \qquad \hat{\boldsymbol{\theta}} = r^{-1}\,\partial_\theta, \qquad \hat{\boldsymbol{\phi}} = (r\,\sin\theta)^{-1}\,\partial_\phi, \tag{4.81}$$

are anholonomic:

$$[\hat{\mathbf{r}}, \hat{\boldsymbol{\theta}}] = -\,r^{-1}\,\hat{\boldsymbol{\theta}}, \qquad [\hat{\mathbf{r}}, \hat{\boldsymbol{\phi}}] = -\,r^{-1}\,\hat{\boldsymbol{\phi}},$$
$$[\hat{\boldsymbol{\theta}}, \hat{\boldsymbol{\phi}}] = -\,r^{-1}\,\cot\theta\,\hat{\boldsymbol{\phi}}. \tag{4.82}$$

The main conceptual difference is that a holonomic basis is integrable, whereas an anholonomic basis is non-integrable.

[a] See section 5.10 for a discussion on *length*.

Example 4.6 Compatibility

A smooth motion (3.12) satisfies the *compatibility condition*

$$\begin{aligned} \alpha_{IJ}{}^{i} &\equiv (\partial_I\partial_J - \partial_J\partial_I)\,\varphi^i \\ &= 0. \end{aligned} \tag{4.83}$$

Equation (4.83) states that partial derivatives of the motion with respect to Lagrangian coordinates $\{X^I\}$ *commute*. For an *incompatible motion*, the elements $\alpha_{IJ}{}^{i}$ are nonzero and define the *incompatibility tensor*. Such a situation involves *material defects* in the form of *dislocations* and *disclinations*, as discussed in example 8.5.

For compatible motion, the Lie bracket of two Eulerian or Lagrangian basis vectors is zero

$$[\mathbf{e}_i, \mathbf{e}_j] = \mathbf{0}, \qquad [\mathbf{e}_I, \mathbf{e}_J] = \mathbf{0}, \tag{4.84}$$

due to the commutativity of partial derivatives:

$$\partial_i\partial_j - \partial_j\partial_i = 0, \qquad \partial_I\partial_J - \partial_J\partial_I = 0. \tag{4.85}$$

Chapter 5

TENSORS

Nothing in life is to be feared, it is only to be understood.
Now is the time to understand more, so that we may fear less.
—MARIE CURIE

In this chapter, we extend the concepts of vectors and one-forms introduced in chapter 4 to *tensors*, which include scalars, vectors, and one-forms as exceptional cases. Tensors are fundamental to the language of physics and form a vital part of its mathematical structure. Like vectors and one-forms, tensors are invariant under coordinate transformations, and their magnitude remains constant regardless of the coordinate system used. To preserve this invariance, tensor components transform in a manner that maintains the tensor's magnitude. While tensor-transformation rules are often used to define tensors, we adopt a different perspective in this book. We consider tensors as geometrical objects associated with spatial points, $\mathbb{S}$, independent of their coordinates. This perspective is in line with the modern mathematical approach that emphasizes a formulation free from coordinates. It is important to note that tensors represent physical quantities that exist independently of the coordinates used to describe them. While we acknowledge the significance of coordinates in quantitative analyses, it is essential to understand that the concept of tensors goes beyond their coordinates.

5.1 Definition

A (p, q) tensor is a mathematical concept representing a multilinear map from a collection of p one-forms and q vectors to $\mathbb{R}$. The term *multilinear* refers to the property that the tensor acts linearly on each of its arguments. For instance, given a $(1, 1)$ tensor $\mathbf{T}$, vectors $\mathbf{u}$ and $\mathbf{w}$, one-forms ω and η, and real numbers a, b, c, d, we write

$$\begin{aligned}\mathbf{T}(a\,\omega + b\,\eta, c\,\mathbf{u} + d\,\mathbf{w}) &= \mathbf{T}(a\,\omega, c\,\mathbf{u}) + \mathbf{T}(a\,\omega, d\,\mathbf{w}) \\ &\quad + \mathbf{T}(b\,\eta, c\,\mathbf{u}) + \mathbf{T}(b\,\eta, d\,\mathbf{w})\end{aligned}$$

$$= a\,c\,\mathbf{T}(\boldsymbol{\omega},\mathbf{u}) + a\,d\,\mathbf{T}(\boldsymbol{\omega},\mathbf{w})$$
$$+ b\,c\,\mathbf{T}(\boldsymbol{\eta},\mathbf{u}) + b\,d\,\mathbf{T}(\boldsymbol{\eta},\mathbf{w}); \tag{5.1}$$

note the similarity to equation (4.32). The $(1,1)$ tensor in equation (5.1) acts on one one-form and one vector to produce—as all tensors do—a real number. In view of duality (4.31), a one-form is a $(0,1)$ tensor and a vector is a $(1,0)$ tensor; notably, a scalar is a $(0,0)$ tensor. Following Misner et al. (2017), one can think of a (p,q) tensor as a "machine" with p one-form slots and q vector slots in any order. When this machine is fed p one-forms and q vectors, it returns a real number. These p one-form slots and q vector slots may appear in any order.

In a coordinate system, $x = \{x^i\}$, we write a tensor by acting on the bases of its vectors and one-forms. Thus—invoking p one-form bases and q vector bases—we write the components of a (p,q) tensor as

$$T^{i_1\ldots i_p}{}_{j_1\ldots j_q} = \mathbf{T}(\mathbf{e}^{i_1},\ldots,\mathbf{e}^{i_p},\mathbf{e}_{j_1},\ldots,\mathbf{e}_{j_q}). \tag{5.2}$$

The order of indices is important since the tensor need not act on each argument in the same manner. We define the components of a tensor by inserting basis vectors and basis one-form elements into the appropriate slots. Thus, tensor (5.2) has p one-form slots followed by q vector slots; a tensor with components $T^{ij}{}_k{}^\ell$ has two one-form slots, followed by one vector slot, followed by another one-form slot.

Again, we emphasize that tensors are defined independent of coordinate systems. Tensor *components*, on the other hand, are tied to a particular choice of coordinates, $T^{i_1\ldots i_p}{}_{j_1\ldots j_q}(x)$. At a given spatial point, the tensor can be also a function of time, $\mathbf{T}(\mathcal{S},t)$; in such a case, its components are $T^{i_1\ldots i_p}{}_{j_1\ldots j_q}(x,t)$.

In a coordinate-free manner, we can write tensor (5.2) as

$$\mathbf{T}(\underbrace{\boldsymbol{\omega}^1,\ldots,\boldsymbol{\omega}^p}_{p \text{ one-form slots}},\overbrace{\mathbf{u}_1,\ldots,\mathbf{u}_q}^{q \text{ vector slots}}) \mapsto \text{real-valued number}, \tag{5.3}$$

where $\boldsymbol{\omega}^1,\ldots,\boldsymbol{\omega}^p$ are p one-forms and $\mathbf{u}_1,\ldots,\mathbf{u}_q$ are q vectors. Following the nomenclature explained below, after equation (5.90), we say that $\mathbf{T}$ is *covariant* of rank p and *contravariant* of rank q. Also, (p,q) tensor is referred to as a *tensor of valence* (p,q) and (p,q)*-valent tensor*. In terms of components, the action of $\mathbf{T}$ on p one-forms and q vectors is

$$\omega^1_{i_1}\cdots\omega^p_{i_p}\,T^{i_1\cdots i_p}{}_{j_1\cdots j_q}\,u^{j_1}_1\cdots u^{j_q}_q. \tag{5.4}$$

Example 5.1 Curvature Tensor

In example 3.3, we introduced spacetime coordinates $\{x^\mu\}$ capturing events $\mathcal{S}$ in a spacetime manifold. In example 4.4, we introduced a corresponding vector basis $\mathbf{e}_\mu = \partial_\mu$ in the tangent space at $\mathcal{S}$, complemented by a one-form basis $\mathbf{e}^\mu = \mathrm{d}x^\mu$ in the cotangent space at $\mathcal{S}$. We have the spacetime duality $\mathbf{e}^\mu(\mathbf{e}_\nu) = \delta^\mu{}_\nu$ and $\mathbf{e}_\mu(\mathbf{e}^\nu) = \delta_\mu{}^\nu$.

The rank-4 *curvature tensor* or *Riemann tensor* **r** with three vector slots followed by a one-form slot returns

$$\mathbf{r}(\mathbf{u},\mathbf{v},\mathbf{w},\boldsymbol{\omega}) = r_{\mu\nu\tau}{}^{\sigma}\, u^{\mu}\, v^{\nu}\, w^{\tau}\, \omega_{\sigma} \tag{5.5}$$

when fed three four-vectors $\mathbf{u} = u^{\mu}\,\mathbf{e}_{\mu}$, $\mathbf{v} = v^{\nu}\,\mathbf{e}_{\nu}$, and $\mathbf{w} = w^{\tau}\,\mathbf{e}_{\tau}$, and one one-form $\boldsymbol{\omega} = \omega_{\sigma}\,\mathbf{e}^{\sigma}$.

The components of the curvature tensor may be obtained by inserting the appropriate basis vectors and one-forms:

$$\mathbf{r}(\mathbf{e}_{\mu},\mathbf{e}_{\nu},\mathbf{e}_{\tau},\mathbf{e}^{\sigma}) = r_{\mu\nu\tau}{}^{\sigma}. \tag{5.6}$$

As a generalization of scalar and vector fields, we consider *tensor fields*. Similar to the scalar and vector cases, these fields can be time-dependent. Thus, a tensor field is a rule that assigns a time-dependent (p,q) tensor to every point $\mathcal{S}$; the components of such a tensor are $T^{i_1\cdots i_p}{}_{j_1\cdots j_q}(x,t)$.

As a generalization of vector equations, we consider *tensor equations*. Similar to the case of vectors, the *free upper indices* and *free lower indices* must match on both sides of the equation; for instance, $\sigma^{ij} = c^{ijk\ell}\,\varepsilon_{k\ell}$. Indices k and ℓ are not free; they are *summation indices* or *dummy indices*, where we follow the rule of summation for each repetition of an upper and lower index on the same side of an equation. Note that the need for matching indices on both sides of an equation applies to entities with several indices, regardless of whether they obey the tensor-transformation rule. For instance, the Christoffel symbol, which is a three-index entity introduced below, is subject to the aforementioned index matching, even though it does not follow the tensor-transformation rule.

5.2 Operations on Tensors

5.2.1 Addition

Similar to vectors, whose sum is defined only for vectors of the same dimension, the sum of tensors is defined only for tensors of the same type. For example, a $(2,3)$ tensor may be added only to another $(2,3)$ tensor. Again, similar to vectors, the resulting tensor is of the same type as its summands. Thus, if two tensors have components $T^{ij}{}_{k\ell m}$ and $S^{ij}{}_{k\ell m}$, their sum has components $R^{ij}{}_{k\ell m} = T^{ij}{}_{k\ell m} + S^{ij}{}_{k\ell m}$.

Since adding two tensors results in a tensor of the same type, and the multiplication of a (p,q) tensor by a scalar results in a (p,q) tensor, tensors of a given type constitute a linear space: another similarity with vectors and vector fields, which are linear spaces, as discussed in chapter 2.

5.2.2 Tensor Product

We can multiply a (p,q) tensor, $\mathbf{T}$, and an (r,s) tensor, $\mathbf{S}$, to produce a $(p+r, q+s)$ tensor denoted by $\mathbf{T}\otimes\mathbf{S}$, where $\otimes$ denotes the *tensor product*; it is also referred to as the *outer product*.

It is common to refer to the lower indices as covariant and the upper as contravariant. Thus, in general, a tensor product increases both the covariant and contravariant ranks of a tensor. For example, the tensor product of two one-forms, $\boldsymbol{\omega} = \omega_i \, \mathbf{e}^i$ and $\boldsymbol{\eta} = \eta_j \, \mathbf{e}^j$, is a $(0,2)$ tensor, $\mathbf{T} = \boldsymbol{\omega} \otimes \boldsymbol{\eta} = T_{ij} \, \mathbf{e}^i \otimes \mathbf{e}^j$, whose components, T_{ij}, are $\omega_i \, \eta_j$.

Using the tensor product, condition (4.33) and considering equation (5.2), we write a (p,q) tensor in terms of its components as

$$\mathbf{T} = T^{i_1 \cdots i_p}{}_{j_1 \cdots j_q} \, \mathbf{e}_{i_1} \otimes \cdots \otimes \mathbf{e}_{i_p} \otimes \mathbf{e}^{j_1} \otimes \cdots \otimes \mathbf{e}^{j_q}. \tag{5.7}$$

Example 5.2 Curvature Tensor in Components

In general relativity, the curvature tensor $\mathbf{r}$ may be expressed in terms of its components (5.6) as

$$\mathbf{r} = r_{\mu\nu\tau}{}^{\sigma} \, \mathbf{e}^{\mu} \otimes \mathbf{e}^{\nu} \otimes \mathbf{e}^{\tau} \otimes \mathbf{e}_{\sigma}. \tag{5.8}$$

Thus, the tensor product of $\mathbf{T}$, whose components are $T^{i_1 \cdots i_p}{}_{j_1 \cdots j_q}$, and $\mathbf{S}$, whose components are $S^{k_1 \cdots k_r}{}_{\ell_1 \cdots \ell_s}$, is a tensor whose components are

$$(\mathbf{T} \otimes \mathbf{S})^{i_1 \cdots i_p}{}_{j_1 \cdots j_q}{}^{k_1 \cdots k_r}{}_{\ell_1 \cdots \ell_s}. \tag{5.9}$$

If we restrict our interests to tensors of the same type at $\mathcal{S}$, addition and multiplication satisfy the commutative,[1] associative,[2] and distributive[3] laws. However, the tensor product is not commutative.

We emphasize that—independent of their type—it is not possible to add, multiply, or perform other algebraic operations on tensors corresponding to different points, $\mathcal{S}$, of a manifold. As a method of relating tensors at different points, we extend the standard concept of differentiation to covariant, Lie, and exterior derivatives, discussed in chapter 8.

5.2.3 Contraction

The contravariant and covariant ranks of a tensor may both be reduced by using an operation called *contraction*, which involves setting one of the contravariant indices, say i_k, equal to one of the covariant indices, say j_ℓ, and summing over the repeated upper and lower indices; in other words,

$$T^{i_1 \cdots i_k \cdots i_p}{}_{j_1 \cdots j_\ell \cdots j_q} \rightarrow T^{i_1 \cdots \overset{k\text{th contravariant index}}{\overset{\downarrow}{m}} \cdots i_p}{}_{j_1 \cdots \underset{\ell\text{th covariant index}}{\underset{\uparrow}{m}} \cdots j_q}, \tag{5.10}$$

[1] For two (p,q) tensors $\mathbf{S}$ and $\mathbf{T}$, we have $\mathbf{S} + \mathbf{T} = \mathbf{T} + \mathbf{S}$.

[2] For three (p,q) tensors $\mathbf{S}$, $\mathbf{T}$, and $\mathbf{U}$, we have $(\mathbf{S} + \mathbf{T}) + \mathbf{U} = \mathbf{S} + (\mathbf{T} + \mathbf{U})$.

[3] For three (p,q) tensors $\mathbf{S}$, $\mathbf{T}$, and $\mathbf{U}$, we have $(\mathbf{S} + \mathbf{T}) \otimes \mathbf{U} = \mathbf{S} \otimes \mathbf{U} + \mathbf{T} \otimes \mathbf{U}$.

which is a $(p-1, q-1)$ tensor. Thus, contraction is the operation that produces a $(p-1, q-1)$ tensor from a (p, q) tensor. In accordance with index summation, discussed in section 5.1, upper indices can be contracted only with lower indices and vice versa.

Example 5.3 Ricci Tensor

In general relativity, the components of the *Ricci tensor* are defined in terms of a contraction of the components of the curvature tensor:

$$r_{\nu\tau} \equiv r_{\sigma\nu\tau}{}^{\sigma}. \tag{5.11}$$

It is conventional to use the same symbol r for the elements of the Ricci and curvature tensors and to distinguish them based on the number of indices.

Similar to the addition and the product of two tensors, a contraction of a tensor entails a tensor, which means that the result satisfies the required transformation rule. If the rule is satisfied by the original tensor, it is satisfied by its contracted counterpart, since we get a term $\lambda^i{}_k \Lambda^k{}_j$, which, as a consequence of equation (4.18), is the Kronecker delta, $\delta^i{}_j$: a tensorial quantity in its own right.

The contraction of a $(1, 1)$ tensor $\mathbf{T}$ produces its *trace*,

$$\text{tr}(\mathbf{T}) \equiv T^i{}_i, \tag{5.12}$$

which is the summation over the repeated upper and lower indices. For instance, in $\mathbb{R}^3$, $\delta^i{}_i = 3$.

In particular, we define the contraction between vector $\mathbf{u}$ and one-form $\boldsymbol{\omega}$ as

$$\boldsymbol{\omega} \cdot \mathbf{u} = \boldsymbol{\omega}(\mathbf{u}) = \mathbf{u}(\boldsymbol{\omega}) = \langle \boldsymbol{\omega}, \mathbf{u} \rangle = \omega_i\, u^i, \tag{5.13}$$

which is called the *dot product* or *Euclidean inner product*.

Let us comment on the notation: $\langle \boldsymbol{\omega}, \mathbf{u} \rangle$ is a coordinate-independent expression, and $\omega_i\, u^i$ refers to components with respect to a particular basis. However, since both $\boldsymbol{\omega}$ and $\mathbf{u}$ are geometrical entities independent of the coordinates in which they are expressed, the value of $\omega_i\, u^i$ does not depend on the choice of basis. Subscripts and superscripts are abstract markers to express algebraic operations.

The inner product between tensors $\mathbf{T}$ and $\mathbf{S}$ is defined only if the last slot of $\mathbf{T}$ takes a vector and the first slot of $\mathbf{S}$ takes a one-form, or vice versa. For example, if a $(1, 1)$ tensor $\mathbf{T}$ has components $T^i{}_j$ and a $(2, 0)$ tensor $\mathbf{S}$ has components S^{ij}, their inner product is a $(2, 0)$ tensor with components $(\mathbf{T} \cdot \mathbf{S})^{ik} = T^i{}_j\, S^{jk}$. More generally, we define the inner product between a (p, q) tensor $\mathbf{T}$ and an (r, s) tensor $\mathbf{S}$ as a $(p+r-1, q+s-1)$ tensor whose components are

$$(\mathbf{T} \cdot \mathbf{S})^{i_1 \cdots i_p}{}_{j_1 \cdots j_{q-1}}{}^{k_2 \cdots k_r}{}_{\ell_1 \cdots \ell_s} = T^{i_1 \cdots i_p}{}_{j_1 \cdots j_{q-1} \underset{\substack{\uparrow \\ \text{last covariant index of } \mathbf{T}}}{m}}\, S^{\overset{\text{first contravariant index of } \mathbf{S}}{\overset{\downarrow}{m}}\, k_2 \cdots k_r}{}_{\ell_1 \cdots \ell_s}, \tag{5.14}$$

where the summation convention implies a sum over the repeated upper and lower indices m.

For a $(2,0)$ tensor, $\mathbf{T}$, and a $(0,2)$ tensor, $\mathbf{S}$, we define the *double-dot product*,

$$\mathbf{T}:\mathbf{S} = T^{ij}\, S_{ij} = \mathbf{S}:\mathbf{T} = S_{ij}\, T^{ij}, \tag{5.15}$$

which is a $(0,0)$ tensor: a scalar.[4] This definition can be extended to (p,q) tensors, where $p+q=2$. We write

$$\mathbf{T}:\mathbf{S} = T^{i}{}_{j}\, S_{i}{}^{j} = \mathbf{S}:\mathbf{T} = S_{i}{}^{j}\, T^{i}{}_{j}, \tag{5.16}$$

which is also a scalar. The double-dot product is also called the *double-inner product* and the *Frobenius inner product*. It induces the *Frobenius norm*, whose geometrical interpretation is a distance in the space of second-rank tensors.

Example 5.4 Hooke's Law

In seismology, Hooke's law involves a linear relationship between the *Cauchy stress tensor* $\boldsymbol{\sigma}$ and the *infinitesimal strain tensor* $\boldsymbol{\varepsilon}$. It is based on a contraction between the rank-4 *elastic tensor* $\mathbf{c}$ and the infinitesimal strain tensor, such that

$$\boldsymbol{\sigma} = \mathbf{c}:\boldsymbol{\varepsilon}. \tag{5.17}$$

In components, we have

$$\sigma^{ij} = c^{ijk\ell}\, \varepsilon_{k\ell}. \tag{5.18}$$

Hooke's law (5.18) defines the most general linear relationship between the elements of the infinitesimal strain $\varepsilon_{k\ell}$ and the Cauchy stress σ^{ij}.

5.2.4 Transpose of (2,0) and (0,2) Tensors

Given two one-forms, $\boldsymbol{\omega}$ and $\boldsymbol{\eta}$, and two vectors, $\mathbf{u}$ and $\mathbf{w}$, the *transpose* of a $(2,0)$ tensor and the transpose of a $(0,2)$ tensor are defined by

$$\mathbf{T}^{t}(\boldsymbol{\omega},\boldsymbol{\eta}) \equiv \mathbf{T}(\boldsymbol{\eta},\boldsymbol{\omega}) \quad \text{and} \quad \mathbf{T}^{t}(\mathbf{u},\mathbf{w}) \equiv \mathbf{T}(\mathbf{w},\mathbf{u}), \tag{5.19}$$

respectively.

Symmetric $(2,0)$ and $(0,2)$ tensors are defined by

$$\mathbf{T}^{t}(\boldsymbol{\omega},\boldsymbol{\eta}) \equiv \mathbf{T}(\boldsymbol{\omega},\boldsymbol{\eta}) \quad \text{and} \quad \mathbf{T}^{t}(\mathbf{u},\mathbf{w}) \equiv \mathbf{T}(\mathbf{u},\mathbf{w}), \tag{5.20}$$

respectively; in components, we write

$$T^{ij} = T^{ji} \quad \text{and} \quad T_{ij} = T_{ji}. \tag{5.21}$$

[4] Other authors (e.g., Malvern, 1969) use $\mathbf{T}\cdot\cdot\,\mathbf{S} \equiv T^{ij}\, S_{ji}$.

In contrast, *antisymmetric* $(2,0)$ and $(0,2)$ tensors are defined by

$$\mathbf{T}^t(\boldsymbol{\omega},\boldsymbol{\eta}) \equiv -\mathbf{T}(\boldsymbol{\omega},\boldsymbol{\eta}) \quad \text{and} \quad \mathbf{T}^t(\mathbf{u},\mathbf{w}) \equiv -\mathbf{T}(\mathbf{u},\mathbf{w}), \tag{5.22}$$

respectively; in components, we write

$$T^{ij} = -T^{ji} \quad \text{and} \quad T_{ij} = -T_{ji}. \tag{5.23}$$

For $(2,0)$ and $(0,2)$ tensors, we define their *symmetric* and *antisymmetric* parts by

$$\widehat{\mathbf{T}} \equiv \tfrac{1}{2}\,(\mathbf{T}+\mathbf{T}^t), \qquad \widetilde{\mathbf{T}} \equiv \tfrac{1}{2}\,(\mathbf{T}-\mathbf{T}^t). \tag{5.24}$$

In components

$$\widehat{T}^{ij} \equiv \tfrac{1}{2}\,(T^{ij}+T^{ji}), \qquad \widetilde{T}^{ij} \equiv \tfrac{1}{2}\,(T^{ij}-T^{ji}), \tag{5.25}$$

$$\widehat{T}_{ij} \equiv \tfrac{1}{2}\,(T_{ij}+T_{ji}), \qquad \widetilde{T}_{ij} \equiv \tfrac{1}{2}\,(T_{ij}-T_{ji}). \tag{5.26}$$

Example 5.5 Moment Tensor

An example of a symmetric $(0,2)$ tensor in seismology is the *moment tensor*, $\mathbf{M}=\mathbf{M}^t$, which is the standard mathematical representation of an earthquake *source mechanism*. Earthquake source mechanisms are graphically depicted using a *beach ball representation*. Let points in the surface of the unit *focal hemisphere* surrounding the earthquake *hypocenter*[a] be identified by vectors $\mathbf{p}$ in the tangent space at the hypocenter. These vectors identify the take-off directions of compressional seismic (P) waves leaving the hypocenter. The P-wave *radiation pattern* is proportional to the scalar field $r \equiv \mathbf{p}\cdot\mathbf{M}\cdot\mathbf{p}$. This radiation pattern is customarily plotted on the lower focal hemisphere using either a stereographic or equal-area projection. The regions where $-1 \le r < 0$ are typically painted black, while the regions where $0 < r \le 1$ are painted white, resulting in the traditional beach ball representation of an earthquake.

The components of a rank-2 tensor are sometimes displayed in matrix form. For example, the spherical polar components (4.76) of the moment tensor may be arranged in the following matrix:

$$(M_{ij}) = \begin{pmatrix} M_{rr} & M_{r\theta} & M_{r\phi} \\ M_{\theta r} & M_{\theta\theta} & M_{\theta\phi} \\ M_{\phi r} & M_{\phi\theta} & M_{\phi\phi} \end{pmatrix}, \tag{5.27}$$

where, because of its symmetry, $M_{r\theta}=M_{\theta r}$, $M_{r\phi}=M_{\phi r}$, and $M_{\theta\phi}=M_{\phi\theta}$. A pictorial glossary of elementary *double-couple* and *non-double-couple* source mechanisms (5.27) and their corresponding beach balls is given in table 5.1.

[a] A hypocenter is the point source location of an earthquake, combining its epicenter and depth.

MOMENT TENSOR	BEACH BALL	MOMENT TENSOR	BEACH BALL
$\frac{1}{\sqrt{3}}\begin{pmatrix}1&0&0\\0&1&0\\0&0&1\end{pmatrix}$		$-\frac{1}{\sqrt{3}}\begin{pmatrix}1&0&0\\0&1&0\\0&0&1\end{pmatrix}$	
$-\frac{1}{\sqrt{2}}\begin{pmatrix}0&0&0\\0&0&1\\0&1&0\end{pmatrix}$		$\frac{1}{\sqrt{2}}\begin{pmatrix}0&0&0\\0&1&0\\0&0&-1\end{pmatrix}$	
$\frac{1}{\sqrt{2}}\begin{pmatrix}0&1&0\\1&0&0\\0&0&0\end{pmatrix}$		$\frac{1}{\sqrt{2}}\begin{pmatrix}0&0&1\\0&0&0\\1&0&0\end{pmatrix}$	
$\frac{1}{\sqrt{2}}\begin{pmatrix}1&0&0\\0&-1&0\\0&0&0\end{pmatrix}$		$\frac{1}{\sqrt{2}}\begin{pmatrix}1&0&0\\0&0&0\\0&0&-1\end{pmatrix}$	
$\frac{1}{\sqrt{6}}\begin{pmatrix}1&0&0\\0&1&0\\0&0&-2\end{pmatrix}$		$\frac{1}{\sqrt{6}}\begin{pmatrix}1&0&0\\0&-2&0\\0&0&1\end{pmatrix}$	
$\frac{1}{\sqrt{6}}\begin{pmatrix}-2&0&0\\0&1&0\\0&0&1\end{pmatrix}$		$-\frac{1}{\sqrt{6}}\begin{pmatrix}-2&0&0\\0&1&0\\0&0&1\end{pmatrix}$	

Table 5.1: In seismology, an earthquake is represented by a symmetric rank-2 moment tensor $\mathbf{M}=\mathbf{M}^t$. The components of such a rank-2 tensor may be arranged in the matrix form (5.27) (*columns 1 and 3*). Moment tensors may be visualized by plotting the scalar field $r=\mathbf{p}\cdot\mathbf{M}\cdot\mathbf{p}$, where the vector field $\mathbf{p}$ in the tangent space at the earthquake hypocenter identifies points in the unit focal sphere centered at the hypocenter (*columns 2 and 4*). The lower-hemisphere projection of this scalar field defines a *beach ball*, in which regions where $-1\leq r<0$ are painted black, while the regions where $0<r\leq 1$ are painted white. Shown are unit source mechanisms, such that $\mathbf{M}:\mathbf{M}=1$. The two sources in the top row are a pure explosion $\mathbf{M}=\mathbf{I}/\sqrt{3}$, with an entirely black beach ball, and a pure implosion $\mathbf{M}=-\mathbf{I}/\sqrt{3}$, with an entirely white beach ball. The next three rows show several double couples, corresponding to vertical strike-slip faults (*second from top*), vertical dip-slip faults (*third from top*), and 45°-dip thrust faults (*fourth from top*). The sources in the fifth and sixth rows are pure compensated linear vector dipoles; the lowermost right entry is an idealized "eyeball" or "fried-egg." Except for the explosion and implosion, all of these sources are purely deviatoric: $\operatorname{tr}(\mathbf{M})=0$. Reproduced from *Dahlen & Tromp (1998).*

5.2.5 Transpose of a (1,1) Tensor

Given a one-form, ω, and a vector, $\mathbf{u}$, the *transpose* of a $(1,1)$ tensor is defined by

$$\mathbf{T}^t(\mathbf{u},\omega)\equiv\mathbf{T}(\omega,\mathbf{u}), \tag{5.28}$$

In components we have

$$u^i\,(T^t)_i{}^j\,\omega_j=\omega_i\,T^i{}_j\,u^j, \tag{5.29}$$

which implies that

$$(T^t)_j{}^i = T^i{}_j. \tag{5.30}$$

Alternatively, the definition of the transpose may be based on the duality product (4.31):

$$\langle \mathbf{T}^t \cdot \boldsymbol{\omega}, \mathbf{u} \rangle \equiv \langle \boldsymbol{\omega}, \mathbf{T} \cdot \mathbf{u} \rangle. \tag{5.31}$$

Written out in components we have

$$(T^t)_i{}^j \, \omega_j \, u^i = \omega_i \, T^i{}_j \, u^j, \tag{5.32}$$

from which we also obtain (5.30).

5.3 Transformations

In view of the definition of a tensor discussed in section 5.1 and the transformation rules of vectors and one-forms discussed in sections 4.1 and 4.2, respectively, we can formulate the transformation rule for tensors. We let $x = \{x^i\}$ and $x' = \{x^{i'}\}$ denote two coordinate systems on a manifold, and we express a tensor in terms of its components in these two coordinate systems as

$$\begin{aligned} \mathbf{T} &= T^{i_1 \cdots i_p}{}_{j_1 \cdots j_q} \, \mathbf{e}_{i_1} \otimes \cdots \otimes \mathbf{e}_{i_p} \otimes \mathbf{e}^{j_1} \otimes \cdots \otimes \mathbf{e}^{j_q} \\ &= T^{i'_1 \cdots i'_p}{}_{j'_1 \cdots j'_q} \, \mathbf{e}_{i'_1} \otimes \cdots \otimes \mathbf{e}_{i'_p} \otimes \mathbf{e}^{j'_1} \otimes \cdots \otimes \mathbf{e}^{j'_q}. \end{aligned} \tag{5.33}$$

Invoking equations (4.16) and (4.44), we write the rule for transforming tensors as

$$T^{i'_1 \dots i'_p}{}_{j'_1 \dots j'_q} = \Lambda^{i'_1}{}_{i_1} \, \cdots \, \Lambda^{i'_p}{}_{i_p} \, T^{i_1 \cdots i_p}{}_{j_1 \cdots j_q} \, \lambda^{j_1}{}_{j'_1} \, \cdots \, \lambda^{j_q}{}_{j'_q}, \tag{5.34}$$

with inverse

$$T^{i_1 \dots i_p}{}_{j_1 \dots j_q} = \lambda^{i_1}{}_{i'_1} \, \cdots \, \lambda^{i_p}{}_{i'_p} \, T^{i'_1 \cdots i'_p}{}_{j'_1 \cdots j'_q} \, \Lambda^{j'_1}{}_{j_1} \, \cdots \, \Lambda^{j'_q}{}_{j_q}. \tag{5.35}$$

Example 5.6 Transformations of the Curvature Tensor

In general relativity, the components of the curvature tensor (5.6) may be transformed from coordinates $\{x^\mu\}$ to coordinates $\{x^{\mu'}\}$ based on four-dimensional versions of (5.34) and (5.35), namely,

$$r_{\mu'\nu'\tau'}{}^{\sigma'} = \lambda^{\mu}{}_{\mu'} \, \lambda^{\nu}{}_{\nu'} \, \lambda^{\tau}{}_{\tau'} \, \Lambda^{\sigma'}{}_{\sigma} \, r_{\mu\nu\tau}{}^{\sigma}, \tag{5.36}$$

and

$$r_{\mu\nu\tau}{}^{\sigma} = \Lambda^{\mu'}{}_{\mu} \, \Lambda^{\nu'}{}_{\nu} \, \Lambda^{\tau'}{}_{\tau} \, \lambda^{\sigma}{}_{\sigma'} \, r_{\mu'\nu'\tau'}{}^{\sigma'}. \tag{5.37}$$

Example 5.7 Transformations of the Stress Tensor

In continuum mechanics, the components of the Cauchy stress tensor (5.18) may be transformed from Eulerian coordinates $\{r^i\}$ to Lagrangian coordinates $\{X^I\}$ based on the deformation gradient (4.27) and its inverse (4.26):

$$t^I{}_J = (F^{-1})^I{}_i \, F^j{}_J \, t^i{}_j. \tag{5.38}$$

The simplest case of equation (5.34) is for a scalar function f, which is a $(0, 0)$ tensor. The rule for a coordinate change is

$$f'(x') = f(x), \tag{5.39}$$

which states that the value of f at point $\mathcal{S}$, namely, f($\mathcal{S}$), is independent of the coordinate system. This is because a $(0, 0)$ tensor is defined by its component, and preserving its essential properties is equivalent to preserving its single component. For example, if all components of a tensor are zero in one coordinate system, they must be zero in all coordinate systems to preserve the property $\mathbf{T} = \mathbf{0}$. Therefore, the function f, which corresponds to its single component, must remain zero in all coordinate systems.

As mentioned in the introductory comments of this chapter, rule (5.34) may be used as a definition of tensors (e.g., Lovelock & Rund, 1989, Section 3.2). However, as commented by Schutz (1980, p. 60),

> The behavior of a tensor's components under a change of basis is at the heart of the older definition of a tensor. It has been replaced more recently by the definition we have used here in terms of linear functions . . . This is not to say that these transformations are unimportant. Most practical calculations involving tensors involve working with their components, and an understanding of their transformation properties is essential.

The existence of a tensor is not defined by its transformation rule. Instead, equation (5.34) is used to ensure that a given entity exhibits the required tensorial properties. In other words, rule (5.34) is the condition for an entity to be a tensor, although it can be argued that this is a subtle distinction between the condition and definition. It is important to note that rule (5.34), which specifies the transformation, is the main condition. Other operations such as contraction and transposition, discussed in section 5.2, do not require rule (5.34) and can be performed on nontensorial entities.

Tensors are geometrical entities that exist independently of a coordinate system. It is better to think of them as a unique multilinear map from vectors and one-forms to real numbers, rather than objects that follow a given transformation rule. In simpler terms, we can view tensors as linear functions that take one-forms and vectors as input and give real numbers as output.

5.3.1 Tetrad Formalism

In general relativity and continuum mechanics, the *tetrad formalism* generalizes the basis of the tangent space from a coordinate basis $\{\partial_i\}$ to a set of non-coordinate local vector fields $\{\mathbf{e}_i\}$, called a *tetrad*, *frame*, *Dreibein* (3D), *Vierbein* (4D), or *Vielbein* (nD), and the basis of the cotangent space from a coordinate basis $\{\mathrm{d}x^i\}$ to a locally defined set of one-form fields, called a *coframe*, $\{\mathbf{e}^i\}$, as illustrated in figure 5.1. Frames and coframes are duals in the sense $\mathbf{e}^i(\mathbf{e}_j) = \delta^i{}_j$ and $\mathbf{e}_i(\mathbf{e}^j) = \delta_i{}^j$. In physics, tetrads are often chosen to be orthogonal, thereby defining an orthogonal frame, but this is not required.

Frames may be obtained based on transformations of the coordinate basis:

$$\mathbf{e}_{i'} = e_{i'}{}^{i}\,\partial_i, \qquad \partial_i = e_i{}^{i'}\,\mathbf{e}_{i'}, \tag{5.40}$$

where $e_{i'}{}^{i}$ denotes the components of a transformation matrix with inverse $e_i{}^{i'}$. In this case, these matrices are *not* defined in terms of the partial derivatives (4.17) of the local diffeomorphic transformations (3.11). The non-coordinate basis vectors are generally anholonomic, because their Lie bracket does not vanish:

$$[\mathbf{e}_{i'}, \mathbf{e}_{j'}] = \tau^{k'}_{i'j'}\,\mathbf{e}_{k'}, \tag{5.41}$$

where $\tau^{k'}_{i'j'}$ denote the structure coefficients

$$\begin{aligned}\tau^{k'}_{i'j'} &= e_j{}^{k'}\left(e_{i'}{}^{i}\,\partial_i e_{j'}{}^{j} - e_{j'}{}^{i}\,\partial_i e_{i'}{}^{j}\right)\\ &= -\,\tau^{k'}_{j'i'}.\end{aligned} \tag{5.42}$$

Similarly, coframes may be obtained based on transformations of the coordinate one-forms:

$$\mathbf{e}^{i'} = e^{i'}{}_{i}\,\mathrm{d}x^i, \qquad \mathrm{d}x^i = e^{i}{}_{i'}\,\mathbf{e}^{i'}, \tag{5.43}$$

and we have the duality

$$e^{i'}(e_{j'}) = e^{i'}{}_{i}\,e_{j'}{}^{i} = \delta^{i'}{}_{j'}, \qquad e_{i'}(e^{j'}) = e_{i'}{}^{i}\,e^{j'}{}_{i} = \delta_{i'}{}^{j'}. \tag{5.44}$$

A vector $\mathbf{u}$ may be expressed in either a coordinate basis or a frame as

$$\mathbf{u} = u^i\,\partial_i = u^{i'}\,\mathbf{e}_{i'}, \tag{5.45}$$

where

$$u^{i'} = e^{i'}{}_{i}\,u^i, \qquad u^i = e^{i}{}_{i'}\,u^{i'}. \tag{5.46}$$

We have been using the identifications $\mathbf{e}_i \equiv \partial_i$ and $\mathbf{e}^i \equiv \mathrm{d}x^i$ for basis vectors and one-forms based on coordinates $\{x^i\}$. Our notation for frames $\mathbf{e}_i$ and coframes $\mathbf{e}^i$ is chosen such that most tensor expressions in the book are valid in both coordinate and non-coordinate bases.

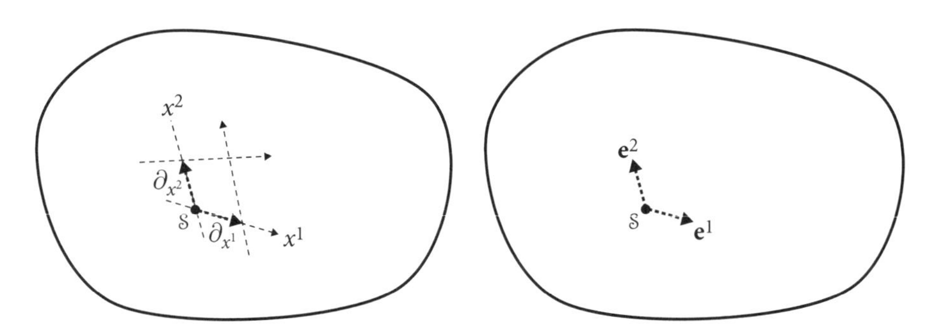

Figure 5.1: *Left*: A point $\mathcal{S}$ in a manifold may be identified with a set of coordinates $\{x^i\}$. These coordinates induce a set of basis vectors $\{\partial_{x^i}\}$ in the tangent space at $\mathcal{S}$. *Right*: Tetrads are basis vectors $\{\mathbf{e}_i\}$ in the tangent space at $\mathcal{S}$ not tied to coordinates. Such "frames" $\{\mathbf{e}_i\}$ define a non-coordinate basis for vectors.

Example 5.8 Tetrads in General Relativity

In general relativity, using the four-dimensional spacetime parametrization introduced in example 5.1, the tetrad formalism is based on the transformations[a]

$$e_a = e_a{}^{\mu}\,\partial_{\mu}, \qquad \partial_{\mu} = e_{\mu}{}^{a}\,e_a, \tag{5.47}$$

where $e_a{}^{\mu}\,e_{\mu}{}^{b} = \delta_a{}^{b}$. In this case, the transformation matrices are *not* defined in terms of the partial derivatives (4.17) of the local diffeomorphic transformations (3.11). A tetrad is usually specified by its coefficients $e_a{}^{\mu}$ with respect to a coordinate basis. The non-coordinate basis is anholonomic:

$$[e_a, e_b] = \tau^{c}_{ab}\,e_c, \tag{5.48}$$

where τ^{c}_{ab} denote the structure coefficients

$$\begin{aligned}\tau^{c}_{ab} &= e_{\nu}{}^{c}\,(e_a{}^{\mu}\,\partial_{\mu}e_b{}^{\nu} - e_b{}^{\mu}\,\partial_{\mu}e_a{}^{\nu})\\ &= -\,\tau^{c}_{ba}.\end{aligned} \tag{5.49}$$

Similarly, the basis of the cotangent space is generalized from a coordinate basis $\{dx^{\mu}\}$ to a locally defined coframe, $\{e^a\}$, based on the transformations

$$e^a \equiv e^a{}_{\mu}\,dx^{\mu}, \qquad dx^{\mu} = e^{\mu}{}_a\,e^a. \tag{5.50}$$

Frames and coframes are duals in the sense

$$e^a(e_b) = e^a{}_{\mu}\,e_b{}^{\mu} = \delta^a{}_b, \qquad e_a(e^b) = e_a{}^{\mu}\,e^b{}_{\mu} = \delta_a{}^b. \tag{5.51}$$

Tetrads are often chosen to be orthogonal, thereby defining an orthogonal frame, but this is not required.

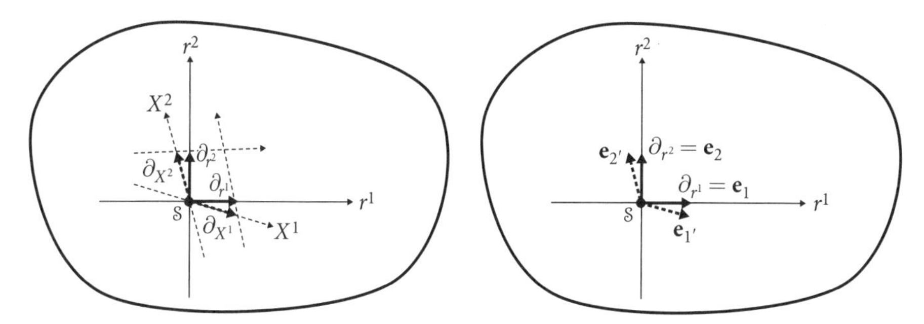

Figure 5.2: *Left*: In continuum mechanics, a point S in a spatial manifold may be identified with either a set of Eulerian coordinates $\{r^i\}$ or a set of local convected or Lagrangian coordinates $\{X^I\}$. These two sets of coordinates are related via the Galilean motion $r^i = \varphi^i(X, T)$ at time $t = T$, as discussed in example 3.4. These coordinate sets induce a set of Eulerian basis vectors $\{\partial_{r^i}\}$ (black arrows) or Lagrangian basis vectors $\{\partial_{X^I}\}$ (dotted black arrows) in the tangent space at S. The Eulerian coordinates are chosen to be Cartesian in this 2D illustration, but this is not required. *Right*: Tetrad formalism in continuum mechanics. Eulerian coordinates $\{r^i\}$ are unaffected by motion and induce a well-behaved vector basis $\{\partial_{r^i}\} = \{\mathbf{e}_i\}$ (black arrows) in the tangent space at S. Lagrangian basis vectors are no longer tied to Lagrangian coordinates $\{X^I\}$. Instead, we use a non-orthonormal frame $\{\mathbf{e}_I\}$ (dotted black arrows) as a non-coordinate basis for vectors, analogous to the tetrad or Vierbein formulation of GR.

The generalized vector and one-form bases may be used to express any tensor in the spacetime manifold. Thus, for example, the four-velocity of a particle, u, may be expressed in a frame basis as

$$u = u^a\, e_a. \tag{5.52}$$

[a]Here, we introduce a notation commonly used to describe tetrads, using lowercase Roman indices at the start of the alphabet, that is, $a, b, c, \ldots$, to denote the elements of the tetrad and not bolding frames, coframes, and tensors.

Example 5.9 Tetrads in Continuum Mechanics

In continuum mechanics, Eulerian coordinates $\{r^i\}$ induce a vector basis $\{\partial_{r^i}\}$ and a one-form basis $\{\mathrm{d}r^i\}$ in the tangent and cotangent bundles of the spatial manifold, whereas Lagrangian coordinates $\{X^I\}$ induce a vector basis $\{\partial_{X^I}\}$ and a one-form basis $\{\mathrm{d}X^I\}$, as illustrated in figure 5.2 (*left*).

In the tetrad formalism of continuum mechanics, Eulerian coordinates $\{r^i\}$ are unaffected by the motion and continue to induce a well-behaved vector basis $\{\partial_{r^i}\} = \{\mathbf{e}_i\}$ in the tangent space of the spatial manifold, but the Lagrangian vector basis is no longer tied to Lagrangian coordinates $\{X^I\}$, as illustrated in figure 5.2 (*right*). We use a non-orthonormal Lagrangian frame $\{\mathbf{e}_I\}$ as a non-coordinate basis for vectors,

$$\mathbf{e}_I = e_I{}^i\, \partial_{r^i}, \tag{5.53}$$

and frames

$$\mathbf{e}^I = e^I{}_i\, \mathrm{d}X^I. \tag{5.54}$$

The coefficients $e_I{}^i$ specify the frame $\mathbf{e}_I$, and the coefficients $e^I{}_i$ specify the coframe $\mathbf{e}^I$, with the duality $\mathbf{e}^I(\mathbf{e}_J) = e^I{}_i\, e_J{}^i = \delta_J{}^I$. We have been using the identifications $\mathbf{e}_I \equiv \partial_I$ and $\mathbf{e}^I \equiv \mathrm{d}X^I$ for Lagrangian basis vectors and one-forms based on coordinates $\{X^I\}$, thereby naturally accommodating frames.

5.3.2 Pseudotensors

A *pseudotensor* transforms according to a modified version of rule (5.34), namely,

$$T^{i'_1\cdots i'_p}{}_{j'_1\cdots j'_q} = \mathrm{sgn}(\lambda)\, \Lambda^{i'_1}{}_{i_1} \cdots \Lambda^{i'_p}{}_{i_p}\, T^{i_1\cdots i_p}{}_{j_1\cdots j_q}\, \lambda^{j_1}{}_{j'_1} \cdots \lambda^{j_q}{}_{j'_q}, \tag{5.55}$$

where λ is the *determinant* of the coordinate transformation,[5]

$$\lambda = \det\left(\partial_{i'}\varphi^i\right), \tag{5.56}$$

and where $\mathrm{sgn}(\lambda) = \pm 1$ denotes the sign of the determinant of the transformation matrix. As an example, reflections generally involve pseudotensors.

5.4 Kronecker or Identity Tensor

The Kronecker delta defined by equation (4.19) transforms as

$$\Lambda^{i'}{}_i\, \delta^i{}_j\, \lambda^j{}_{j'} = \Lambda^{i'}{}_i\, \lambda^i{}_{j'} = \delta^{i'}{}_{j'}, \tag{5.57}$$

where we used the fact that $\Lambda^{i'}{}_i$ and $\lambda^j{}_{j'}$ are inverses of one another, as stated in equation (4.18). We see that the Kronecker delta transforms according to rule (5.34), which defines the components of a tensor. We denote it by $\mathbf{I}$ and call it the *Kronecker tensor* or the *identity tensor*.

5.5 Logarithms and Exponentials of (1,1) Tensors

Let us discuss operations on a $(1,1)$ tensor that are extensions of logarithms and exponentials acting on real numbers. As in $\mathbb{R}$, these operations are inverses of one another; the limitation of their application to $(1,1)$ tensors becomes apparent as we proceed.

[5] See section 5.8 for a definition of the determinant in terms of alternating symbols.

Following the analogy of standard expressions for logarithms and exponentials expressed in terms of the power series, we write

$$\log \mathbf{T} = \log(\mathbf{I} + \mathbf{S}) = \sum_{k=1}^{\infty} \frac{(-1)^{k-1}}{k} \mathbf{S}^k = \mathbf{S} - \frac{1}{2}\mathbf{S}^2 + \frac{1}{3}\mathbf{S}^3 - \frac{1}{4}\mathbf{S}^4 + \cdots, \tag{5.58}$$

and

$$\exp \mathbf{T} = \sum_{k=0}^{\infty} \frac{1}{k!} \mathbf{T}^k = \mathbf{I} + \mathbf{T} + \frac{1}{2}\mathbf{T}^2 + \frac{1}{3!}\mathbf{T}^3 + \frac{1}{4!}\mathbf{T}^4 + \cdots, \tag{5.59}$$

where **I**, **S**, and **T** are dimensionless $(1,1)$ tensors.[6] Series (5.58) is the *Mercator series*, which is a special case of the Taylor series that relates to the natural logarithm; series (5.59) is the Maclaurin series. To ensure convergence of series (5.58), we require that the *Frobenius norm*, which is the square root of the sum of the squared components of **S**, be less than unity; series (5.59) is convergent for any **T**. In section 6.3, we encounter an example of the logarithm of a tensor in the form of the *Hencky* strain tensor.

5.6 Tensor Densities and Capacities

We encounter objects whose components transform according to a modification of rule (5.34), namely[7]

$$\overline{T}^{i'_1\cdots i'_p}{}_{j'_1\cdots j'_q} = \lambda^w \, \Lambda^{i'_1}{}_{i_1} \cdots \Lambda^{i'_p}{}_{i_p} \, \overline{T}^{i_1\cdots i_p}{}_{j_1\cdots j_q} \, \lambda^{j_1}{}_{j'_1} \cdots \lambda^{j_q}{}_{j'_q}, \tag{5.60}$$

where λ is the determinant of the coordinate transformation (5.56), and where w is a positive integer. For an orientation-reversing coordinate transformation, the determinant can be negative. Such an object is a (p,q) *tensor density of weight* w. Tensor densities are also referred to as *relative tensors*. Tensor densities explicitly involve the determinant associated with the map φ, and hence a coordinate system, making them not "pure" tensors, which, by definition, are strictly geometrical objects.

Similarly, a (p,q) *tensor capacity of weight* w, where w is a positive integer, transforms according to[8]

$$\underline{T}^{i'_1\cdots i'_p}{}_{j'_1\cdots j'_q} = \frac{1}{\lambda^w} \, \Lambda^{i'_1}{}_{i_1} \cdots \Lambda^{i'_p}{}_{i_p} \, \underline{T}^{i_1\cdots i_p}{}_{j_1\cdots j_q} \, \lambda^{j_1}{}_{j'_1} \cdots \lambda^{j_q}{}_{j'_q}. \tag{5.61}$$

For a function f, which is a $(0,0)$ tensor, we refer to *scalar densities of weight* w if

$$\bar{f}' = \lambda^w \bar{f}, \tag{5.62}$$

and *scalar capacities of weight* w if

[6]We use log, not ln, to denote the natural logarithm.

[7]Throughout this book we use an "overline" to denote tensor densities of weight one. On rare occasions, we use two overlines to denote a density of weight two. This notation becomes cumbersome for weights greater than two.

[8]Throughout this book we use an "underline" to denote tensor capacities of weight one.

$$\underline{f}' = \frac{1}{\lambda^w} \underline{f}. \tag{5.63}$$

5.6.1 Pseudotensor Densities and Capacities

Pseudotensor densities and *pseudotensor capacities* transform according to a modification of rules (5.60) and (5.61) involving an additional sign change under an orientation-reversing coordinate transformation, namely,

$$\overline{T}^{i'_1 \cdots i'_p}{}_{j'_1 \cdots j'_q} = \operatorname{sgn}(\lambda)\, \lambda^w\, \Lambda^{i'_1}{}_{i_1} \cdots \Lambda^{i'_p}{}_{i_p}\, \overline{T}^{i_1 \cdots i_p}{}_{j_1 \cdots j_q}\, \lambda^{j_1}{}_{j'_1} \cdots \lambda^{j_q}{}_{j'_q}, \tag{5.64}$$

and

$$\underline{T}^{i'_1 \cdots i'_p}{}_{j'_1 \cdots j'_q} = \operatorname{sgn}(\lambda)\, \frac{1}{\lambda^w}\, \Lambda^{i'_1}{}_{i_1} \cdots \Lambda^{i'_p}{}_{i_p}\, \underline{T}^{i_1 \cdots i_p}{}_{j_1 \cdots j_q}\, \lambda^{j_1}{}_{j'_1} \cdots \lambda^{j_q}{}_{j'_q}, \tag{5.65}$$

respectively.

5.7 Levi-Civita Density and Capacity

The *Levi-Civita symbol*, also known as the *alternating symbol*, defines an important tensor density/capacity. There are two different Levi-Civita symbols with all lower/upper indices: $\underline{\epsilon}_{ijk\cdots}$ and $\overline{\epsilon}^{ijk\cdots}$. The number of indices they contain equals the dimension of the space. For instance, in spaces of dimensions 2, 3, and 4, the Levi-Civita capacities are, respectively, $\underline{\epsilon}_{ij}$, $\underline{\epsilon}_{ijk}$, and $\underline{\epsilon}_{ijkl}$. In general, they are defined by

$$\underline{\epsilon}_{ijk\cdots} = \begin{cases} +1 & \text{if } i, j, \ldots \text{ is an } \textit{even} \text{ permutation of } 1, 2, \ldots, \\ 0 & \text{if any indices are identical,} \\ -1 & \text{if } i, j, \ldots \text{ is an } \textit{odd} \text{ permutation of } 1, 2, \ldots, \end{cases} \tag{5.66}$$

and, similarly,

$$\overline{\epsilon}^{ijk\cdots} = \begin{cases} +1 & \text{if } i, j, \ldots \text{ is an } \textit{even} \text{ permutation of } 1, 2, \ldots, \\ 0 & \text{if any indices are identical,} \\ -1 & \text{if } i, j, \ldots \text{ is an } \textit{odd} \text{ permutation of } 1, 2, \ldots. \end{cases} \tag{5.67}$$

There is an important relation between Levi-Civita symbols and determinants. For example, the determinant of transformation (5.56) may be expressed in terms of the Levi-Civita density and capacity as

$$\lambda = \frac{1}{n!}\, \underline{\epsilon}_{ijk\cdots}\, \frac{\partial \varphi^i}{\partial x^{i'}}\, \frac{\partial \varphi^j}{\partial x^{j'}}\, \frac{\partial \varphi^k}{\partial x^{k'}} \cdots \overline{\epsilon}^{i'j'k'\cdots}. \tag{5.68}$$

This definition of the determinant of the coordinate transformation is more satisfying than equation (5.56), because it does not invoke an operation from matrix algebra, that is, it does not involve regarding the components of $\partial_{i'} \varphi^i$ as a matrix and taking the determinant of this matrix. We define the determinant of a $p + q = 2$ tensor in section 5.8, where we discover that the determinants are either pure scalars or scalar densities/capacities of weight two.

We write Levi-Civita symbols with overlines and underlines, which suggests that these objects must be more than just "symbols." To see that this is indeed the case, consider the transformation

$$\underline{\epsilon}'_{i'j'k'\cdots} = \underline{\epsilon}_{ijk\cdots}\, \lambda^{i}{}_{i'}\, \lambda^{j}{}_{j'}\, \lambda^{k}{}_{k'} \cdots = \underline{\epsilon}_{ijk\cdots} \frac{\partial \varphi^i}{\partial x^{i'}} \frac{\partial \varphi^j}{\partial x^{j'}} \frac{\partial \varphi^k}{\partial x^{k'}} \cdots, \tag{5.69}$$

where we seek to determine the nature of $\underline{\epsilon}'_{i'j'k'\cdots}$: could this transformation define a tensor? To find out, we multiply this equation by $\overline{\epsilon}^{\,i'j'k'\cdots}$ to obtain a scalar:

$$\overline{\epsilon}^{\,i'j'k'\cdots}\, \underline{\epsilon}'_{i'j'k'\cdots} = \underline{\epsilon}_{ijk\cdots} \frac{\partial \varphi^i}{\partial x^{i'}} \frac{\partial \varphi^j}{\partial x^{j'}} \frac{\partial \varphi^k}{\partial x^{k'}} \cdots \overline{\epsilon}^{\,i'j'k'\cdots} = n!\, \lambda, \tag{5.70}$$

where—in the last equality—we used determinant (5.68). Using the fact that

$$\overline{\epsilon}^{\,i'j'k'\cdots}\, \underline{\epsilon}_{i'j'k'\cdots} = n!, \tag{5.71}$$

we observe that

$$\underline{\epsilon}'_{i'j'k'\cdots} = \lambda\, \underline{\epsilon}_{i'j'k'\cdots}, \tag{5.72}$$

which implies the transformation rule:

$$\underline{\epsilon}_{i'j'k'\cdots} = \frac{1}{\lambda}\, \underline{\epsilon}_{ijk\cdots}\, \lambda^{i}{}_{i'}\, \lambda^{j}{}_{j'}\, \lambda^{k}{}_{k'} \cdots. \tag{5.73}$$

Using a similar approach, it follows that $\overline{\epsilon}^{\,ijk\cdots}$ transforms according to

$$\overline{\epsilon}^{\,i'j'k'\cdots} = \lambda\, \Lambda^{i'}{}_{i}\, \Lambda^{j'}{}_{j}\, \Lambda^{k'}{}_{k} \cdots \overline{\epsilon}^{\,ijk\cdots}. \tag{5.74}$$

We conclude that, in fact, $\underline{\epsilon}_{ijk\cdots}$ and $\overline{\epsilon}^{\,ijk\cdots}$ are more than "symbols": they are, respectively, a capacity and a density, in the sense that, if changing coordinates, we compute the new components of the Levi-Civita capacity and density using rules (5.60) and (5.61).

Example 5.10 Cross Product

In classical tensor calculus in $\mathbb{R}^3$, the Levi-Civita symbol is used to represent the components of the cross product of two vectors $\mathbf{u}$ and $\mathbf{v}$, as $(\mathbf{u} \times \mathbf{v})_i \equiv \underline{\epsilon}_{ijk}\, u^j\, v^k$. We see that the cross product of two vectors *does not produce another vector*, but a $(0, 1)$ tensor capacity. Similarly, $\mathbf{u} \cdot (\mathbf{v} \times \mathbf{w})$ is a scalar capacity given in terms of components by $\underline{\epsilon}_{ijk}\, u^i\, v^j\, w^k$.

5.8 Determinant of Rank-2 Tensors

We would like to avoid using notions from matrix algebra to define determinants of $p+q=2$ tensors, that is, $(2,0)$, $(1,1)$, and $(0,2)$ tensors. For example, arranging the components of such tensors, that is, T_{ij}, $T^i{}_j$, and T^{ij}, as a matrix—regardless of the placement of the indices—one may calculate determinants, but is the determinant $\det(T_{ij})$ the same kind of scalar as the determinant $\det(T^i{}_j)$?

To bring the construction of a determinant within the realm of tensor algebra, the determinant of a $(1,1)$ tensor $\mathbf{T}$ in n dimensions is defined by

$$\det\mathbf{T} \equiv \frac{1}{n!}\,\underline{\epsilon}_{i_1\cdots i_n}\,T^{i_1}{}_{j_1}\,\cdots\,T^{i_n}{}_{j_n}\,\overline{\epsilon}^{j_1\cdots j_n}, \tag{5.75}$$

where $\underline{\epsilon}_{i_1\cdots i_n}$ and $\overline{\epsilon}^{i_1\cdots i_n}$ are, respectively, the Levi-Civita capacity and density discussed in section 5.7. Using the rules for the transformation of tensors (5.34), tensor densities (5.60), and tensor capacities (5.61), we deduce that the determinant of a $(1,1)$ tensor is a pure scalar; in other words, it is transformation-invariant. Equation (5.75) illustrates the power of the overline/underline notation for densities and capacities, clarifying that the product of a density with a capacity leads to a pure tensor, in this instance a pure scalar. Note that based on equation (5.71), equation (5.75) implies that

$$\underline{\epsilon}_{i_1\cdots i_n}\,T^{i_1}{}_{j_1}\,\cdots\,T^{i_n}{}_{j_n} = \det\mathbf{T}\,\underline{\epsilon}_{j_1\cdots j_n}. \tag{5.76}$$

Following definition (5.75), determinants of $(2,0)$ and $(0,2)$ tensors may also be described in terms of products of two Levi-Civita capacities or densities, respectively. Specifically, in three dimensions, the determinant of a $(0,2)$ tensor is

$$\overline{\overline{\det\mathbf{T}}} = \frac{1}{3!}\,\overline{\epsilon}^{ijk}\,T_{i\ell}\,T_{jm}\,T_{kn}\,\overline{\epsilon}^{\ell mn}, \tag{5.77}$$

and the determinant of a $(2,0)$ tensor is

$$\underline{\underline{\det\mathbf{T}}} = \frac{1}{3!}\,\underline{\epsilon}_{ijk}\,T^{i\ell}\,T^{jm}\,T^{kn}\,\underline{\epsilon}_{\ell mn}. \tag{5.78}$$

Note that (5.77) defines a scalar density of weight two, as indicated by the double overline, whereas (5.78) defines a scalar capacity of weight two, as indicated by the double underline.

5.9 Inverse of Rank-2 Tensors

Provided its determinant is nonzero, the *inverse* of a $(0,2)$ tensor, $\mathbf{T}$, is a $(2,0)$ tensor, $\mathbf{T}^{-1}$, such that

$$T_{ik}\,(T^{-1})^{kj} = \delta^j{}_i, \qquad (T^{-1})^{ik}\,T_{kj} = \delta^i{}_j. \tag{5.79}$$

Similarly, the inverse of a $(2,0)$ tensor, $\mathbf{T}$, is a $(0,2)$ tensor, $\mathbf{T}^{-1}$, such that

$$T^{ik}\,(T^{-1})_{kj} = \delta^i{}_j, \qquad (T^{-1})_{ik}\,T^{kj} = \delta^j{}_i. \tag{5.80}$$

Finally, the inverse of a $(1,1)$ tensor, $\mathbf{T}$, is a $(1,1)$ tensor, $\mathbf{T}^{-1}$, such that

$$T^i{}_k\,(T^{-1})^k{}_j = (T^{-1})^i{}_k\,T^k{}_j = \delta^i{}_j. \tag{5.81}$$

Results (5.80)–(5.81) may be succinctly written in the tensor notation

$$\mathbf{T}\cdot\mathbf{T}^{-1} = \mathbf{T}^{-1}\cdot\mathbf{T} = \mathbf{I}, \tag{5.82}$$

where $\mathbf{I}$ denotes the identity tensor introduced in section 5.4.

5.10 Metric Tensor

The concept of *distance* is not intrinsic to a manifold. It has to be introduced in the context of the *metric tensor*, which we discuss in this chapter. Notably, a differentiable manifold whose tangent spaces are endowed with a positive-definite metric is called a *Riemannian manifold*.

5.10.1 Formulation

Consider two points in proximity to one another with coordinates $\{x^i\}$ and $\{x^i + dx^i\}$, respectively.[9] In classical tensor calculus, the squared distance, ds^2, between these points is expressed in terms of the elements of the metric tensor, g_{ij}, as

$$ds^2 \equiv g_{ij}\,dx^i\,dx^j. \tag{5.83}$$

One can regard this expression as the *definition* of the components of the metric tensor. It follows from this definition that the metric tensor is a symmetric $(0,2)$ tensor in the sense that

$$g_{ij} = g_{ji}. \tag{5.84}$$

In Cartesian components, the squared distance between two points with coordinates $\{z^i\}$ and $\{z^i + dz^i\}$ may be expressed as

$$ds^2 = \delta_{ij}\,dz^i\,dz^j, \tag{5.85}$$

which implies that the components of the metric tensor in Cartesian coordinates are δ_{ij}.[10]

In tensor calculus, the metric tensor, $\mathbf{g}$, is expressed in terms of its components as

$$\mathbf{g} = g_{ij}\,\mathrm{d}x^i \otimes \mathrm{d}x^j = g_{ij}\,\mathbf{e}^i \otimes \mathbf{e}^j. \tag{5.86}$$

[9] Note the use of the italic d, for small coordinate "displacements," to distinguish it from the Roman d, used for differentials.

[10] Only in Cartesian coordinates do we use the notation δ_{ij} with two lower indices, with the usual convection that the symbol is 1 when $i = j$ and zero otherwise.

The metric tensor is a positive-definite $(0,2)$ tensor: a symmetric machine with two slots that accept vectors. It is symmetric in the sense (5.20), as required by positive definiteness: $\mathbf{g}(\mathbf{u},\mathbf{u})>0$, for nonzero vectors $\mathbf{u}$. Notably, this is the definition of a *Riemannian metric*. In this book, we assume that the metric tensor is *nondegenerate*: if $\mathbf{g}(\mathbf{u},\mathbf{w})=0$ for all $\mathbf{w}$, then $\mathbf{u}=\mathbf{0}$. Thus, for two vectors, $\mathbf{u}$ and $\mathbf{w}$, we have

$$\mathbf{g}(\mathbf{u},\mathbf{w})=\mathbf{g}(\mathbf{w},\mathbf{u})=g_{ij}\,u^i\,w^j=g_{ij}\,w^i\,u^j. \tag{5.87}$$

In particular, if we feed the metric tensor a small vector, $dx^i\,\mathbf{e}_i$, in both slots, then it returns

$$\mathbf{g}(dx^i\,\mathbf{e}_i,dx^j\,\mathbf{e}_j)=dx^i\,dx^j\,\mathbf{g}(\mathbf{e}_i,\mathbf{e}_j)=g_{ij}\,dx^i\,dx^j=ds^2, \tag{5.88}$$

that is, the squared distance ds^2, as in definition (5.83). In section 5.2.3 we introduced the dot product (5.13) between a one-form and a vector. The existence of a metric enables us to define the *inner product* between two vectors, $\mathbf{u}$ and $\mathbf{w}$, as

$$(\mathbf{u},\mathbf{w})\equiv\mathbf{g}(\mathbf{u},\mathbf{w})=g_{ij}\,u^i\,w^j. \tag{5.89}$$

Since the metric tensor is a $(0,2)$ tensor, its components transform as

$$g_{i'j'}=g_{ij}\,\lambda^i{}_{i'}\,\lambda^j{}_{j'}. \tag{5.90}$$

The reason one-forms are referred to as "covariant vectors" in classical tensor analysis is rooted in the importance of the metric tensor and the fact that their components transform like the components of the metric tensor; in other words, they are covariant to each other, as opposed to vectors, which are contravariant to the metric-tensor components.

In particular, the transformation from Cartesian coordinates $\{z^i\}$ to general spatial coordinates $\{x^{i'}\}$ is given by

$$g_{i'j'}=\delta_{ij}\,\lambda^i{}_{i'}\,\lambda^j{}_{j'}. \tag{5.91}$$

Given a local map from Cartesian coordinates z to general spatial coordinates x, that is, $z=\varphi(x)$, equation (5.91) provides us with a recipe for the calculation of the components of the metric tensor in a general spatial coordinate system.

Upon calculating the determinant of $(0,2)$ tensor (5.90), based on expression (5.77), and taking the square root of the result, we find

$$\bar{g}'=\lambda\,\bar{g}, \tag{5.92}$$

where λ is given by equation (5.68), and where we defined[11]

$$\bar{g}\equiv\left(\overline{\overline{\det\mathbf{g}}}\right)^{1/2}. \tag{5.93}$$

[11] In general relativity, it is conventional to use a metric tensor with a negative determinant. In that case, we define $\bar{g}\equiv\left(-\overline{\overline{\det\mathbf{g}}}\right)^{1/2}$.

Thus, we deduce from expression (5.62) that $\overline{g}$ transforms like a scalar density of weight one, and that the determinant of the metric tensor, $\overline{\overline{\det \mathbf{g}}}$, transforms like a scalar density of weight two.

To formulate the *inverse* of the metric tensor, in the sense discussed in section 5.9, we verify that its determinant is nonzero. Consider $g_{ij}\, v^i\, u^j = 0$. Upon setting $v^i = 1$ and the remaining components to zero, we obtain a system of n linear equations, $g_{ij}\, u^j = 0$, whose solution is $\mathbf{u} = \mathbf{0}$. Thus, it follows that $\overline{\overline{\det \mathbf{g}}} \neq 0$, as claimed.

Since $\overline{\overline{\det \mathbf{g}}}$ is nonzero, the metric tensor has an inverse $\mathbf{g}^{-1}$ with elements conventionally denoted by g^{ij}, such that

$$g^{ij}\, g_{jk} = \delta^i{}_k \quad \text{and} \quad g_{ij}\, g^{jk} = \delta^k{}_i. \tag{5.94}$$

Thus, the inverse of the metric tensor $\mathbf{g}^{-1}$ is a $(2, 0)$ tensor: two slots that accept one-forms. The square root of the determinant of the inverse of the metric tensor is a scalar capacity of weight one:

$$\underline{g} \equiv \left(\underline{\underline{\det \mathbf{g}^{-1}}} \right)^{1/2} = 1/\overline{g}. \tag{5.95}$$

Example 5.11 Metric Tensor in Continuum Mechanics

In continuum mechanics, the metric tensor may be expressed in either inertial Eulerian or comoving Lagrangian components as

$$\mathbf{g} = g_{ij}\, \mathbf{e}^i \otimes \mathbf{e}^j = g_{IJ}\, \mathbf{e}^I \otimes \mathbf{e}^J. \tag{5.96}$$

The two sets of components are related via the deformation gradient (4.26) and its inverse (4.27) as

$$g_{ij} = (F^{-1})^I{}_i\, (F^{-1})^J{}_j\, g_{IJ}, \tag{5.97}$$

$$g_{IJ} = F^i{}_I\, F^j{}_J\, g_{ij}. \tag{5.98}$$

Whereas the Eulerian components of the metric are stationary, its Lagrangian components are time-dependent. In classical continuum mechanics (see, e.g., Malvern, 1969), the combination of terms on the right-hand side of equation (5.98) is called the *right Cauchy–Green deformation tensor* or *Green deformation tensor*, with components commonly denoted as $C_{IJ} = F^i{}_I\, F^j{}_J\, g_{ij}$. In example 6.3, we further explore the Cauchy–Green tensor.

Upon calculating the determinant of the metric tensor in Lagrangian coordinates (5.98) based on expression (5.77), using the rule $\det(AB) = \det(A)\det(B)$ for the product of two matrices A and B and taking the square root of the result, we find

$$\overline{G} = F\overline{g}. \tag{5.99}$$

Here F is the determinant of the deformation gradient (4.26) defined by

$$F = \frac{1}{3!}\,\underline{\epsilon}_{ijk}\,F^i{}_I\,F^j{}_J\,F^k{}_K\,\overline{\epsilon}^{IJK}. \tag{5.100}$$

We also defined the Lagrangian and Eulerian square roots of the determinant of the metric as

$$\overline{G} \equiv \left(\tfrac{1}{3!}\,\overline{\epsilon}^{IJK}\,g_{IL}\,g_{JM}\,g_{KN}\,\overline{\epsilon}^{LMN}\right)^{1/2}, \tag{5.101}$$

and

$$\overline{g} \equiv \left(\tfrac{1}{3!}\,\overline{\epsilon}^{ijk}\,g_{i\ell}\,g_{jm}\,g_{kn}\,\overline{\epsilon}^{\ell mn}\right)^{1/2}, \tag{5.102}$$

respectively. We conclude from (5.99) that the square root of the determinant of the metric $\overline{g}$ transforms like a scalar density of weight one.

Example 5.12 Metric Tensor in Relativity

In the theory of relativity, the value of the magnitude of an observer's four-velocity $\|\mathbf{u}\|$, introduced in example 4.4, may be expressed in terms of the metric tensor as

$$\|\mathbf{u}\|^2 = g_{\mu\nu}\,u^\mu\,u^\nu = \pm c^2, \tag{5.103}$$

where c denotes the speed of light. Thus, the magnitude of the four-velocity is *always* $\pm c$, where the sign depends on the choice of *metric signature*.

An event $\mathcal{S}$ in *Minkowski spacetime* has coordinates $\{x^\mu(\mathcal{S})\} = \{ct, x, y, z\}$ which are constrained by

$$-c^2t^2 + x^2 + y^2 + z^2 = \text{constant}. \tag{5.104}$$

This equation expresses the *second postulate* of special relativity, which states that the speed of light in free space has the same value, c, in all inertial frames of reference.[a] The *Minkowski metric* of *special relativity*, with components denoted by $\eta_{\mu\nu}$, is defined by

$$\begin{aligned} \mathbf{g} &= \eta_{\mu\nu}\,\mathrm{d}x^\mu \otimes \mathrm{d}x^\nu \\ &\equiv -c^2\,\mathrm{d}t \otimes \mathrm{d}t + \mathrm{d}x \otimes \mathrm{d}x + \mathrm{d}y \otimes \mathrm{d}y + \mathrm{d}z \otimes \mathrm{d}z, \end{aligned} \tag{5.105}$$

reflecting the *metric signature* $(-,+,+,+)$.

The components of the four-velocity, introduced in example 4.4, along a world-line $\{x^\mu(\tau)\}$ are given by

$$u^\mu = \frac{dx^\mu}{d\tau}. \tag{5.106}$$

For an observer at rest in an inertial frame of reference, its four-velocity is parallel to the direction of the time coordinate: $u^0 = c$. For an observer moving at a constant speed v, the four-velocity takes the form $\{u^\mu\} = \gamma\,\{c, v^i\}$, where $v^i = dx^i/dt$, and

where

$$\gamma \equiv \frac{dt}{d\tau} = \frac{1}{\sqrt{1 - v^2/c^2}} \tag{5.107}$$

denotes the *Lorentz factor*, which captures *time dilation*.

[a] The *first postulate* of special relativity states that the laws of physics take the same form in all inertial frames of reference.

Lowering an index and *raising an index* on a component of a tensor may be accomplished using the metric tensor and its inverse, respectively. For instance,

$$T^i{}_j = T^{ik} g_{kj}, \qquad T^{ij} = T^i{}_k g^{kj}. \tag{5.108}$$

Thus, the metric tensor may be used to change the nature of a tensor slot from accepting a vector to accepting a one-form, and vice versa.

The introduction of the metric tensor enabled us to generalize the *dot product* (5.13) to vectors based on definition (5.89):

$$\mathbf{u} \cdot \mathbf{w} \equiv g_{ij}\, u^i\, w^j. \tag{5.109}$$

Similarly, the *double-dot product* between a $(2,0)$ tensor and a $(0,2)$ tensor, given by equation (5.15), can be generalized to include (p,q) tensors with $p+q=2$:

$$\mathbf{T} : \mathbf{S} \equiv T^{ij} S_{ij} = T_{ij} S^{ij} = T^i{}_k S_i{}^k = T_i{}^j S^i{}_j. \tag{5.110}$$

The metric tensor may also be used to raise and lower indices of nontensorial entities, such as the Christoffel symbol, as shown in section 7.1.8.

5.10.2 Geometrical Meaning

A metric tensor endows a manifold with a structure that provides a geometrical meaning: the concept of distance. The importance of such a structure is indicated by distinguishing manifolds endowed with positive-definite metric tensors by referring to them as *Riemannian manifolds*. For a Riemannian manifold, we require $\mathbf{g}$ to be a smooth $(0,2)$ tensor field, which allows us to obtain a unique value at each point of tangency and to integrate the square root of equation (5.83) between two points on the manifold to obtain the length of a curve connecting these points. Specifically, if $\mathbf{u}$ is a tangent to the curve $x(\lambda)$, such that $\mathrm{d}x^i/\mathrm{d}\lambda = u^i$, then the length of the curve between $\lambda = \lambda_0$ and $\lambda = \lambda_1$ is given by

$$l = \int ds = \int_{\lambda_0}^{\lambda_1} \sqrt{\mathbf{g}(\mathbf{u},\mathbf{u})}\, d\lambda = \int_{\lambda_0}^{\lambda_1} \sqrt{g_{ij} \frac{\mathrm{d}x^i}{\mathrm{d}\lambda} \frac{\mathrm{d}x^j}{\mathrm{d}\lambda}}\, d\lambda. \tag{5.111}$$

Expression (5.83) is a statement of the *Pythagorean theorem* for infinitesimal considerations, which is not valid for finite intervals on a manifold. Below, using the metric, we can formulate a notion of lengths of vectors and one-forms.

5.10.3 Norm of Vectors and One-Forms

The existence of a metric tensor endows the manifold with a structure that enables us to define the notion of the *norm*, which is a *length*, of a vector. It is obtained by feeding the metric the same vector $\mathbf{u}$ in both slots:

$$||\mathbf{u}||^2 \equiv (\mathbf{u}, \mathbf{u}) = \mathbf{g}(\mathbf{u}, \mathbf{u}) = g_{ij}\, u^i\, u^j = \mathbf{u} \cdot \mathbf{u}, \tag{5.112}$$

where in the last equality we used definition (5.109) of the dot product between two vectors. A *unit vector* has unit length and is obtained by scaling it by its length:

$$\hat{\mathbf{u}} \equiv \frac{\mathbf{u}}{||\mathbf{u}||}. \tag{5.113}$$

We can also define an angle, θ, between two vectors, $\mathbf{u}$ and $\mathbf{w}$:

$$\cos\theta \equiv \frac{\mathbf{g}(\mathbf{u}, \mathbf{w})}{||\mathbf{u}||\; ||\mathbf{w}||}. \tag{5.114}$$

We say that two vectors are *orthogonal* to one another if $\mathbf{g}(\mathbf{u}, \mathbf{w}) = 0$.

Using the inverse metric, we can define similar concepts for a one-form, $\boldsymbol{\omega}$, namely

$$||\boldsymbol{\omega}||^2 \equiv \mathbf{g}^{-1}(\boldsymbol{\omega}, \boldsymbol{\omega}) = g^{ij}\, \omega_i\, \omega_j, \tag{5.115}$$

and

$$\hat{\boldsymbol{\omega}} \equiv \frac{\boldsymbol{\omega}}{||\boldsymbol{\omega}||}. \tag{5.116}$$

For two one-forms, $\boldsymbol{\omega}$ and $\boldsymbol{\eta}$, we define

$$\cos\theta \equiv \frac{\mathbf{g}^{-1}(\boldsymbol{\omega}, \boldsymbol{\eta})}{||\boldsymbol{\omega}||\; ||\boldsymbol{\eta}||}. \tag{5.117}$$

5.10.4 Metric in Tetrads

Using the tetrads introduced in section 5.3.1, we can find an expression of the metric in a tetrad coframe basis:

$$\begin{aligned} \mathbf{g} &= g_{ij}\, \mathrm{d}x^i \otimes \mathrm{d}x^j \\ &= g_{ij}\, e^i{}_{i'}\, e^j{}_{j'}\, \mathbf{e}^{i'} \otimes \mathbf{e}^{j'} \\ &= g_{i'j'}\, \mathbf{e}^{i'} \otimes \mathbf{e}^{j'}, \end{aligned} \tag{5.118}$$

where

$$g_{i'j'} = g_{ij}\, e^i{}_{i'}\, e^j{}_{j'}. \tag{5.119}$$

We now have an opportunity to select *orthogonal* coframes by choosing

$$g_{i'j'} = \delta_{i'j'}, \tag{5.120}$$

which implies that

$$g_{ij} = e_i{}^{i'}\, e_j{}^{j'}\, \delta_{i'j'}. \tag{5.121}$$

Thus, by calculating the determinant of this expression,

$$\bar{g} = \det(e_i{}^{i'}), \tag{5.122}$$

we may interpret $e_i{}^{i'}$ as the "square root of the metric."

5.11 Adjoint of a (1,1) Tensor

The *adjoint* of a $(1,1)$ tensor $\mathbf{T}$ is denoted by $\mathbf{T}^\dagger$ and is defined in terms of the dot product (5.89) as

$$(\mathbf{T}^\dagger \cdot \mathbf{u}, \mathbf{w}) \equiv (\mathbf{u}, \mathbf{T} \cdot \mathbf{w}). \tag{5.123}$$

Written out in components, we have

$$g_{ij}\, (T^\dagger)^i{}_k\, u^k\, w^j = g_{ij}\, u^i\, T^j{}_k\, w^k, \tag{5.124}$$

implying that

$$(T^\dagger)^i{}_j = g^{i\ell}\, T^k{}_\ell\, g_{kj}. \tag{5.125}$$

Notice that the definition of the adjoint (5.123) involves the inner product and therefore the metric, whereas the definition of the transpose (5.31) involves the duality product but not the metric.

Alternatively, the adjoint of a $(1,1)$ tensor may be defined in terms of the duality product as

$$\begin{aligned} \langle \mathbf{g} \cdot \mathbf{T}^\dagger \cdot \mathbf{u}, \mathbf{w} \rangle &\equiv \langle \mathbf{g} \cdot \mathbf{u}, \mathbf{T} \cdot \mathbf{w} \rangle \\ &= \langle \mathbf{T}^t \cdot \mathbf{g} \cdot \mathbf{u}, \mathbf{w} \rangle, \end{aligned} \tag{5.126}$$

where in the second equality we used the definition of the transpose (5.31). Thus, the adjoint $\mathbf{T}^\dagger$ and the transpose $\mathbf{T}^t$ are related via

$$\mathbf{T}^\dagger = \mathbf{g}^{-1} \cdot \mathbf{T}^t \cdot \mathbf{g}, \tag{5.127}$$

or in components

$$(T^\dagger)^i{}_j = g^{i\ell}\, (T^t)_\ell{}^k\, g_{kj}. \tag{5.128}$$

Upon combining expressions (5.128) and (5.125) and lowering and raising indices with the metric, we find that

$$(T^{\dagger})^{i}{}_{j} = (T^{t})^{i}{}_{j} = T_{j}{}^{i}. \tag{5.129}$$

The first equality implies that the adjoint and the transpose of a $(1,1)$ tensor are the same. If we permit ourselves to think of a $(1,1)$ tensor as a matrix, then the last equality in expression (5.129) implies exchanging its rows and columns to obtain its adjoint or transpose.

The $(1,1)$ tensors considered in this book are *endomorphisms*, mapping a linear space onto itself. More generally, one can consider linear maps between *distinct* manifolds with distinct dimensions, as illustrated in figure 3.3. In such a case, the transpose of the mapping $T^{i}{}_{j'}$ is defined by

$$(T^{t})_{j'}{}^{i} = T^{i}{}_{j'}, \tag{5.130}$$

and the relationship between the transpose and adjoint of the mapping $T^{i}{}_{j'}$ may be expressed in the form

$$(T^{\dagger})^{i}{}_{j'} = g^{i\ell}\,(T^{t})_{\ell}{}^{k'}\,g'_{k'j'}, \tag{5.131}$$

where an index i denotes components in one space, with metric g_{ij}, whereas a primed index i' denotes components in the other space, with metric $g'_{i'j'}$. Such is the case in the formulation of *inverse problems* (Tarantola, 2005).

5.12 Tensor Densities and Capacities Revisited

We have seen that in a Riemannian manifold, according to expression (5.92), the square root of the determinant of the metric tensor transforms like a scalar density of weight one: $\overline{g}' = \lambda\,\overline{g}$. Accordingly, tensor densities and capacities of weight w may be defined in terms of an "ordinary" tensor, with elements $T^{i_1\cdots i_p}{}_{j_1\cdots j_q}$, as

$$\overline{T}^{i_1\cdots i_p}{}_{j_1\cdots j_q} = \overline{g}^{w}\,T^{i_1\cdots i_p}{}_{j_1\cdots j_q}, \tag{5.132}$$

and

$$\underline{T}^{i_1\cdots i_p}{}_{j_1\cdots j_q} = \underline{g}^{w}\,T^{i_1\cdots i_p}{}_{j_1\cdots j_q}, \tag{5.133}$$

respectively. These objects transform according to rules (5.60) and (5.61). In general relativity, tensor densities are commonly expressed in a Fraktur typeface: $\mathfrak{T}^{i_1\cdots i_p}{}_{j_1\cdots j_q}$.

5.13 Levi-Civita Pseudotensor

In section 5.7, we defined the Levi-Civita density and capacity, which take the values of $\{+1,0,-1\}$, in any coordinate system. Tensors may be obtained as a result of the canonical bijections:

$$\epsilon_{ijk\cdots} \equiv \overline{g}\,\underline{\epsilon}_{ijk\cdots} \quad \text{and} \quad \epsilon^{ijk\cdots} \equiv \underline{g}\,\overline{\epsilon}^{ijk\cdots}. \tag{5.134}$$

Instead of $\{+1,0,-1\}$, these tensor elements take values of $\{+\overline{g},0,-\overline{g}\}$ and $\{+\underline{g},0,-\underline{g}\}$, respectively.

In three dimensions, one may identify the elements ϵ_{ijk} with the *volume form* $\boldsymbol{\epsilon}$, an antisymmetric $(0,3)$ pseudoform discussed extensively in section 8.5:

$$\boldsymbol{\epsilon} = \epsilon_{ijk}\, \mathbf{e}^i \otimes \mathbf{e}^j \otimes \mathbf{e}^k. \tag{5.135}$$

The sign of ϵ_{ijk} depends on the orientation of the manifold. We must compensate for the sign change in the transformation to keep the integration volume element positive. Thus, the volume form or *Levi-Civita pseudotensor* transforms according to the modified rules discussed in section 5.3.2, making it a *pseudotensor* (Frankel, 2004).

An *oriented surface element* or *surface patch* may be obtained by feeding the volume form (5.135) two nonparallel vectors, $\mathbf{u}$ and $\mathbf{w}$, which together define a local plane:

$$\hat{\mathbf{n}}\, dS \equiv \boldsymbol{\epsilon}(\,\cdot\,, \mathbf{u}, \mathbf{w}). \tag{5.136}$$

The surface element $\hat{\mathbf{n}}\, dS$, a one-form, is *oriented* because exchanging the vectors $\mathbf{u}$ and $\mathbf{w}$ changes the sign of the surface element. In section 8.6 we discuss surfaces as forms and clarify the notation for a surface element.

Example 5.13 Volume Form in Continuum Mechanics

In continuum mechanics, the *volume form* $\boldsymbol{\epsilon}$ may be expressed in either inertial Eulerian or comoving Lagrangian components as

$$\boldsymbol{\epsilon} = \epsilon_{ijk}\, \mathbf{e}^i \otimes \mathbf{e}^j \otimes \mathbf{e}^k = \epsilon_{IJK}\, \mathbf{e}^I \otimes \mathbf{e}^J \otimes \mathbf{e}^K, \tag{5.137}$$

where

$$\epsilon_{ijk} \equiv \overline{g}\, \underline{\epsilon}_{ijk} \qquad \text{and} \qquad \epsilon_{IJK} \equiv \overline{G}\, \underline{\epsilon}_{IJK}. \tag{5.138}$$

The square roots of the determinants of the Eulerian and Lagrangian components of the metric tensor are related via (5.99). The two sets of components are related via the deformation gradient (4.26) and its inverse (4.27) as

$$\epsilon_{ijk} = (F^{-1})^I{}_i\, (F^{-1})^J{}_j\, (F^{-1})^K{}_k\, \epsilon_{IJK}, \tag{5.139}$$

$$\epsilon_{IJK} = F^i{}_I\, F^j{}_J\, F^k{}_K\, \epsilon_{ijk}. \tag{5.140}$$

Whereas the Eulerian components of the volume form are stationary, its Lagrangian components are time-dependent.

Example 5.14 Surface One-Form in Continuum Mechanics

In continuum mechanics, a surface element (5.136) has Eulerian components

$$\hat{n}_i\, dS = \epsilon_{ijk}\, u^j\, w^k, \tag{5.141}$$

and Lagrangian components

$$\hat{n}_I \, dS = \epsilon_{IJK} \, u^J \, w^K. \tag{5.142}$$

The *volume transport* by a vector field $\mathbf{v}$ through an oriented surface element $\hat{\mathbf{n}} \, dS$ spanned by two vectors $\mathbf{u}$ and $\mathbf{w}$ may now be expressed as

$$\mathbf{v} \cdot \hat{\mathbf{n}} \, dS = \boldsymbol{\epsilon}(\mathbf{v}, \mathbf{u}, \mathbf{w}) = v^i \, \hat{n}_i \, dS = v^I \, \hat{n}_I \, dS. \tag{5.143}$$

The metric tensor $\mathbf{g}$ and the Levi-Civita pseudotensor $\boldsymbol{\epsilon}$ are the two most important entities in the structure of a Riemannian manifold.

Example 5.15 Cross Product Revisited

The implication for the cross product between two vectors, $\mathbf{u} \times \mathbf{v}$, is that its components are given by $\epsilon_{ijk} \, u^j \, v^k$, which are the components of a pseudo one-form. The proper volume spanned by three vectors $\mathbf{u}$, $\mathbf{v}$, and $\mathbf{w}$ is $\boldsymbol{\epsilon}(\mathbf{u}, \mathbf{v}, \mathbf{w}) = \epsilon_{ijk} \, u^i \, v^j \, w^k$, which is a *pseudoscalar*, that is, a scalar that changes sign under coordinate reflections. Here we see the status of the Levi-Civita pseudotensor as the measurer of volume: it is a machine that takes three vectors and returns the volume spanned by the three vectors. In other words, it counts how many times the unit volume fits in the volume defined by the vectors.

5.14 Kronecker Determinants

On an n-dimensional manifold, the Levi-Civita tensor has n indices. For any nonnegative integer $p \leq n$, consider an integer q such that $p + q = n$. The following property holds:

$$\epsilon^{i_1 \ldots i_p s_1 \ldots s_q} \, \epsilon_{j_1 \ldots j_p s_1 \ldots s_q} = q! \, \det \begin{pmatrix} \delta^{i_1}{}_{j_1} & \delta^{i_1}{}_{j_2} & \cdots & \delta^{i_1}{}_{j_p} \\ \delta^{i_2}{}_{j_1} & \delta^{i_2}{}_{j_2} & \cdots & \delta^{i_2}{}_{j_p} \\ \vdots & \vdots & \ddots & \vdots \\ \delta^{i_p}{}_{j_1} & \delta^{i_p}{}_{j_2} & \cdots & \delta^{i_p}{}_{j_p} \end{pmatrix}. \tag{5.144}$$

The determinant on the right-hand side is called the *Kronecker determinant* or the *generalized Kronecker delta* and is denoted by

$$\delta^{i_1 i_2 \ldots i_p}{}_{j_1 j_2 \ldots j_p} \equiv \det \begin{pmatrix} \delta^{i_1}{}_{j_1} & \delta^{i_1}{}_{j_2} & \cdots & \delta^{i_1}{}_{j_p} \\ \delta^{i_2}{}_{j_1} & \delta^{i_2}{}_{j_2} & \cdots & \delta^{i_2}{}_{j_p} \\ \vdots & \vdots & \ddots & \vdots \\ \delta^{i_p}{}_{j_1} & \delta^{i_p}{}_{j_2} & \cdots & \delta^{i_p}{}_{j_p} \end{pmatrix}. \tag{5.145}$$

Because the Kronecker determinant is defined as a product of Levi-Civita tensors, it is itself a tensor. It generalizes the definition of the Kronecker tensor $\delta^i{}_j$, because it has the following properties:

$$\delta^{i_1 i_2 \ldots i_m}{}_{j_1 j_2 \ldots j_m} = \begin{cases} +1 & i_1, \ldots, i_m \text{ even permutation of } j_1, \ldots, j_m, \\ -1 & i_1, \ldots, i_m \text{ odd permutation of } j_1, \ldots, j_m, \\ 0 & \text{two } i\text{'s or } j\text{'s are the same}, \\ 0 & \text{sets } \{i_1, \ldots, i_m\} \text{ and } \{j_1, \ldots, j_m\} \text{ differ.} \end{cases}$$

Since applying the same permutation to the indices of the two Levi-Civita tensors in equation (5.144) does not change the overall sign of the expression, we have

$$\begin{aligned} \epsilon^{i_1 \ldots i_p s_1 \ldots s_q} \epsilon_{j_1 \ldots j_p s_1 \ldots s_q} &= \epsilon^{s_1 \ldots s_q i_1 \ldots i_p} \epsilon_{s_1 \ldots s_q j_1 \ldots j_p} \\ &= q! \, \delta^{i_1 i_2 \ldots i_p}{}_{j_1 j_2 \ldots j_p}, \end{aligned} \tag{5.146}$$

but if we perform only a permutation in one of the Levi-Civita tensors, then we must take care of the sign of the permutation to obtain

$$\begin{aligned} \epsilon^{i_1 \ldots i_p s_1 \ldots s_q} \epsilon_{s_1 \ldots s_q j_1 \ldots j_p} &= \epsilon^{s_1 \ldots s_q i_1 \ldots i_p} \epsilon_{j_1 \ldots j_p s_1 \ldots s_q} \\ &= (-1)^{pq} \, q! \, \delta^{i_1 i_2 \ldots i_p}{}_{j_1 j_2 \ldots j_p}. \end{aligned} \tag{5.147}$$

Such a change of sign happens only in spaces with even dimension ($n = 2, 4, \ldots$), because in spaces with odd dimension ($n = 3, 5, \ldots$) the condition $p + q = n$ implies that pq is an even number, and $(-1)^{pq} = +1$.

We note the following two properties of the generalized Kronecker delta:

$$\frac{1}{q!} \delta^{i_1 \ldots i_p j_1 \ldots j_q}{}_{k_1 \ldots k_p \ell_1 \ldots \ell_q} \, \delta^{m_1 \ldots m_q}{}_{j_1 \ldots j_q} = \delta^{i_1 \ldots i_p m_1 \ldots m_q}{}_{k_1 \ldots k_p \ell_1 \ldots \ell_q}, \tag{5.148}$$

and

$$\frac{1}{q!} \epsilon_{i_1 \ldots i_p k_1 \ldots k_q} \, \delta^{k_1 \ldots k_q}{}_{j_1 \ldots j_q} = \epsilon_{i_1 \ldots i_p j_1 \ldots j_q}. \tag{5.149}$$

There are a number of important relations between the Kronecker and Levi-Civita tensors. For example, on a two-dimensional manifold, we have

$$\delta^{ij}{}_{k\ell} = \epsilon^{ij} \epsilon_{k\ell} = \delta^i{}_k \delta^j{}_\ell - \delta^i{}_\ell \delta^j{}_k, \tag{5.150}$$

$$\delta^j{}_k = \epsilon^{ij} \epsilon_{ik} = \delta^j{}_k, \tag{5.151}$$

$$\delta = \tfrac{1}{2} \epsilon^{ij} \epsilon_{ij} = 1. \tag{5.152}$$

Perhaps more interestingly, on a three-dimensional manifold, we have

$$\delta^{ijk}{}_{\ell mn} = \epsilon^{ijk}\,\epsilon_{\ell mn} = \delta^i{}_\ell\,\delta^j{}_m\,\delta^k{}_n + \delta^i{}_m\,\delta^j{}_n\,\delta^k{}_\ell + \delta^i{}_n\,\delta^j{}_\ell\,\delta^k{}_m \\ - \delta^i{}_\ell\,\delta^j{}_n\,\delta^k{}_m - \delta^i{}_n\,\delta^j{}_m\,\delta^k{}_\ell - \delta^i{}_m\,\delta^j{}_\ell\,\delta^k{}_n, \tag{5.153}$$

$$\delta^{jk}{}_{\ell m} = \epsilon^{ijk}\,\epsilon_{i\ell m} = \delta^j{}_\ell\,\delta^k{}_m - \delta^j{}_m\,\delta^k{}_\ell, \tag{5.154}$$

$$\delta^k{}_\ell = \tfrac{1}{2}\,\epsilon^{ijk}\,\epsilon_{ij\ell} = \delta^k{}_\ell, \tag{5.155}$$

$$\delta = \tfrac{1}{6}\,\epsilon^{ijk}\,\epsilon_{ijk} = 1. \tag{5.156}$$

And finally, for relativistic readers, on a four-dimensional manifold, we have

$$\begin{aligned} \delta^{ijk\ell}{}_{mnpq} &= \epsilon^{ijk\ell}\,\epsilon_{mnpq} \\ &= \delta^i{}_m\,\delta^j{}_n\,\delta^k{}_p\,\delta^\ell{}_q + \delta^i{}_m\,\delta^j{}_p\,\delta^k{}_q\,\delta^\ell{}_n + \delta^i{}_m\,\delta^j{}_q\,\delta^k{}_n\,\delta^\ell{}_p \\ &\quad + \delta^i{}_n\,\delta^j{}_q\,\delta^k{}_p\,\delta^\ell{}_m + \delta^i{}_n\,\delta^j{}_p\,\delta^k{}_m\,\delta^\ell{}_q + \delta^i{}_n\,\delta^j{}_m\,\delta^k{}_q\,\delta^\ell{}_p \\ &\quad + \delta^i{}_p\,\delta^j{}_q\,\delta^k{}_m\,\delta^\ell{}_n + \delta^i{}_p\,\delta^j{}_m\,\delta^k{}_n\,\delta^\ell{}_q + \delta^i{}_p\,\delta^j{}_n\,\delta^k{}_q\,\delta^\ell{}_m \\ &\quad + \delta^i{}_q\,\delta^j{}_m\,\delta^k{}_p\,\delta^\ell{}_n + \delta^i{}_q\,\delta^j{}_n\,\delta^k{}_m\,\delta^\ell{}_p + \delta^i{}_q\,\delta^j{}_p\,\delta^k{}_n\,\delta^\ell{}_m \\ &\quad - \delta^i{}_m\,\delta^j{}_n\,\delta^k{}_q\,\delta^\ell{}_p - \delta^i{}_m\,\delta^j{}_p\,\delta^k{}_n\,\delta^\ell{}_q - \delta^i{}_m\,\delta^j{}_q\,\delta^k{}_p\,\delta^\ell{}_n \\ &\quad - \delta^i{}_n\,\delta^j{}_p\,\delta^k{}_q\,\delta^\ell{}_m - \delta^i{}_n\,\delta^j{}_q\,\delta^k{}_m\,\delta^\ell{}_p - \delta^i{}_n\,\delta^j{}_m\,\delta^k{}_p\,\delta^\ell{}_q \\ &\quad - \delta^i{}_p\,\delta^j{}_q\,\delta^k{}_n\,\delta^\ell{}_m - \delta^i{}_p\,\delta^j{}_m\,\delta^k{}_q\,\delta^\ell{}_n - \delta^i{}_p\,\delta^j{}_n\,\delta^k{}_m\,\delta^\ell{}_q \\ &\quad - \delta^i{}_q\,\delta^j{}_m\,\delta^k{}_n\,\delta^\ell{}_p - \delta^i{}_q\,\delta^j{}_n\,\delta^k{}_p\,\delta^\ell{}_m - \delta^i{}_q\,\delta^j{}_p\,\delta^k{}_m\,\delta^\ell{}_n, \end{aligned} \tag{5.157}$$

$$\begin{aligned} \delta^{jk\ell}{}_{mnp} &= \epsilon^{ijk\ell}\,\epsilon_{imnp} \\ &= \delta^j{}_m\,\delta^k{}_n\,\delta^\ell{}_p + \delta^j{}_n\,\delta^k{}_p\,\delta^\ell{}_m + \delta^j{}_p\,\delta^k{}_m\,\delta^\ell{}_n \\ &\quad - \delta^j{}_m\,\delta^k{}_p\,\delta^\ell{}_n - \delta^j{}_n\,\delta^k{}_m\,\delta^\ell{}_p - \delta^j{}_p\,\delta^k{}_n\,\delta^\ell{}_m, \end{aligned} \tag{5.158}$$

$$\delta^{k\ell}{}_{mn} = \tfrac{1}{2}\,\epsilon^{ijk\ell}\,\epsilon_{ijmn} = (\delta^k{}_m\,\delta^\ell{}_n - \delta^k{}_n\,\delta^\ell{}_m), \tag{5.159}$$

$$\delta^\ell{}_m = \tfrac{1}{6}\,\epsilon^{ijk\ell}\,\epsilon_{ijkm} = \delta^\ell{}_m, \tag{5.160}$$

$$\delta = \tfrac{1}{24}\,\epsilon^{ijk\ell}\,\epsilon_{ijk\ell} = 1. \tag{5.161}$$

Example 5.16 Cramer's Rule

In this example, we use expression (5.76) for a $(1,1)$ tensor, $\mathbf{T}$, that is,

$$\underline{\epsilon}_{ijk}\,T^i{}_\ell\,T^j{}_m\,T^k{}_n = \det \mathbf{T}\,\underline{\epsilon}_{\ell mn}, \tag{5.162}$$

to obtain *Cramer's rule* for its inverse, $\mathbf{T}^{-1}$. Upon contracting (5.162) with $\bar{\epsilon}^{pmn}$, using property (5.155), we find

$$\bar{\epsilon}^{pmn}\,\underline{\epsilon}_{ijk}\,T^{i}{}_{\ell}\,T^{j}{}_{m}\,T^{k}{}_{n}=2\det\mathbf{T}\,\delta^{p}{}_{\ell}. \tag{5.163}$$

Upon multiplying this result with $(T^{-1})^{\ell}{}_{q}$, we find

$$(T^{-1})^{p}{}_{q}=\tfrac{1}{2}\,(\det\mathbf{T})^{-1}\,\bar{\epsilon}^{pmn}\,\underline{\epsilon}_{qjk}\,T^{j}{}_{m}\,T^{k}{}_{n}. \tag{5.164}$$

Cramer's rule (5.164) gives the elements of the inverse $\mathbf{T}^{-1}$ in terms of products of the elements of $\mathbf{T}$.

For the Lagrangian components of the three-dimensional metric tensor we have

$$\bar{\epsilon}^{IJK}\,g_{PL}\,g_{JM}\,g_{KM}\,\bar{\epsilon}^{LMN}=2\,\overline{G}^{2}\,\delta^{I}{}_{P}. \tag{5.165}$$

and Cramer's rule takes the form

$$g^{PQ}=\tfrac{1}{2}\,\underline{G}^{2}\,\bar{\epsilon}^{PJK}\,\bar{\epsilon}^{QMN}\,g_{JM}\,g_{KN}, \tag{5.166}$$

thereby providing an explicit expression for the elements of its inverse.

In four-dimensional general relativity, we have

$$g^{\mu\nu}=\tfrac{1}{6}\,\underline{g}^{2}\,\bar{\epsilon}^{\mu\tau\sigma\alpha}\,\bar{\epsilon}^{\nu\beta\eta\zeta}\,g_{\tau\beta}\,g_{\sigma\eta}\,g_{\alpha\zeta}, \tag{5.167}$$

where in this case we used property (5.160).

5.15 Rotations

Rotations in n dimensions may be represented in terms of a $(1,1)$ *rotation tensor*, $\mathbf{Q}$. The inverse of a rotation tensor, which for a general $(1,1)$ tensor is discussed in section 5.9, has the important property

$$\mathbf{Q}^{-1}=\mathbf{Q}^{t}, \tag{5.168}$$

which means that its inverse is equal to its transpose. Thus, in components, rotations have the properties

$$Q^{i}{}_{k}\,(Q^{t})^{k}{}_{j}=\delta^{i}{}_{j}, \qquad (Q^{t})^{i}{}_{k}\,Q^{k}{}_{j}=\delta^{i}{}_{j}, \tag{5.169}$$

where, according to expression (5.129), the elements of the transpose are determined in terms of the elements of the original matrix by $(Q^{t})^{i}{}_{j}=Q_{j}{}^{i}$. The determinant of a rotation matrix is one:

$$\det\mathbf{Q}=1. \tag{5.170}$$

The set of all rotation forms a group: the *special orthogonal group*, SO(n), which is a subgroup of the *orthogonal group*, O(n); the latter also includes reflections.

We can conveniently envision the components of a $(1,1)$ tensor in $\mathbb{R}^n$ as an $n \times n$ matrix, which for $\mathbf{Q}$ we refer to as a *rotation matrix*. Such a notation facilitates clarification of the nomenclature, namely, the *orthogonal* group, which is a group of *orthogonal transformations*. The dot product of any two columns or any two rows is zero: they are orthogonal to each other. Furthermore, the norm of any column and any row is equal to one; we can say that they are orthonormal.

To gain further insight into $\mathbf{Q}$, let us consider its geometrical meaning by examining its components in three spatial dimensions, which we write as

$$\mathbf{Q} = \begin{pmatrix} Q^1{}_1 & Q^1{}_2 & Q^1{}_3 \\ Q^2{}_1 & Q^2{}_2 & Q^2{}_3 \\ Q^3{}_1 & Q^3{}_2 & Q^3{}_3 \end{pmatrix}. \tag{5.171}$$

5.15.1 Euler Angles

Given a Cartesian coordinate system, we can represent rotations about the x- , y-, and z-axes as

$$\mathbf{Q}_x(\phi) = \begin{pmatrix} 1 & 0 & 0 \\ 0 & \cos\phi & \sin\phi \\ 0 & -\sin\phi & \cos\phi \end{pmatrix}, \tag{5.172}$$

$$\mathbf{Q}_y(\phi) = \begin{pmatrix} \cos\phi & 0 & -\sin\phi \\ 0 & 1 & 0 \\ \sin\phi & 0 & \cos\phi \end{pmatrix}, \tag{5.173}$$

and

$$\mathbf{Q}_z(\phi) = \begin{pmatrix} \cos\phi & \sin\phi & 0 \\ -\sin\phi & \cos\phi & 0 \\ 0 & 0 & 1 \end{pmatrix}, \tag{5.174}$$

respectively. A general rotation may be expressed as a product of these three $(1,1)$ tensors based on the three *Euler angles*. Following the convention of Edmonds (1960), we specify a rotation from an unprimed coordinate system x, y, z to a primed system x', y', z' by the following sequence (see figure 5.3).

1. A rotation about the z-axis through an angle $0 \leq \alpha \leq 2\pi$ leading to an intermediate coordinate system ξ, η, $\zeta = z$. This rotation is represented by the matrix

$$\mathbf{Q}_z(\alpha) = \begin{pmatrix} \cos\alpha & \sin\alpha & 0 \\ -\sin\alpha & \cos\alpha & 0 \\ 0 & 0 & 1 \end{pmatrix}. \tag{5.175}$$

2. A rotation about the η-axis through an angle $0 \leq \beta \leq \pi$, leading to a second intermediate coordinate system ξ', $\eta' = \eta$, ζ'. This rotation about the so-called *line of nodes* is

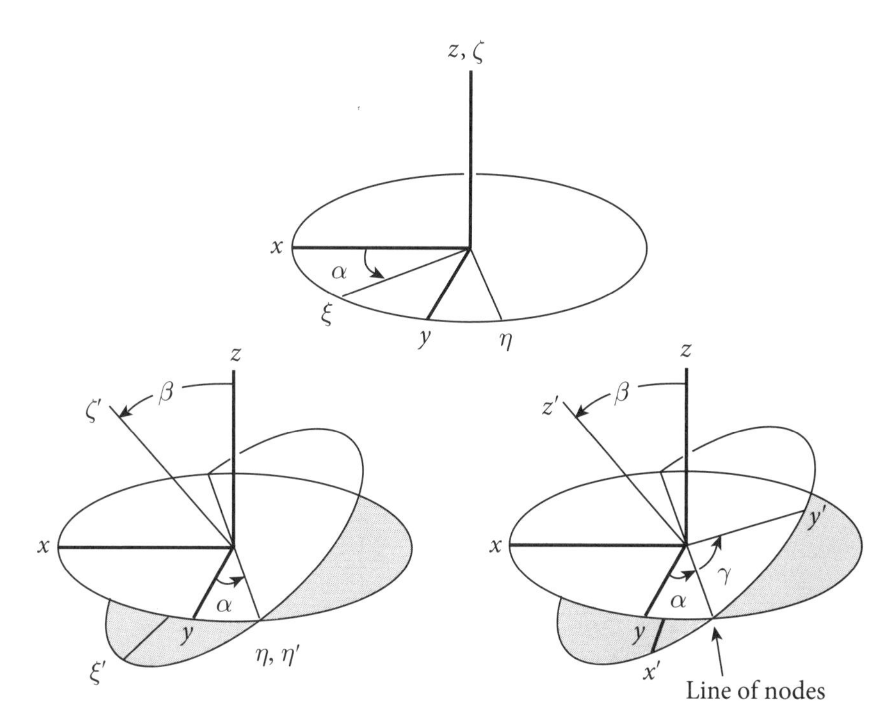

Figure 5.3: Ordered sequence of axial rotations defining the three Euler angles: (*top*) rotation $\mathbf{Q}_z(\alpha)$ through an angle α about the z-axis; (*bottom left*) rotation $\mathbf{Q}_y(\beta)$ through an angle β about the line of nodes η; (*bottom right*) rotation $\mathbf{Q}_z(\gamma)$ through an angle γ about the ζ'-axis. The coordinates of a spatial point $\mathcal{S}$ are transformed from $\{x, y, z\}$ to $\{x', y', z'\}$. Redrawn from Dahlen & Tromp (1998).

represented by the matrix

$$\mathbf{Q}_y(\beta) = \begin{pmatrix} \cos\beta & 0 & -\sin\beta \\ 0 & 1 & 0 \\ \sin\beta & 0 & \cos\beta \end{pmatrix}. \tag{5.176}$$

3. A rotation about the ζ'-axis through an angle $0 \leq \gamma \leq 2\pi$, leading to the final coordinate system $x', y', z' = \zeta'$. The matrix describing this third rotation is

$$\mathbf{Q}_z(\gamma) = \begin{pmatrix} \cos\gamma & \sin\gamma & 0 \\ -\sin\gamma & \cos\gamma & 0 \\ 0 & 0 & 1 \end{pmatrix}. \tag{5.177}$$

The final effect of this sequence of axial rotations is the product,

$$\mathbf{Q} = \mathbf{Q}_z(\gamma)\, \mathbf{Q}_y(\beta)\, \mathbf{Q}_z(\alpha), \tag{5.178}$$

of matrices (5.175)–(5.177). Every rotation in three dimensions is characterized in terms of an *axis of rotation*, which remains fixed during rotation, and an *angle of rotation* about this axis. The eigenvalues of a 3×3 rotation tensor are 1, and $\cos\phi \pm i \sin\phi = e^{\pm i\phi}$, where the eigenvector associated with eigenvalue 1 corresponds to the axis of rotation, and ϕ, determined by the remaining two eigenvalues, corresponds to the associated angle of rotation.

5.15.2 Rodrigues's Formula

Rotations are closely related to antisymmetric tensors, namely, for any antisymmetric tensor $\mathbf{q} = -\mathbf{q}^t$,

$$\mathbf{Q} = \exp \mathbf{q} = \sum_{k=0}^{\infty} \frac{1}{k!}\mathbf{q}^k = \mathbf{I} + \mathbf{q} + \ldots + \frac{1}{k!}\mathbf{q}^k + \ldots \tag{5.179}$$

is a rotation tensor. Conversely, for any rotation tensor $\mathbf{Q}$,

$$\mathbf{q} = \log \mathbf{Q} = \sum_{k=1}^{\infty} \frac{(-1)^{k+1}}{k}(\mathbf{Q} - \mathbf{I})^k \tag{5.180}$$

is an antisymmetric tensor, where $\mathbf{I}$ denotes the Kronecker or identity tensor.

The basis of antisymmetric tensors consists of the three $(1, 1)$ tensors

$$\mathbf{q}_x = \begin{pmatrix} 0 & 0 & 0 \\ 0 & 0 & -1 \\ 0 & 1 & 0 \end{pmatrix}, \quad \mathbf{q}_y = \begin{pmatrix} 0 & 0 & 1 \\ 0 & 0 & 0 \\ -1 & 0 & 0 \end{pmatrix},$$

$$\mathbf{q}_z = \begin{pmatrix} 0 & -1 & 0 \\ 1 & 0 & 0 \\ 0 & 0 & 0 \end{pmatrix}, \tag{5.181}$$

These entities are related by the Lie bracket; $[\mathbf{q}_x, \mathbf{q}_y] = \mathbf{q}_z$, and so on, such that a general antisymmetric tensor $\mathbf{q}$ may be expressed as

$$\mathbf{q} = q^x\,\mathbf{q}_x + q^y\,\mathbf{q}_y + q^z\mathbf{q}_z = \begin{pmatrix} 0 & -q^z & q^y \\ q^z & 0 & -q^x \\ -q^y & q^x & 0 \end{pmatrix}. \tag{5.182}$$

This means that any vector in $\mathbb{R}^3$, such as (q^x, q^y, q^z), can be identified with an antisymmetric matrix; there is an analogy between such a matrix and the cross-product operation.

Let us consider SO(3). It may be shown that

$$\mathbf{q}^3 = -q^2\,\mathbf{q}, \tag{5.183}$$

where $q^2 = (\mathbf{q} : \mathbf{q})/2$, with ":" denoting the double-dot product introduced in equation (5.16); hence, $\mathbf{q} : \mathbf{q}$ is the negative of trace of $\mathbf{q}^2$. After three multiplications of $\mathbf{q}$ with itself, we recover a multiple of $\mathbf{q}$. Note that if

$$\mathbf{q} = \begin{pmatrix} 0 & q^3 & -q^2 \\ -q^3 & 0 & q^1 \\ q^2 & -q^1 & 0 \end{pmatrix}, \tag{5.184}$$

then

$$q = \left[(q^1)^2 + (q^2)^2 + (q^3)^2\right]^{1/2}. \tag{5.185}$$

Property (5.183) may be used to obtain an explicit expression for the rotation tensor, $\mathbf{Q}$, from equation (5.179), namely,

$$\mathbf{Q} = \exp \mathbf{q} = \mathbf{I} + \frac{\sin q}{q}\,\mathbf{q} + \frac{1 - \cos q}{q^2}\,\mathbf{q}^2. \tag{5.186}$$

This remarkable result is known as *Rodrigues's formula* (Rodrigues, 1840).

Transposing equation (5.186), we find

$$\mathbf{Q}^t = \mathbf{I} - \frac{\sin q}{q}\,\mathbf{q} + \frac{1 - \cos q}{q^2}\,\mathbf{q}^2. \tag{5.187}$$

Subtracting equation (5.187) from (5.186), we obtain

$$\mathbf{Q} - \mathbf{Q}^t = 2\,\frac{\sin q}{q}\,\mathbf{q}, \tag{5.188}$$

which implies that the complement to equation (5.186) is

$$\mathbf{q} = \tfrac{1}{2}\,\frac{q}{\sin q}\,(\mathbf{Q} - \mathbf{Q}^t). \tag{5.189}$$

Calculating the trace of equation (5.186), using $\operatorname{tr}\mathbf{q} = 0$ and $\operatorname{tr}\mathbf{q}^2 = -\mathbf{q} : \mathbf{q} = -2\,q^2$, we find

$$\operatorname{tr}\mathbf{Q} = 3 - 2\,(1 - \cos q), \tag{5.190}$$

which implies that, in terms of $\mathbf{Q}$,

$$\cos q = \tfrac{1}{2}(\operatorname{tr}\mathbf{Q} - 1). \tag{5.191}$$

Thus, equation (5.189) is an explicit expression for $\mathbf{q}$ in terms of $\mathbf{Q}$, just as equation (5.186) is an explicit expression for $\mathbf{Q}$ in terms of $\mathbf{q}$.

Example 5.17 Elastic Tensor

In example 5.4, we introduced Hooke's law, which linearly relates the Cauchy stress, $\boldsymbol{\sigma}$, and the infinitesimal strain, $\boldsymbol{\varepsilon}$, via the elastic tensor, $\mathbf{c}$, and in this example, we investigate symmetries of the elastic tensor under rotations and reflections. In seismology, the *strain energy density* U, measured per unit mass, is quadratic in the infinitesimal strain tensor ε:

$$\rho U = \tfrac{1}{2}\, \varepsilon : \mathbf{c} : \varepsilon, \tag{5.192}$$

where ρ denotes the mass density. The Cauchy stress tensor is defined in terms of the internal energy density as

$$\sigma \equiv \rho \frac{\partial U}{\partial \varepsilon} = \mathbf{c} : \varepsilon, \tag{5.193}$$

which is Hooke's law (5.17).

Thanks to the symmetry of the strain tensor, $\varepsilon = \varepsilon^t$, and the quadratic form of the strain energy density (5.192), the elastic tensor $\mathbf{c}$ has the symmetries

$$c^{ijk\ell} = c^{jik\ell}, \quad c^{ijk\ell} = c^{ij\ell k}, \quad c^{ijk\ell} = c^{k\ell ij}, \tag{5.194}$$

which reduce the number of independent elements from 81 to 21. Any symmetries of the material may reduce the number of independent elements of the elastic tensor further. Let $Q^{i'}{}_i$ denote the elements of an orthogonal matrix that transforms the Cartesian frame $\{x^i\}$ to a Cartesian frame $\{x^{i'}\}$, such that $Q^{-1} = Q^t$ and $\det(Q) = 1$ (for rotations) or $\det(Q) = -1$ (for reflections). In the transformed coordinate system the elastic tensor takes the form

$$c^{i'j'k'\ell'} = Q^{i'}{}_i\, Q^{j'}{}_j\, Q^{k'}{}_k\, Q^{\ell'}{}_\ell\, c^{ijk\ell}. \tag{5.195}$$

A transformation that leaves the elements of the elastic tensor unchanged reduces the number of independent elements. Common symmetry classes are as follows (see, e.g., Bona et al., 2004).

Triclinic: Fully anisotropic; 21 independent elements.

Monoclinic: One symmetry plane; 13 independent elements.

Orthotropic: Three mutually orthogonal planes of reflection symmetry; 9 independent elements.

Transversely isotropic: Three mutually orthogonal planes of reflection symmetry and axial symmetry about one axis; 5 independent elements.

Cubic: Three mutually orthogonal planes of reflection symmetry and 90° rotation symmetry with respect to those planes; 3 independent elements.

Isotropic: Complete rotation and reflection symmetry; 2 independent elements.

The *isotropic elastic tensor* takes the form

$$c^{ijk\ell} = (\kappa - \tfrac{2}{3}\,\mu)\, g^{ij} g^{k\ell} + \mu\, (g^{ik} g^{j\ell} + g^{i\ell} g^{jk}), \tag{5.196}$$

where κ denotes the *bulk modulus* or *incompressibility* and μ denotes the *shear modulus* or *rigidity*. In an isotropic medium, using the elastic tensor (5.196), Hooke's law (5.17) takes the form

$$\boldsymbol{\sigma} = \kappa \, \mathrm{tr}(\boldsymbol{\varepsilon}) \, \mathbf{I} + 2\,\mu\,\mathbf{d}, \tag{5.197}$$

where $\mathbf{d}$ denotes the *strain deviator*

$$\mathbf{d} \equiv \boldsymbol{\varepsilon} - \tfrac{1}{3}\,\mathrm{tr}(\boldsymbol{\varepsilon})\,\mathbf{I}. \tag{5.198}$$

The elastic tensor, $\mathbf{c}$, relates a given strain to stress, and the complementary *compliance tensor*, $\mathbf{c}^{-1}$, relates a given stress to strain:

$$\boldsymbol{\epsilon} = \mathbf{c}^{-1} : \boldsymbol{\sigma}. \tag{5.199}$$

The isotropic compliance tensor is

$$c^{-1}_{ijk\ell} = \left(\frac{1}{9\,\kappa} - \frac{1}{6\,\mu} \right) g_{ij}\, g_{k\ell} + \frac{1}{4\,\mu} \left(g_{ik}\, g_{j\ell} + g_{i\ell}\, g_{jk} \right). \tag{5.200}$$

The isotropic elastic tensor (5.196) and the isotropic compliance tensor (5.200) are orthogonal in the sense

$$c^{ijmn}\, c^{-1}_{mnk\ell} = \tfrac{1}{2} \left(\delta^i{}_k\, \delta^j{}_\ell + \delta^i{}_\ell\, \delta^j{}_k \right). \tag{5.201}$$

In gases and liquids, the shear modulus vanishes, $\mu = 0$, and the isotropic bulk modulus κ governs the physics. Thus, the stress tensor takes the isotropic form,

$$\boldsymbol{\sigma} = -p\,\mathbf{I}, \tag{5.202}$$

where $p = -\kappa\,\mathrm{tr}(\boldsymbol{\varepsilon})$ denotes the induced pressure.

Chapter 6

MAPS BETWEEN MANIFOLDS

Mathematics is a place where you can do things which you can't do in the real world.
—MARCUS DU SAUTOY

This chapter concerns the mapping of tensors between manifolds. We provide a brief and general overview of the different types of maps used in this book and their relationships. Then, we focus on maps between manifolds of different dimensions, including limitations in mapping functions, vectors, one-forms, and tensors between such manifolds. To avoid these limitations, we restrict the discussion to maps between two points on the same manifold.

6.1 Maps

In general, as discussed by Lovelock & Rund (1989), a map from set M to N is a rule that assigns to each element of M an element of N; $M \to N$. We might refer to M as a domain of N. In general, many elements of N can be assigned to a single element of M. Also, even though each element of M must have an assignment, not all elements of N must be involved.

If all elements of N are involved, a map is *surjective*; it is also referred to as *onto*. In other words, each element of N has a corresponding element of M.

If each element of N that is involved corresponds to at most one element of M, the map is *injective*; it is also referred to as *one-to-one*. We emphasize that—unlike for the surjective mapping—not all elements of N need to be involved. However, those that are must have only one corresponding element of M.

If a map is both surjective and injective, it is called a *bijection*. In other words, all the elements of both sets are involved in such a manner that each element of one set corresponds to exactly one element of the other set. It follows that such a map has an inverse.

We note that the one-to-oneness is necessary but not sufficient to ensure the inverse, in general, since not involving all elements of N, but only its subset, leads to only a *partially defined map*. To exemplify this issue, consider $f\colon M \to N$, such that $f(x) = x^2$, and the following four cases. First, $M = N = \mathbb{R} \equiv (-\infty, +\infty)$; second, $M \in (0, +\infty)$ and $N \in (-\infty, +\infty)$; third, $M \in (-\infty, +\infty)$ and $N \in (0, +\infty)$; fourth, $M \in (0, +\infty)$ and $N \in (0, +\infty)$. These four cases are different from each other. The first is neither injective nor surjective, the second is injective but not surjective, the third is surjective but not injective, and the fourth is bijective. Only the last case has an inverse, which is, by definition, the square-root function.

A bijection for which both the map and its inverse are continuous is called a *homeomorphism*. If they are not only continuous but also smooth, in the sense of being infinitely differentiable, it is a *diffeomorphism*. Two manifolds related by a diffeomorphism are called *diffeomorphic*. From the viewpoint of differential geometry, diffeomorphic manifolds are indistinguishable from each other.

6.2 Maps between Manifolds of Different Dimensions

Consider an n-dimensional manifold N and an m-dimensional manifold M, with coordinate systems $x = \{x^1, \ldots, x^n\}$ and $x' = \{x^{1'}, \ldots, x^{m'}\}$. We assume that there exists a map $\varphi : M \to N$ between points $\mathcal{S}'(x') \in M$ and points $\mathcal{S}(x) \in N$:

$$\mathcal{S} = \varphi(\mathcal{S}'). \tag{6.1}$$

In chapters 4 and 5, x' and x denote transformed and original coordinates, respectively, *in the same manifold*. Presently, we do not assume that these coordinate systems have the same dimensions, nor do we assume φ to have an inverse.

6.2.1 Pullback

Let there be a function in the manifold N such that $\mathrm{f} : N \to \mathbb{R}$, that is, f maps points $\mathcal{S}$ in N to the real numbers. We wish to bring this function to manifold M. A natural way to do it is to take a composition $M \overset{\varphi}{\to} N \overset{\mathrm{f}}{\to} \mathbb{R}$; that is, to consider $\varphi^*\mathrm{f}(\mathcal{S}') \equiv \mathrm{f}(\varphi(\mathcal{S}'))$, as illustrated in figure 6.1. We refer to composition $\mathrm{f} \circ \varphi$ as a *pullback* of f from N to M, and denote it by $\varphi^*\mathrm{f}$.

The term "pullback" suggests a temporal sequence: a function in N that could have been originally associated with M and is brought back from N to M. Such an interpretation is particularly insightful in the context of deformations: N represents a deformation of M, and we might wish to examine how the deformed function compares to the original one; hence, we bring it back. We give such an example in section 6.3.

The pullback can be applied to functions, which are zero-forms. It can be also applied to one-forms and general $(0, q)$ tensors. To become familiar with the concept, let us consider a $(0, 2)$ tensor and invoke equation (4.17) to define

$$(\varphi^*\boldsymbol{\omega})_{i'j'} \equiv \omega_{ij}\, \lambda^i{}_{i'}\, \lambda^j{}_{j'}, \tag{6.2}$$

where

$$\lambda^i{}_{i'} = \partial_{i'}\varphi^i. \tag{6.3}$$

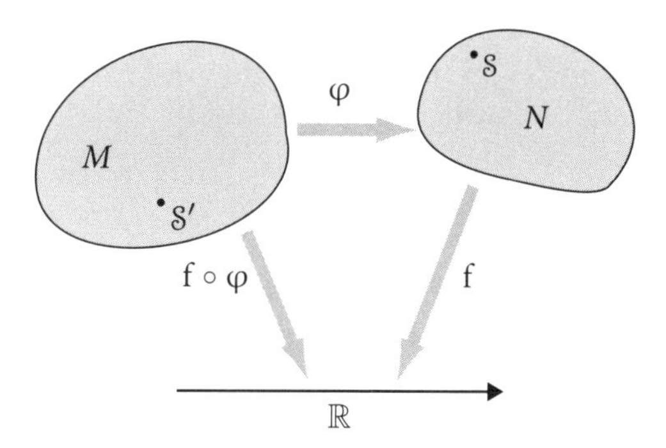

Figure 6.1: Pullback of a function f, which resides in manifold N, to manifold M, using the map, $\mathcal{S} = \varphi(\mathcal{S}')$. Explicitly, we compose $\mathrm{f}: N \to \mathbb{R}$ and $\varphi : M \to N$ to obtain the pullback: $\varphi^* \mathrm{f} = \mathrm{f} \circ \varphi : M \to \mathbb{R}$.

Equation (6.2) is akin to a change of coordinates. This is not a coincidence: if φ possesses an inverse, we have the transformation used in chapters 4 and 5. Herein, however, the index i' ranges from 1 to m, whereas the index i ranges from 1 to n.

The pullback of a function f is given by $\varphi^* \mathrm{f} = \mathrm{f} \circ \varphi$. Consider the exterior derivative of the function f, which is the one-form df. The pullback of this one-form is

$$(\varphi^* \mathrm{df})_{i'} \equiv (\partial_i f)\, \lambda^i{}_{i'} = \partial_{i'} f', \tag{6.4}$$

which is the transformation rule—effectively the chain rule—for the partial derivatives of the function f.

Example 6.1 Metric on the Sphere

Let M be the two-dimensional unit sphere with coordinates $x' = \{\theta, \phi\}$, embedded at the origin of N: a three-dimensional space described by Cartesian coordinates, $z = \{x, y, z\}$. We write $\varphi : M \to N$ as

$$\begin{aligned} x &= x^1 = \varphi^1(\theta, \phi) = \sin\theta \cos\phi, \\ y &= x^2 = \varphi^2(\theta, \phi) = \sin\theta \sin\phi, \\ z &= x^3 = \varphi^3(\theta, \phi) = \cos\theta. \end{aligned} \tag{6.5}$$

Even though this map is one-to-one (i.e., injective), it is not *bijective*,[a] since the image is the unit sphere in $\mathbb{R}^3$, not the entire space; hence, there is no inverse map. Consider the Cartesian metric. Following expression (5.83), we write $g_{ij} = \delta_{ij}$,[b] which is a $(0, 2)$ tensor. Let us use the pullback to obtain the corresponding metric in spherical coordinates. To obtain the pullback $\varphi^* \mathbf{g}$, we apply transformation rule (5.34), with the explicit formula stated in the first rule of expression (4.17), to write

$$(\varphi^* \mathbf{g})_{i'j'} = g_{ij}\, \lambda^i{}_{i'}\, \lambda^j{}_{j'} = \delta_{ij}\, \lambda^i{}_{i'}\, \lambda^j{}_{j'}, \tag{6.6}$$

where the second equality results from N being endowed with the Euclidean metric. Written explicitly, the transformation matrix is

$$(\lambda^{i}{}_{i'}) = \begin{pmatrix} \partial_{1'}\varphi^1 & \partial_{1'}\varphi^2 & \partial_{1'}\varphi^3 \\ \partial_{2'}\varphi^1 & \partial_{2'}\varphi^2 & \partial_{2'}\varphi^3 \end{pmatrix}$$
$$= \begin{pmatrix} \cos\theta\cos\phi & \cos\theta\sin\phi & -\sin\theta \\ -\sin\theta\sin\phi & \sin\theta\cos\phi & 0 \end{pmatrix}. \tag{6.7}$$

Hence,

$$\varphi^*\mathbf{g} = \begin{pmatrix} 1 & 0 \\ 0 & \sin^2\theta \end{pmatrix}, \tag{6.8}$$

that is, the Euclidean metric, $g_{ij} = \delta_{ij}$ in TN, pulled back to the cotangent space of the sphere, TM.

[a] A map $\varphi : M \to N$ is bijective if it is both one-to-one (injective) and onto (surjective).
[b] Note that we are using a slightly different version of the Kronecker delta symbol here, one with two lower indices. Its value is still one if the indices are the same, and zero otherwise.

Extending the pattern, we define the pullback of any $(0, q)$ tensor as

$$\overset{\text{pullback to } m \leq n\text{-dimensional cotangent space of } M}{\overset{\downarrow}{(\varphi^*\mathbf{T})_{i'_1 \dots i'_q}}} = \underset{\text{in } n\text{-dimensional cotangent space of } N}{\underset{\uparrow}{T_{i_1 \dots i_q}}} \lambda^{i_1}{}_{i'_1} \cdots \lambda^{i_q}{}_{i'_q}. \tag{6.9}$$

Because the transformation matrices $\lambda^{i_k}{}_{i'_k}$ are well-defined, that is, they just involve the map φ and not its inverse, pulling back a $(0, q)$ tensor is a well-defined operation. In particular, one can always pull back a k-form, that is, a skewsymmetric $(0, k)$ tensor discussed in chapter 8.

6.2.2 Pushforward

Given the function $\mathrm{g} : M \to \mathbb{R}$, we cannot compose it with φ to obtain a function $N \to \mathbb{R}$, since to do so, we need to write $N \overset{\varphi^{-1}}{\to} M \overset{\mathrm{g}}{\to} \mathbb{R}$, but, in general, φ is not invertible. However, we can *pushforward* vectors in TN to obtain vectors in TM, and—more generally—$(p, 0)$ tensors. To circumvent the lack of an inverse, φ^{-1}, consider a vector $\mathbf{u}$ in the tangent space of M, TM, which acts on a one-form in the cotangent space of M, T^*M. Let us choose the pullback of the exterior derivative of a function f in N, $\varphi^*\mathrm{df}$, to be this one-form; it resides in T^*M, and is defined by f, which resides in N. Now we can evaluate $\mathbf{u}(\varphi^*\mathrm{df})$, which is the action of the vector $\mathbf{u}$ in TM on the one-form $\varphi^*\mathrm{df}$ in T^*M. Thus, we obtain the definition of the operation reciprocal to pullback: it acts in TM and involves a function from N.

We refer to this operation as the *pushforward* from TM to TN, and denote it by $\varphi_*\mathbf{u}$:

$$\varphi_*\mathbf{u}(\mathrm{df}) \equiv \mathbf{u}(\varphi^*\mathrm{df}). \tag{6.10}$$

The asterisk $*$ used as a superscript and as a subscript in definition (6.10) follows a convention akin to the one used for *adjoint operators*. It is required for consistency of the order of composition of functions. In general, the subscript asterisk does not change the order of operations: $(\varphi_1\,\varphi_2)_* = (\varphi_1)_*\,(\varphi_2)_*$, the superscript asterisk does: $(\varphi_1\,\varphi_2)^* = (\varphi_2)^*\,(\varphi_1)^*$, akin to, say, an inverse: $(\varphi_1\,\varphi_2)^{-1} = (\varphi_2)^{-1}\,(\varphi_1)^{-1}$.

Definition (6.10) is valid for individual vectors. It does not allow us, in general, to push forward vector fields, since φ is neither one-to-one nor onto. We may push forward vector fields if φ is invertible. By contrast, one-form fields may be pulled back by arbitrary maps φ.

Using the definition of the action of a vector on a one-form (4.31), which has the component expression (4.36), we may write expression (6.10) in components as

$$(\varphi_*\mathbf{u})^i\,(\mathrm{df})_i = u^{i'}\,(\varphi^*\mathrm{df})_{i'}. \tag{6.11}$$

Note how the left-hand side of this expression involves the components of the pushforward $\varphi_*\mathbf{u}$ in TN and df in T^*N, whereas the right-hand side involves the components of $\mathbf{u}$ in TM and $\varphi^*\mathrm{df}$ in T^*M. Using the components df and the expression for the pullback of a one-form (6.4), we write

$$(\varphi_*\mathbf{u})^i\,\partial_i f = u^{i'}\,(\partial_i f\,)\,\lambda^i{}_{i'}. \tag{6.12}$$

This result holds for any function f, and thus we obtain the pushforward operator

$$(\varphi_*\mathbf{u})^i = \lambda^i{}_{i'}\,u^{i'}. \tag{6.13}$$

Again, we see the appearance of the transformation matrix $\lambda^i{}_{i'}$, but not of its inverse: in general, we can push forward a vector.

One-forms and, in general, k-forms can be pulled back but not pushed forward; the opposite is true for vectors and, in general, k-vectors. For a differentiable map φ we may perform the operations $(\varphi^*\omega)_{i'} = \omega_i\,\lambda^i{}_{i'}$ and $(\varphi_*\mathbf{u})^i = \lambda^i{}_{i'}\,u^{i'}$ without danger, because the transformation matrix $\lambda^i{}_{i'}$ is well-behaved. An important consequence is that $(0,k)$ tensors—in particular k-forms—may be integrated, because a change of coordinates involves the elements $\lambda^i{}_{i'}$ and the Jacobian $\lambda = \det(\lambda^i{}_{i'})$, which are well-defined.

The pullback of a form and pushforward of a vector are related by

$$\varphi^*\boldsymbol{\omega}(\mathbf{u}) = \boldsymbol{\omega}(\varphi_*\mathbf{u}), \tag{6.14}$$

which is valid for all $\boldsymbol{\omega}$ in T^*N and all $\mathbf{u}$ in TM. Since a $(0,q)$ tensor is a linear map of q vectors to $\mathbb{R}$, we generalize this expression to

$$\varphi^*\mathbf{T}(\mathbf{u}^1,\ldots,\mathbf{u}^q) = \mathbf{T}(\varphi_*\mathbf{u}^1,\ldots,\varphi_*\mathbf{u}^q), \tag{6.15}$$

which is equivalent to expression (6.9), written in components. Because of the duality of vectors and one-forms, for $(p,0)$ tensors, we write

$$\varphi_* \mathbf{T}(\omega^1, \ldots, \omega^p) = \mathbf{T}(\varphi^* \omega^1, \ldots, \varphi^* \omega^p), \tag{6.16}$$

which, expressed in components, is equivalent to

$$\begin{array}{c}\text{pushforward to } n\text{-dimensional tangent space of } N \\ \downarrow \\ (\varphi_* \mathbf{T})^{i_1 \ldots i_q} = \lambda^{i_1}{}_{i'_1} \cdots \lambda^{i_q}{}_{i'_q} \, T^{i'_1 \ldots i'_q}. \\ \uparrow \\ \text{in } m \leq n\text{-dimensional tangent space of } M\end{array} \tag{6.17}$$

In general, a (p, q) tensor cannot be pulled back or pushed forward between manifolds of different dimensions. Let us consider the case of manifolds of the same dimension.

6.3 Maps between Manifolds of the Same Dimensions

Let us revisit the formulation presented in section 6.2, but constraining our interest to manifolds of the same dimension, which is the necessary condition for the invertibility of the map φ. Notably, this scenario includes bijective maps between charts of the same manifold, thereby providing an alternative perspective on local coordinate changes, discussed in section 3.3.

Following expressions (6.15) and (6.16) and letting $\varphi^{-1} = \Phi$, we write the pushforward as

$$\varphi_* \mathbf{T}(\omega^1, \ldots, \omega^p, \mathbf{u}^1, \ldots, \mathbf{u}^q) = \mathbf{T}(\varphi^* \omega^1, \ldots, \varphi^* \omega^p, \Phi_* \mathbf{u}^1, \ldots, \Phi_* \mathbf{u}^q). \tag{6.18}$$

In principle, the arguments of expression (6.18) would require $\varphi^* \mathbf{u}$, which is not defined, in general. Since we assume that φ is invertible, we may replace $\varphi^* \mathbf{u}$ with $(\varphi^{-1})_* \equiv \Phi_*$. In components,

$$(\varphi_* \mathbf{T})^{i_1 \ldots i_p}{}_{j_1 \ldots j_q} = \lambda^{i_1}{}_{i'_1} \cdots \lambda^{i_p}{}_{i'_p} \, T^{i'_1 \ldots i'_p}{}_{j'_1 \ldots j'_q} \, \Lambda^{j'_1}{}_{j_1} \cdots \Lambda^{j'_q}{}_{j_q}. \tag{6.19}$$

These expressions state explicitly the pushforward of a (p, q) tensor. Since the pushforward and pullback are related by φ and Φ, the pullback expression analogous to expression (6.18) is

$$\varphi^* \mathbf{T}(\omega^1, \ldots, \omega^p, \mathbf{u}^1, \ldots, \mathbf{u}^q) = \mathbf{T}(\Phi^* \omega^1, \ldots, \Phi^* \omega^p, \varphi_* \mathbf{u}^1, \ldots, \varphi_* \mathbf{u}^q), \tag{6.20}$$

where $\Phi^* \equiv (\varphi^{-1})^*$ is the substitute for φ_*. In components,

$$(\varphi^* \mathbf{T})^{i'_1 \ldots i'_p}{}_{j'_1 \ldots j'_q} = \Lambda^{i'_1}{}_{i_1} \cdots \Lambda^{i'_p}{}_{i_p} \, T^{i_1 \ldots i_p}{}_{j_1 \ldots j_q} \, \lambda^{j_1}{}_{j'_1} \cdots \lambda^{j_q}{}_{j'_q}. \tag{6.21}$$

This transformation has the same form as the rule for the transformation of the components of a general tensor, namely (5.34). In other words, we have rediscovered the tensor-transformation rule, derived in this chapter by considering the concepts of pullbacks and pushforwards.

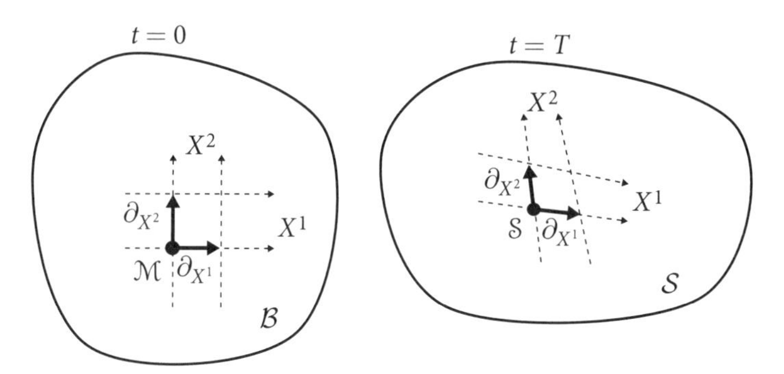

Figure 6.2: *Left*: *Referential state* of the continuum at time $t = 0$ captured by the *referential manifold* $\mathcal{B}$. A *material point* $\mathcal{M}$ may be identified with *referential coordinates* $\{X^I\}$, which define a *chart* in the referential manifold. A local vector basis in the tangent space of the referential manifold at $\mathcal{M}$ is denoted by the partial derivatives $\{\partial_{X^I}\}$. For convenience, the referential coordinates are chosen to be Cartesian, but this is not required. *Right*: Deformed state of the continuum at time $t = T$ captured by the *spatial manifold* $\mathcal{S}$. A *spatial point* $\mathcal{S}$, *not* tied to a specific element of the continuum, may be identified with the comoving Lagrangian coordinates $\{X^I\}$ of the material particle that happens to occupy its location at time $t = T$. Thus, Lagrangian coordinates define an evolving local chart in the spatial manifold at time $t = T$. A local non-orthonormal vector basis in the tangent space of the spatial manifold at $\mathcal{S}$ at time $t = T$ is denoted by the partial derivatives $\{\partial_{X^I}\}$. Importantly, Lagrangian coordinates in the spatial manifold are chosen such that they are identical to the referential coordinates in the referential manifold at the referential time $t = 0$, when the referential and spatial manifolds capture the same state of the continuum. Thus, the Lagrangian coordinates evolve with the flow of matter.

Example 6.2 Referential Manifold

The notions of *strain* and *stress* in continuum mechanics require the introduction of an equilibrium state at some referential time $t = 0$, for example, the state of the Earth before an earthquake. This state is captured by the *referential manifold* $\mathcal{B}$. As illustrated in figure 6.2 (*left*), an element of the continuum in the referential state is labeled by a *material point* $\mathcal{M}$, which may be identified with a set of *referential coordinates* $\{X^I\}$. These referential coordinates define a *chart* in the *referential manifold* and remain associated with whichever element of the continuum they identify. The local vector basis in the tangent space of the referential manifold at material point $\mathcal{M}$ is identified with the partial derivatives ∂_{X^I}, analogous to the identification of vectors with tangents to curves, as discussed in section 4.1.1.

The state of the continuum at a later time $t = T$ as a result of the motion φ, as discussed in example 3.4, is captured by the *spatial manifold* $\mathcal{S}$, shown in figure 6.2 (*right*). A *spatial point* $\mathcal{S}$ in the spatial manifold is not tied to a specific element of the continuum: it simply denotes a location in inertial space. Thus, whereas a material point $\mathcal{M}$ labels a specific particle in the referential manifold $\mathcal{B}$ at time $t = 0$, a spatial point $\mathcal{S}$ labels a location in the inertial spatial manifold $\mathcal{S}$ not tied to any particular particle or time. As discussed in example 3.2, spatial points $\mathcal{S}$ in the spatial manifold $\mathcal{S}$ may be identified by either a set of inertial Eulerian coordinates $\{r^i\}$ or a set of Lagrangian coordinates $\{X^I\}$, labeling whatever material particle happens to be located at $\mathcal{S}$ at

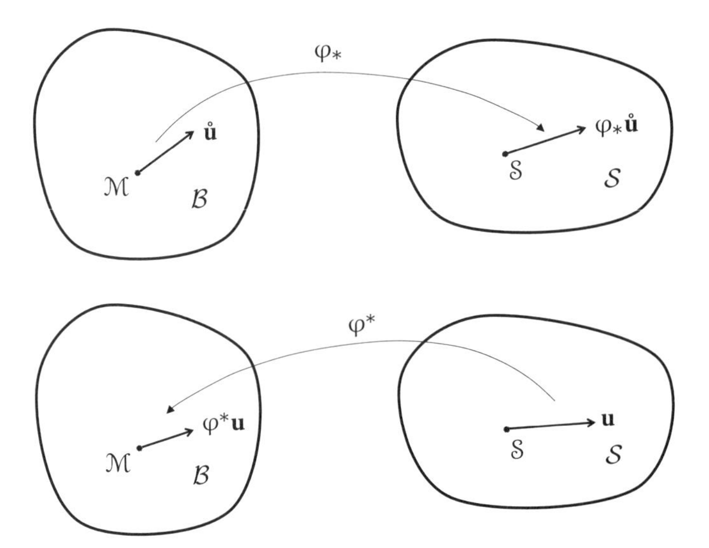

Figure 6.3: *Top*: Let $\mathring{\mathbf{u}} = \mathring{u}^I\, \mathring{\mathbf{e}}_I$ denote a vector in tangent space of the referential manifold $\mathcal{B}$ at material location $\mathcal{M}$. The *pushforward* of $\mathring{\mathbf{u}}$ from the referential manifold $\mathcal{B}$ to the spatial manifold $\mathcal{S}$ with the motion φ is defined as $\varphi_*\mathring{\mathbf{u}} \equiv \mathring{u}^I\, F^i{}_I\, \mathbf{e}_i$, which defines a vector in the tangent space of the spatial manifold $\mathcal{S}$ at the current location of the particle $\mathcal{S}$. *Bottom*: Conversely, let $\mathbf{u} = u^i\, \mathbf{e}_i$ denote a vector in the tangent space of the spatial manifold $\mathcal{S}$ at location $\mathcal{S}$. The *pullback* of $\mathbf{u}$ from the spatial manifold $\mathcal{S}$ to the material manifold $\mathcal{B}$ with the motion φ is defined as $\varphi^*\mathbf{u} \equiv u^i\, (F^{-1})^I{}_i\, \mathring{\mathbf{e}}_I = u^I\, \mathring{\mathbf{e}}_I$. If the referential vector $\mathring{\mathbf{u}}$ is the pullback of a spatial vector $\mathbf{u}$, that is, $\mathring{\mathbf{u}} = \varphi^*\mathbf{u}$, then $\mathring{u}^I = u^I$; in other words, the referential components of the referential vector equal the Lagrangian components of the spatial vector.

time $t = T$. Thus, Lagrangian coordinates define an evolving local chart in the spatial manifold $\mathcal{S}$. Lagrangian coordinates in the spatial manifold, $\{X^I\}$, are chosen such that they are identical to the Cartesian referential coordinates in the referential manifold $\mathcal{B}$ at the referential time $t = 0$, when the referential and spatial manifolds capture the same state of the continuum. In other words, at times $t \geq 0$ the referential coordinates $\{X^I\}$, *comove* or *convect* with the material to evolve into a set of *Lagrangian* or *convected* or *comoving* coordinates.

To introduce the concepts of stress and strain, we need to be able to compare tensors in the spatial manifold $\mathcal{S}$ at time $t = T$ with tensors in the referential manifold $\mathcal{B}$ at time $t = 0$, and vice versa. This is accomplished based on pushforwards and pullbacks associated with the motion φ, which are concepts illustrated in figure 6.3.

To clearly distinguish tensors in the referential manifold $\mathcal{B}$ from tensors in the spatial manifold $\mathcal{S}$, we use a "$\circ$" for their identification, and we use $\mathring{\mathbf{e}}_I$ and $\mathring{\mathbf{e}}^I$ to denote referential basis vectors and one-forms, respectively. Thus, the metric tensor in the referential manifold is denoted by

$$\mathring{\mathbf{g}} = \mathring{g}_{IJ}\, \mathring{\mathbf{e}}^I \otimes \mathring{\mathbf{e}}^J, \tag{6.22}$$

where

$$\mathring{g}_{IJ}(X) \equiv g_{IJ}(X,0). \tag{6.23}$$

The pushforward of the metric in the referential manifold to the spatial manifold is defined by

$$\begin{aligned} \varphi_* \mathring{\mathbf{g}} &\equiv \mathring{g}_{IJ}\,(F^{-1})^I{}_i\,(F^{-1})^J{}_j\,\mathbf{e}^i \otimes \mathbf{e}^j \\ &= \mathring{g}_{IJ}\,\mathbf{e}^I \otimes \mathbf{e}^J, \end{aligned} \tag{6.24}$$

where in the second equality we expressed the pushforward in Lagrangian components in the spatial manifold. Because referential coordinates in the referential manifold coincide with Lagrangian coordinates in the spatial manifold at time $t=0$, the Lagrangian components of the pushforward $\varphi_* \mathring{\mathbf{g}}$ are identical to the referential components of the referential metric tensor, $\mathring{g}_{IJ}$.

Similarly, the pullback of the metric tensor in the spatial manifold to the referential manifold is defined by

$$\begin{aligned} \varphi^* \mathbf{g} &\equiv g_{ij}\,F^i{}_I\,F^j{}_J\,\mathring{\mathbf{e}}^I \otimes \mathring{\mathbf{e}}^J \\ &= g_{IJ}\,\mathring{\mathbf{e}}^I \otimes \mathring{\mathbf{e}}^J. \end{aligned} \tag{6.25}$$

In this case, the referential components of the pullback $\varphi^* \mathbf{g}$ are identical to the Lagrangian components of the spatial metric tensor, g_{IJ}.

Based on the introduction of the volume form in continuum mechanics in example 5.13, we may also wish to establish the *change in volume* when an element of the continuum transitions from the referential manifold to the spatial manifold. In the referential manifold $\mathcal{B}$, the volume element is

$$\begin{aligned} \mathring{\boldsymbol{\epsilon}} &= \mathring{\epsilon}_{IJK}\,\mathring{\mathbf{e}}^I \otimes \mathring{\mathbf{e}}^J \otimes \mathring{\mathbf{e}}^K \\ &= \overline{\mathring{G}}\,\underline{\epsilon}_{IJK}\,\mathring{\mathbf{e}}^I \otimes \mathring{\mathbf{e}}^J \otimes \mathring{\mathbf{e}}^K, \end{aligned} \tag{6.26}$$

where $\overline{\mathring{G}}$ is the square root of the determinant of the referential metric given by (6.22). The pushforward of the referential volume element is

$$\begin{aligned} \varphi_* \mathring{\boldsymbol{\epsilon}} &= \overline{\mathring{G}}\,\underline{\epsilon}_{IJK}\,(F^{-1})^I{}_i\;(F^{-1})^J{}_j\,(F^{-1})^K{}_k\,\mathbf{e}^i \otimes \mathbf{e}^j \otimes \mathbf{e}^k \\ &= \overline{\mathring{G}}\,\underline{G}\,\epsilon_{IJK}\,\mathbf{e}^I \otimes \mathbf{e}^J \otimes \mathbf{e}^K \\ &= F^{-1}\,\overline{\mathring{G}}\,\underline{g}\,\boldsymbol{\epsilon} \\ &= J^{-1}\,\boldsymbol{\epsilon}, \end{aligned} \tag{6.27}$$

where in the third equality, we used relationship (5.99), and where in the last equality we have introduced the *Jacobian of the motion* (see, e.g., Marsden & Hughes, 1983, Proposition 5.3)

$$J \equiv \overline{G}\,\underline{\mathring{G}} = F\,\overline{g}\,\underline{\mathring{G}}. \tag{6.28}$$

Note that at the referential time, we have

$$J(X,0)=1. \tag{6.29}$$

Thus, we have the following relationship between the volume form and the pushforward of the referential volume form:

$$\epsilon = J\,\varphi_*\mathring{\epsilon}. \tag{6.30}$$

This is clearly a measure of the volume change, starting from $J=1$ at the referential time $t=0$. Conversely, in terms of the pullback we have the relationship

$$\varphi^*\epsilon = J\,\mathring{\epsilon}. \tag{6.31}$$

Based on the introduction of the surface one-form in continuum mechanics in example 5.14, we may also wish to establish the *change in an oriented surface* when a surface element of the continuum transitions from the referential manifold to the spatial manifold. In the spatial manifold $\mathcal{S}$, consider two nonparallel vectors $\mathbf{u}$ and $\mathbf{w}$ which define an oriented patch

$$\hat{\mathbf{n}}\,dS \equiv \epsilon(\,\cdot\,,\mathbf{u},\mathbf{w}). \tag{6.32}$$

The pullback of vectors $\mathbf{u}$ and $\mathbf{w}$ from the spatial manifold to the referential manifold yields two vectors $\mathring{\mathbf{u}}$ and $\mathring{\mathbf{w}}$, whose referential components are the same as the Lagrangian components u^I and w^I, as illustrated in figure 6.1. The surface element spanned by these vectors in the referential manifold is

$$\hat{\mathring{\mathbf{n}}}\,d\mathring{S} = \mathring{\epsilon}(\,\mathring{\cdot}\,,\mathring{\mathbf{u}},\mathring{\mathbf{w}}), \tag{6.33}$$

where $\mathring{\epsilon}$ denotes the referential volume form. The pushforward of this surface element is

$$\begin{aligned}\varphi_*(\hat{\mathring{\mathbf{n}}}\,d\mathring{S}) &= \varphi_*\mathring{\epsilon}(\varphi_*\,\mathring{\cdot}\,,\varphi_*\mathring{\mathbf{u}},\varphi_*\mathring{\mathbf{w}})\\ &= J^{-1}\,\epsilon(\cdot\,,\mathbf{u},\mathbf{w})\\ &= J^{-1}\,\hat{\mathbf{n}}\,dS,\end{aligned} \tag{6.34}$$

where in the second equality we used (6.30). We conclude that the pushforward of the referential surface element, $\varphi_*(\hat{\mathring{\mathbf{n}}}\,d\mathring{S})$, and the corresponding surface element in the spatial manifold, $\hat{\mathbf{n}}\,dS$, are related via

$$\hat{\mathbf{n}}\,dS = J\,\varphi_*(\hat{\mathring{\mathbf{n}}}\,d\mathring{S}). \tag{6.35}$$

This expression relating $\hat{\mathbf{n}}\,dS$ to $\varphi_*(\hat{\mathring{\mathbf{n}}}\,d\mathring{S})$ is the areal analogue of the volume relation (6.30). Expressed in component form, we have *Nanson's relation* (Dahlen &

Tromp, 1998, equation 2.47)

$$\hat{n}_i \, dS = J \, \mathring{\hat{n}}_I \, d\mathring{S} \, (F^{-1})^I{}_i. \tag{6.36}$$

In general, the Lagrangian components of the pushforward of a tensor $\mathring{\mathbf{T}}$ from the referential to the spatial manifold, $\varphi_* \mathring{\mathbf{T}}$, are identical to its referential components, and the referential components of the pullback of a tensor $\mathbf{T}$ from the spatial to the referential manifold, $\varphi^* \mathbf{T}$, are identical to its Lagrangian components.

In terms of the inverse map $\Phi : \mathcal{S} \to \mathcal{B}$, we note that the pullback Φ^* is equivalent to the pushforward φ_*, and that the pushforward Φ_* is equivalent to the pullback φ^*.

Example 6.3 Cauchy–Green Tensor

The *Cauchy–Green tensor*, $\mathbf{C}$, is a $(1, 1)$ tensor defined in terms of the pushforward $\varphi_* \mathring{\mathbf{g}}^{-1}$ and the metric $\mathbf{g}$ in the spatial manifold $\mathcal{S}$ via

$$\mathbf{C} \equiv (\varphi_* \mathring{\mathbf{g}}^{-1}) \cdot \mathbf{g}. \tag{6.37}$$

This combination of the metric tensor $\mathbf{g}$ with the pushforward of its inverse $\varphi_* \mathring{\mathbf{g}}^{-1}$ in the referential manifold ensures that (1) the Cauchy–Green tensor is a $(1, 1)$ tensor, and (2) the Cauchy–Green tensor is dimensionless. We can construct functions of dimensionless $(1, 1)$ tensors, such as the square root or logarithm, as discussed in section 5.5 and applied in example 6.4.

Alternatively, the Cauchy–Green tensor may be defined in terms of the pullback of the metric in the spatial manifold, $\varphi^* \mathbf{g}$, as follows:

$$\mathbf{C} = \varphi_* \left(\mathring{\mathbf{g}}^{-1} \cdot \varphi^* \mathbf{g} \right). \tag{6.38}$$

Example 6.4 Definitions of Strain

If we pull back the metric tensor $\mathbf{g}$ in the spatial manifold $\mathcal{S}$ to the referential manifold $\mathcal{B}$, $\varphi^* \mathbf{g}$, we can compare this metric tensor to the referential metric $\mathring{\mathbf{g}}$ in the referential manifold $\mathcal{B}$:

$$\mathring{\mathbf{E}} \equiv \tfrac{1}{2} \, (\varphi^* \mathbf{g} - \mathring{\mathbf{g}}). \tag{6.39}$$

The tensor $\mathring{\mathbf{E}}$ represents a measure of strain in the referential manifold $\mathcal{B}$. If we raise the first index of this measure of strain in the referential manifold $\mathcal{B}$ with the referential inverse metric $\mathring{\mathbf{g}}^{-1}$ and push the result forward from the referential manifold $\mathcal{B}$ to the spatial manifold $\mathcal{S}$, we obtain the *Lagrangian strain tensor*

$$\mathbf{E}^{\mathrm{L}} \equiv \tfrac{1}{2}\left[\varphi_*\left(\mathring{\mathbf{g}}^{-1}\cdot\varphi^*\mathbf{g}\right)-\mathbf{I}\right]=\tfrac{1}{2}\,(\mathbf{C}-\mathbf{I}), \tag{6.40}$$

thanks to relationship (6.38).

Conversely, if we pushforward the metric tensor $\mathring{\mathbf{g}}$ in the referential manifold $\mathcal{B}$ to the spatial manifold $\mathcal{S}$, $\varphi_*\mathring{\mathbf{g}}$, we can compare this metric tensor to the spatial metric $\mathbf{g}$ in the spatial manifold $\mathcal{S}$:

$$\mathbf{E}=\tfrac{1}{2}(\mathbf{g}-\varphi_*\mathring{\mathbf{g}}). \tag{6.41}$$

The tensor $\mathbf{E}$ represents a measure of strain in the spatial manifold $\mathcal{S}$. If we raise the first index of this measure of strain in the spatial manifold $\mathcal{S}$ with the inverse metric $\mathbf{g}^{-1}$, we obtain the *Eulerian strain tensor* or *Almansi strain tensor*

$$\mathbf{E}^{\mathrm{E}} \equiv \tfrac{1}{2}\left(\mathbf{I}-\mathbf{C}^{-1}\right), \tag{6.42}$$

where we used the inverse of the Cauchy–Green tensor (6.37).

As a final example, the *logarithmic*, *Hencky*, or *true* strain tensor (Hencky, 1928; Truesdell & Toupin, 1960) is defined in terms of the Cauchy–Green tensor $\mathbf{C}$ via

$$\mathbf{E}^{\log} \equiv \tfrac{1}{2}\,\log(\mathbf{C}), \tag{6.43}$$

based on expression (5.58) for the logarithm of a $(1,1)$ tensor.

The infinitesimal strain tensor $\boldsymbol{\varepsilon}$, introduced in example 5.4, is the small-strain limit of all the strain tensors discussed in this example.

Example 6.5 Definitions of Stress

In addition to the Cauchy stress tensor introduced in example 5.4, other definitions of stress are commonly used in continuum mechanics. For example, the *second Piola–Kirchhoff stress tensor* is defined in terms of the pullback of the Cauchy stress tensor $\boldsymbol{\sigma}$ from the spatial manifold to the referential manifold as

$$\mathring{\mathbf{S}} \equiv J\,\varphi^*\boldsymbol{\sigma}, \tag{6.44}$$

where J denotes the Jacobian of the motion (6.28). This relationship between the second Piola–Kirchhoff stress and the Cauchy stress is called the *Piola transformation*. In referential components in the referential manifold, we have

$$\mathring{S}_I{}^J = J\,F^i{}_I\,(F^{-1})^J{}_j\,\sigma_i{}^j = J\,\sigma_I{}^J. \tag{6.45}$$

This stress tensor "lives" in the tangent/cotangent bundle of the referential manifold $\mathcal{B}$.

Upon pushing the second Piola–Kirchhoff stress from the referential manifold back to the spatial manifold, we obtain the *Kirchhoff stress tensor*

$$\boldsymbol{\tau} \equiv \varphi_*(\mathring{\mathbf{S}}) = J\boldsymbol{\sigma}. \tag{6.46}$$

This stress tensor is also called the *weighted Cauchy stress tensor*, which “lives” in the tangent/cotangent bundle of the spatial manifold. In Lagrangian components in the spatial manifold $\mathcal{S}$, equation (6.46) takes the form

$$\tau_I{}^J = J\sigma_I{}^J. \tag{6.47}$$

Example 6.6 Two-Point Tensors in Continuum Mechanics

It is common in continuum mechanics to invoke the concept of *two-point tensors*. These are tensors with some “legs” in the referential manifold $\mathcal{B}$ expressed relative to referential frames $\mathring{\mathbf{e}}^I$ and coframes $\mathring{\mathbf{e}}_I$, and their remaining legs in the spatial manifold $\mathcal{S}$ expressed relative to Eulerian frames $\mathbf{e}^i$ and coframes $\mathbf{e}_i$. A prime example is the deformation gradient (4.26), which may be regarded as a “machine,” $\mathbf{F}$, with one spatial one-form slot and one referential vector slot:

$$\mathbf{F}(\boldsymbol{\omega}, \mathring{\mathbf{u}}) \equiv \omega_i \, F^i{}_I \, \mathring{u}^I. \tag{6.48}$$

Here $\boldsymbol{\omega} = \omega_i \, \mathbf{e}^i$ is a one-form in the cotangent space of the spatial manifold $\mathcal{S}$, and $\mathring{\mathbf{u}} = \mathring{u}^I \, \mathring{\mathbf{e}}_I$ is a vector in the tangent space of the referential manifold $\mathcal{B}$. The components of a two-point tensor may be obtained by substituting spatial and preferential frames and coframes in the appropriate slots, for example,

$$F^i{}_I = \mathbf{F}(\mathbf{e}^i, \mathring{\mathbf{e}}_I). \tag{6.49}$$

Another example of a two-point tensor in continuum mechanics involves the *first Piola–Kirchhoff stress*. Its components are defined in terms of the components of the second Piola–Kirchhoff stress (6.45) as

$$P_i{}^I \equiv (F^{-1})^J{}_i \, \mathring{S}_J{}^I. \tag{6.50}$$

Effectively, one “leg” of the second Piola–Kirchhoff stress is pushed to the cotangent space of the spatial manifold. Like the deformation gradient, the first Piola–Kirchhoff stress involves an object, $P_i{}^I$, with mixed Eulerian and referential indices. We may use this object to define the first Piola–Kirchhoff stress two-point tensor, $\mathbf{P}$, with one spatial vector slot and one referential one-form slot:

$$\mathbf{P}(\mathbf{u}, \mathring{\boldsymbol{\omega}}) \equiv u^i \, P_i{}^I \, \mathring{\omega}_I. \tag{6.51}$$

Its components may then be obtained as

$$P_i{}^I = \mathbf{P}(\mathbf{e}_i, \mathring{\mathbf{e}}^I). \tag{6.52}$$

In this book, we eschew the use of two-point tensors, because the author finds such split-personality tensors disturbing and geometrically offensive.

Chapter 7

DIFFERENTIATION ON MANIFOLDS

> Equations are just the boring part of mathematics. I attempt to see things in terms of geometry.
>
> —STEPHEN HAWKING

Tensors can only be compared or combined at a single point on a manifold within the corresponding tangent or cotangent spaces. To compare tensors located at neighboring points on the manifold, they must somehow be "transported" to the same location, but such a procedure affects the components of the transported tensor. In this chapter, we delve into the concept of comparing tensors at two neighboring points of a manifold.

7.1 Covariant Derivative

7.1.1 Formulation

We seek to define a notion of differentiation for a general tensor.[1] We denote this *yet-to-be-defined operation* by $\boldsymbol{\nabla}$, and its action on the tensor $\mathbf{T}$ by $\boldsymbol{\nabla}\mathbf{T}$. The outcome of this action should be such that $\boldsymbol{\nabla}\mathbf{T}$ defines a proper tensor, which is a multilinear function of vectors and one-forms that transforms according to rule (5.34). We call this tensor the *covariant derivative* or the *gradient* of $\mathbf{T}$. Unlike partial differentiation, which is tied to a specific coordinate system, covariant differentiation is a geometrical operation, independent of the notion of coordinates.

Since tensors are linear functions of vectors and one-forms that reside, respectively, in the tangent and cotangent spaces at a specific spatial point $\mathcal{S}$ in a manifold, one cannot take the difference between two tensors at two distinct locations. What form, then, should this derivative take? A natural place to start is the differential or exterior derivative of a scalar

[1] The approach in this section is inspired by Wald (1984).

function (4.39),

$$\mathrm{d}f = \partial_i f \,\mathrm{d}x^i. \tag{7.1}$$

If we feed this one-form—a machine with a single slot that takes vectors—a vector $\mathbf{u}$, it returns[2]

$$\mathrm{d}f(\mathbf{u}) = \partial_i f \,\mathrm{d}x^i(\mathbf{u}) = u^i \,\partial_i f \equiv \langle \boldsymbol{\nabla}\mathrm{f}, \mathbf{u}\rangle, \tag{7.2}$$

where we used expression (4.41). Thus, for a scalar, we identify covariant differentiation with exterior differentiation:

$$\mathrm{df} \equiv \boldsymbol{\nabla}\mathrm{f}, \tag{7.3}$$

which defines a one-form; in terms of the components of this one-form, we may write

$$\partial_i f \equiv (\boldsymbol{\nabla}\mathrm{f})_i. \tag{7.4}$$

Thus, we see that the covariant derivative takes a $(0,0)$ tensor, a scalar function, and turns it into a $(0,1)$ tensor, a one-form. What about the action on a general tensor? If $\mathbf{T}$ is a (p,q) tensor, its covariant derivative $\boldsymbol{\nabla}\mathbf{T}$ is a $(p,q+1)$ tensor of the form

$$\begin{aligned}
\boldsymbol{\nabla}\mathbf{T} &= (\boldsymbol{\nabla}\mathbf{T})^{i_1\cdots i_p}{}_{j_1\cdots j_q m}\,\mathbf{e}^m \otimes \mathbf{e}_{i_1} \otimes \cdots \otimes \mathbf{e}_{i_p} \otimes \mathbf{e}^{j_1} \otimes \cdots \otimes \mathbf{e}^{j_q} \\
&= \nabla_m T^{i_1\cdots i_p}{}_{j_1\cdots j_q}\,\mathbf{e}^m \otimes \mathbf{e}_{i_1} \otimes \cdots \otimes \mathbf{e}_{i_p} \otimes \mathbf{e}^{j_1} \otimes \cdots \otimes \mathbf{e}^{j_q} \\
&= T^{i_1\cdots i_p}{}_{j_1\cdots j_q;m}\,\mathbf{e}^m \otimes \mathbf{e}_{i_1} \otimes \cdots \otimes \mathbf{e}_{i_p} \otimes \mathbf{e}^{j_1} \otimes \cdots \otimes \mathbf{e}^{j_q} \\
&= T^{i_1\cdots i_p}{}_{j_1\cdots j_q|m}\,\mathbf{e}^m \otimes \mathbf{e}_{i_1} \otimes \cdots \otimes \mathbf{e}_{i_p} \otimes \mathbf{e}^{j_1} \otimes \cdots \otimes \mathbf{e}^{j_q}.
\end{aligned} \tag{7.5}$$

With abuse of notation, it is common to denote the components of the covariant derivative by $\nabla_m T^{i_1\cdots i_p}{}_{j_1\cdots j_q}$[3] or, using the so-called *semicolon notation*, by $T^{i_1\cdots i_p}{}_{j_1\cdots j_q;m}$ or, using a vertical bar notation, by $T^{i_1\cdots i_p}{}_{j_1\cdots j_q|m}$. Next, we need to be specific about the properties of the covariant-derivative operator $\boldsymbol{\nabla}$.

The covariant derivative has the following attributes.

- For a scalar function f, equations (7.3) and (7.4) hold: the covariant derivative of a function is its exterior derivative. The directional derivative of the function f with respect to the vector field $\mathbf{u}$ is defined by (7.2). For a scalar function, the covariant derivative is equal to the partial derivative:

$$(\boldsymbol{\nabla}\mathrm{f})_i = \nabla_i f = \partial_i f. \tag{7.6}$$

- For two (p,q) tensors, $\mathbf{S}$ and $\mathbf{T}$, and two real numbers, a and b, we require *linearity*:

$$\boldsymbol{\nabla}(a\,\mathbf{S} + b\,\mathbf{T}) = a\,\boldsymbol{\nabla}\mathbf{S} + b\,\boldsymbol{\nabla}\mathbf{T}. \tag{7.7}$$

---

[2] Note how we are careful with the notation to distinguish the $(0,0)$ tensor f, which is defined at a point $\mathcal{S}$, from the function f, which depends on the coordinates x.

[3] We use this notation in this book.

- For two arbitrary tensors, **S** and **T**, we impose *Leibniz's rule*:

$$\nabla_m (S^{i_1\cdots i_p}{}_{j_1\cdots j_q}\, T^{k_1\cdots k_r}{}_{\ell_1\cdots \ell_s}) = (\nabla_m S^{i_1\cdots i_p}{}_{j_1\cdots j_q})\, T^{k_1\cdots k_r}{}_{\ell_1\cdots \ell_s} + S^{i_1\cdots i_p}{}_{j_1\cdots j_q}\, (\nabla_m T^{k_1\cdots k_r}{}_{\ell_1\cdots \ell_s}). \tag{7.8}$$

- Covariant differentiation and contraction commute. Thus, if **T** is a (p,q) tensor and **S** is a $(p-1,q-1)$ tensor obtained by contracting the i_kth contravariant index of **T** with the j_kth covariant index of **T**, as in expression (5.10), then

$$\nabla_m S^{i_1\cdots i_{k-1}\, i_{k+1}\cdots i_p}{}_{j_1\cdots j_{k-1}\, j_{k+1}\cdots j_q} = \nabla_m T^{i_1\cdots k\cdots i_p}{}_{j_1\cdots k\cdots j_q}, \tag{7.9}$$

is a $(p-1,q)$ tensor.

To find an explicit expression for the covariant derivative of a general tensor, we need to consider its action on basis vectors, $\mathbf{e}_i$, and basis one-forms, $\mathbf{e}^i$. We define the action of the covariant derivative on a basis vector as

$$\nabla_k \mathbf{e}_i \equiv \Gamma^j_{ki}\, \mathbf{e}_j, \tag{7.10}$$

where, as usual, there is a sum over repeated upper and lower indices, and where the coefficients Γ^j_{ki} remain to be determined.[4] Basically, definition (7.10) expresses that the action of the covariant-derivative operator on a basis vector, $\nabla_k \mathbf{e}_i$, produces some linear combination of the basis vectors, $\Gamma^j_{ki}\, \mathbf{e}_j$. In other words, $\nabla_k \mathbf{e}_i$ belongs to the tangent space. In terms of the duality product (4.33), we have

$$\Gamma^j_{ki} = \langle \mathbf{e}^j, \nabla_k \mathbf{e}_i \rangle. \tag{7.11}$$

With the introduction of the coefficients Γ^j_{ki}, using Leibniz's rule (7.8), we may write the covariant derivative of a vector as

$$\begin{aligned} \boldsymbol{\nabla}\mathbf{u} &= \mathbf{e}^k \otimes \nabla_k (u^i\, \mathbf{e}_i) \\ &= \mathbf{e}^k \otimes (\partial_k u^i\, \mathbf{e}_i + u^i\, \nabla_k \mathbf{e}_i) \\ &= (\partial_k u^j + \Gamma^j_{ki}\, u^i)\, \mathbf{e}^k \otimes \mathbf{e}_j, \end{aligned} \tag{7.12}$$

where the components, u^i, are treated as scalar functions. Thus, the components of the covariant derivative of a vector are

$$\nabla_k u^j = \partial_k u^j + \Gamma^j_{ki}\, u^i. \tag{7.13}$$

Next, let us consider the action of the covariant derivative on a one-form basis. We write

$$\nabla_k \mathbf{e}^i = \tilde{\Gamma}^i_{kj}\, \mathbf{e}^j, \tag{7.14}$$

where the coefficients $\tilde{\Gamma}^i_{kj}$ remain to be determined. Analogous to (7.10), we require that the action of the covariant-derivative operator ∇_k on a basis one-form, $\nabla_k \mathbf{e}^i$, may be expressed

[4] The order of the lower indices k and i on Γ^j_{ki} involves a choice, which must be made consistently once and for all.

as a linear combination of basis one-forms, $\tilde{\Gamma}^i_{kj}\,\mathbf{e}^j$. Basically, we demand that $\nabla_k \mathbf{e}^i$ "lives" in the cotangent space. We write the covariant derivative of a one-form as

$$\begin{aligned}\boldsymbol{\nabla}\boldsymbol{\omega} &= \mathbf{e}^k \otimes \nabla_k(\omega_j\,\mathbf{e}^j)\\ &= \mathbf{e}^k \otimes (\partial_k\omega_j\,\mathbf{e}^j + \omega_j\,\nabla_k\mathbf{e}^j)\\ &= (\partial_k\omega_j + \tilde{\Gamma}^i_{kj}\,\omega_i)\,\mathbf{e}^k \otimes \mathbf{e}^j.\end{aligned} \tag{7.15}$$

We anticipate a relationship between coefficients Γ^i_{jk} and $\tilde{\Gamma}^i_{jk}$. To find it, we consider the action of the covariant derivative on the Euclidean inner product (5.13) between a vector and a one-form, which is a scalar. According to Leibniz' rule (7.8) and the commutative rule for contractions (7.9), we have

$$\nabla_k(\omega_j\,u^j) = (\nabla_k\omega_j)\,u^j + \omega_j\,(\nabla_k u^j). \tag{7.16}$$

Using expression (7.13) and (7.15) and the fact that the left-hand side is the gradient of a scalar, we find

$$\begin{aligned}\partial_k(\omega_j\,u^j) &= (\partial_k\omega_j + \tilde{\Gamma}^i_{kj}\,\omega_i)\,u^j + (\partial_k u^j + \Gamma^j_{ki}\,u^i)\,\omega_j\\ &= \partial_k(\omega_j\,u^j) + (\tilde{\Gamma}^i_{kj} + \Gamma^i_{kj})\,\omega_i\,u^j.\end{aligned} \tag{7.17}$$

We conclude from expression (7.17) that the anticipated relationship between Γ^i_{kj} and $\tilde{\Gamma}^i_{kj}$ is

$$\tilde{\Gamma}^i_{kj} = -\Gamma^i_{kj}, \tag{7.18}$$

and thus the covariant derivative of the one-form basis is

$$\nabla_k\mathbf{e}^i = -\,\Gamma^i_{kj}\,\mathbf{e}^j, \tag{7.19}$$

and the covariant derivative of a one-form (7.15) becomes

$$\boldsymbol{\nabla}\boldsymbol{\omega} = (\partial_k\omega_j - \Gamma^i_{kj}\,\omega_i)\,\mathbf{e}^j \otimes \mathbf{e}^k. \tag{7.20}$$

The coefficients Γ^i_{jk} are called the *connection coefficients.*

At this point, we have the required ingredients to obtain the covariant derivative of a tensor. The covariant derivative of a (p,q) tensor is a $(p,q+1)$ tensor whose components are

$$\begin{aligned}(\boldsymbol{\nabla}\mathbf{T})^{i_1 i_2\cdots i_p}{}_{j_1 j_2\cdots j_q m} &\equiv \nabla_m T^{i_1 i_2\cdots i_p}{}_{j_1 j_2\cdots j_q}\\ &= \partial_m T^{i_1 i_2\cdots i_p}{}_{j_1 j_2\cdots j_q}\\ &\quad + \Gamma^{i_1}_{mn}\,T^{n i_2\cdots i_p}{}_{j_1 j_2\cdots j_q} + \Gamma^{i_2}_{mn}\,T^{i_1 n\cdots i_p}{}_{j_1 j_2\cdots j_q} + \cdots\\ &\quad - \Gamma^n_{mj_1}\,T^{i_1 i_2\cdots i_p}{}_{n j_2\cdots j_q} - \Gamma^n_{mj_2}\,T^{i_1 i_2\cdots i_p}{}_{j_1 n\cdots j_q} - \cdots .\end{aligned} \tag{7.21}$$

7.1.2 Transformation of Connection Coefficients

Since expression (7.12) defines a (1,1) tensor, its components must transform according to rule (5.34):

$$\nabla_{j'} u^{k'} = \Lambda^{k'}{}_{k}\, \lambda^{j}{}_{j'}\, \nabla_{j} u^{k}. \tag{7.22}$$

Written in its full glory, using (7.13), this implies that

$$\partial_{j'} u^{k'} + \Gamma^{k'}_{j'i'}\, u^{i'} = \Lambda^{k'}{}_{k}\, \lambda^{j}{}_{j'}\, (\partial_{j} u^{k} + \Gamma^{k}_{ji}\, u^{i}), \tag{7.23}$$

where $\partial_j \equiv \partial/\partial x^j$ and $\partial_{j'} \equiv \partial/\partial x^{j'}$. Since the components of a vector transform according to

$$u^{i'} = \Lambda^{i'}{}_{i}\, u^{i}, \tag{7.24}$$

expression (7.23) becomes

$$\partial_{j'} (\Lambda^{k'}{}_{k}\, u^{k}) + \Gamma^{k'}_{j'i'}\, \Lambda^{i'}{}_{i}\, u^{i} = \Lambda^{k'}{}_{k}\, \lambda^{j}{}_{j'}\, (\partial_{j} u^{k} + \Gamma^{k}_{ji}\, u^{i}). \tag{7.25}$$

Using the chain rule, we may rewrite it as

$$\Lambda^{k'}{}_{k}\, \lambda^{j}{}_{j'}\, \partial_{j} u^{k} + u^{k}\, \lambda^{j}{}_{j'}\, \partial_{j} \Lambda^{k'}{}_{k} + \Gamma^{k'}_{j'i'}\, \Lambda^{i'}{}_{i}\, u^{i} = \Lambda^{k'}{}_{k}\, \lambda^{j}{}_{j'}\, (\partial_{j} u^{k} + \Gamma^{k}_{ji}\, u^{i}), \tag{7.26}$$

which implies that

$$u^{k}\, \lambda^{j}{}_{j'}\, \partial_{j} \Lambda^{k'}{}_{k} + \Gamma^{k'}_{j'i'}\, \Lambda^{i'}{}_{i}\, u^{i} = \Lambda^{k'}{}_{k}\, \lambda^{j}{}_{j'}\, \Gamma^{k}_{ji}\, u^{i}. \tag{7.27}$$

This equation holds for any vector $\mathbf{u}$, and thus we have

$$\lambda^{j}{}_{j'}\, \partial_{j} \Lambda^{k'}{}_{k} + \Gamma^{k'}_{j'i'}\, \Lambda^{i'}{}_{k} = \Lambda^{k'}{}_{\ell}\, \lambda^{j}{}_{j'}\, \Gamma^{\ell}_{jk}. \tag{7.28}$$

Finally, this may be rearranged and relabeled to give the relationship between the transformed and original connection coefficients:

$$\Gamma^{k'}_{j'i'} = \lambda^{i}{}_{i'}\, \Lambda^{k'}{}_{k}\, \lambda^{j}{}_{j'}\, \Gamma^{k}_{ji} - \lambda^{i}{}_{i'}\, \lambda^{j}{}_{j'}\, \partial_{j} \Lambda^{k'}{}_{i}. \tag{7.29}$$

An alternate expression for the relationship between connection coefficients may be obtained by invoking relationships (4.18) to get

$$\Gamma^{k'}_{j'i'} = \lambda^{i}{}_{i'}\, \Lambda^{k'}{}_{k}\, \lambda^{j}{}_{j'}\, \Gamma^{k}_{ji} + \Lambda^{k'}{}_{i}\, \partial_{i'} \lambda^{i}{}_{j'}. \tag{7.30}$$

Note that the connection coefficients *do not define the components of a (1,2) tensor*, since they do not transform according to tensorial rules; notably, the second summands of rules (7.29) and (7.30) contain second derivatives. If Φ is a linear function of x^i, the second derivatives disappear and the transformation is in agreement with the tensor-transformation rule; however, in general, Γ^i_{jk} is not a tensor, even though it might behave as one in this particular case. The nontensorial quality of Γ^i_{jk} is apparent in the notation; if it were a tensor, we would write Γ^i_{jk}.

Example 7.1 Material Velocity Gradient

In example 4.1 we introduced a vector field $\mathbf{v}$ to represent the flow of matter in continuum mechanics. The gradient of the material velocity, $\boldsymbol{\nabla}\mathbf{v}$, is a $(1,1)$ tensor. This tensor may be expressed in either Eulerian or Lagrangian coordinates as

$$\boldsymbol{\nabla}\mathbf{v} = \nabla_i v^j\, \mathbf{e}^i \otimes \mathbf{e}_j = \nabla_I v^J\, \mathbf{e}^I \otimes \mathbf{e}_J, \tag{7.31}$$

subject to transformation rules based on the deformation gradient (4.26) and its inverse (4.27):

$$\nabla_i v^j = (F^{-1})^I{}_i\, F^j{}_J\, \nabla_I v^J, \qquad \nabla_I v^J = F^i{}_I\, (F^{-1})^J{}_j\, \nabla_i v^j. \tag{7.32}$$

The relation between Eulerian and Lagrangian connection coefficients, Γ^k_{ij} and Γ^K_{IJ}, is

$$\Gamma^K_{IJ} = (F^{-1})^K{}_k\, F^i{}_I\, F^j{}_J\, \Gamma^k_{ij} + (F^{-1})^K{}_k\, \partial_I F^k{}_J, \tag{7.33}$$

which is a particular example of relationship (7.30). If the Eulerian coordinates are Cartesian, the connection coefficients Γ^k_{ij} vanish, and the Lagrangian connection coefficients may be expressed in terms of the motion (3.12) and its inverse (3.13) as

$$\Gamma^K_{IJ} = (\partial_k \Phi^K)\, \partial_I \partial_J \varphi^k. \tag{7.34}$$

Example 7.2 Deformation Rate and Vorticity

The gradient of the material velocity, $\boldsymbol{\nabla}\mathbf{v}$, defines the $(1,1)$ tensor

$$\mathbf{G} \equiv (\boldsymbol{\nabla}\mathbf{v})^t, \tag{7.35}$$

where a superscript t denotes the transpose, discussed in section 5.2.5. We can write the material velocity gradient (7.35) as the sum of a symmetric and an antisymmetric tensor, as discussed in sections 5.2.4 and 5.2.5, namely

$$\mathbf{G} = \mathbf{D} + \mathbf{W}, \tag{7.36}$$

where

$$\mathbf{D} \equiv \tfrac{1}{2}\left(\mathbf{G}+\mathbf{G}^t\right) = \widehat{\mathbf{G}} = \tfrac{1}{2}\left[(\boldsymbol{\nabla}\mathbf{v})^t + \boldsymbol{\nabla}\mathbf{v}\right] \tag{7.37}$$

is the *symmetric deformation-rate tensor*,

$$\mathbf{D} = \mathbf{D}^t, \tag{7.38}$$

and

$$\mathbf{W} \equiv \tfrac{1}{2}\left(\mathbf{G}-\mathbf{G}^t\right) = \widetilde{\mathbf{G}} = \tfrac{1}{2}\left[(\boldsymbol{\nabla}\mathbf{v})^t - \boldsymbol{\nabla}\mathbf{v}\right] \tag{7.39}$$

the *antisymmetric vorticity tensor*,

$$\mathbf{W} = -\mathbf{W}^t. \tag{7.40}$$

The deformation-rate and vorticity tensors play an important role in continuum and fluid mechanics.

7.1.3 Divergence

The *divergence* of a tensor field $\operatorname{div}\mathbf{T} = \boldsymbol{\nabla}\cdot\mathbf{T}$ is obtained by contracting the last contravariant (upper) index with the *last* covariant (lower) index of $\boldsymbol{\nabla}\mathbf{T}$ in expression (7.21), thereby reducing the contravariant rank of the tensor by one,

$$\begin{aligned}\operatorname{div}\mathbf{T} &\equiv \boldsymbol{\nabla}\cdot\mathbf{T}\\ &= (\boldsymbol{\nabla}\cdot\mathbf{T})^{i_1\cdots i_{p-1}m}{}_{j_1\cdots j_q m}\,\mathbf{e}_{i_1}\otimes\cdots\otimes\mathbf{e}_{i_{p-1}}\otimes\mathbf{e}^{j_1}\otimes\cdots\otimes\mathbf{e}^{j_q};\end{aligned} \tag{7.41}$$

in components,

$$\begin{aligned}(\boldsymbol{\nabla}\cdot\mathbf{T})^{i_1\cdots i_{p-1}m}{}_{j_1 j_2\cdots j_q m} &= \nabla_m T^{i_1\cdots i_{p-1}m}{}_{j_1\cdots j_q}\\ &= \partial_m T^{i_1\cdots i_{p-1}m}{}_{j_1 j_2\cdots j_q}\\ &\quad + \Gamma^{i_1}_{mn}\,T^{n i_2\cdots i_{p-1}m}{}_{j_1 j_2\cdots j_q} + \Gamma^{i_2}_{mn}\,T^{i_1 n\cdots i_{p-1}m}{}_{j_1 j_2\cdots j_q} + \cdots\\ &\quad - \Gamma^{n}_{m j_1}\,T^{i_1\cdots i_{p-1}m}{}_{n j_2\cdots j_q} - \Gamma^{n}_{m j_2}\,T^{i_1\cdots i_{p-1}m}{}_{j_1 n\cdots j_q} - \cdots.\end{aligned} \tag{7.42}$$

The notation $\boldsymbol{\nabla}\cdot\mathbf{T}$ is not ideal for higher order tensors because the dot " $\cdot$ " suggests contracting the covariant index of the gradient operator with the first contravariant index of the tensor, as in expression (5.14).

7.1.4 Parallel Transport

The notion of parallelism implies constancy; for instance, in Euclidean geometry, a constant vector field involves vectors that are parallel to each other and of equal length. Given a scalar function, f, on an n-manifold, we can measure its rate of change, which is a one-form, df. If

$df = \mathbf{0}$, it follows that f is constant throughout the manifold. Beyond scalar functions, there is no unambiguous notion of constancy, and hence parallelism, on a manifold, even for a vector field. The vanishing of the covariant derivative allows us to define the concept of *constancy* of the vector field.

Analytic Formulation

Given a covariant-derivative operator, we define a notion of *parallel transport*. Suppose we have a curve in a manifold, parameterized in terms of λ, with tangent vector $\mathbf{u} = d/d\lambda$, as illustrated in figure 4.2. A vector $\mathbf{w}$ is said to be parallel transported along the curve defined by vector $\mathbf{u}$ if the equation

$$\mathbf{u} \cdot (\boldsymbol{\nabla}\mathbf{w}) \equiv \boldsymbol{\nabla}_{\mathbf{u}}\mathbf{w} = \mathbf{0} \tag{7.43}$$

holds everywhere along the curve. The derivative $\boldsymbol{\nabla}_{\mathbf{u}}\mathbf{w}$ is called the *directional derivative* of the vector $\mathbf{w}$; this derivative provides a measure of the change of vector field $\mathbf{w}$ in the direction defined by the vector $\mathbf{u}$. Using the component expression for covariant derivative (7.13), we find

$$u^j(\partial_j w^k + \Gamma^k_{ji}\, w^i) = \frac{dw^k}{d\lambda} + \Gamma^k_{ji}\, u^j\, w^i = 0. \tag{7.44}$$

where we identified $\mathbf{u} \equiv u^i\, \partial_i$ with the directional derivative along the curve $d/d\lambda$.

Geometrical Interpretation

To give a geometrical interpretation of parallel transport, let us consider the need for such a concept in the definition of a derivative. We need to answer the following question: how do we transport a vector on a surface such that it remains parallel to itself and tangent to the surface? If a surface is flat, it is easy to accommodate both conditions and hence we can use the standard definition of a derivative

$$\mathbf{w}' \equiv \lim_{\Delta\lambda \to 0} \frac{\mathbf{w}(\,\mathcal{S}(\lambda+\Delta\lambda)\,) - \mathbf{w}(\,\mathcal{S}(\lambda)\,)}{\Delta\lambda}; \tag{7.45}$$

one can compare the vectors at two different points, $\mathcal{S}(\lambda)$ and $\mathcal{S}(\lambda+\Delta\lambda)$. However, such a comparison is not possible in general. Consider a surface embedded in a Euclidean space, $\mathbb{R}^3$. Let a vector be tangent to that surface at the point $\mathcal{S}(\lambda)$. In general, parallel transport of this vector in $\mathbb{R}^3$ to the point $\mathcal{S}(\lambda+\Delta\lambda)$ does not result in it remaining parallel to the surface. A solution to this problem was proposed by Tullio Levi-Civita at the beginning of the twentieth century.

As illustrated in figure 7.1, let $\mathcal{S}(\lambda)$ denote the location of points along a curve parameterized in terms of λ, and let $\mathbf{u} = d/d\lambda$ denote the corresponding tangent vector field. Vector $\mathbf{w}_0 = \mathbf{w}(\mathcal{S}(\lambda_0))$ is *parallel transported* along the curve $\mathcal{S}(\lambda)$ by keeping its orientation fixed relative to the tangent vector $\mathbf{u}(\mathcal{S}(\lambda))$. Covariant differentiation of a vector $\mathbf{w}$ involves parallel transporting the vector at location $\mathcal{S}(\lambda+\Delta\lambda)$ back to location $\mathcal{S}(\lambda)$, and then taking the difference between the two vectors at $\mathcal{S}(\lambda)$ in the limit $\Delta\lambda \to 0$:

$$\boldsymbol{\nabla}_{\mathbf{u}}\mathbf{w} = \lim_{\Delta\lambda \to 0} \frac{\mathbf{w}(\,\mathcal{S}(\lambda+\Delta\lambda)\,)_{\text{parallel transported to } \mathcal{S}(\lambda)} - \mathbf{w}(\,\mathcal{S}(\lambda)\,)}{\Delta\lambda}. \tag{7.46}$$

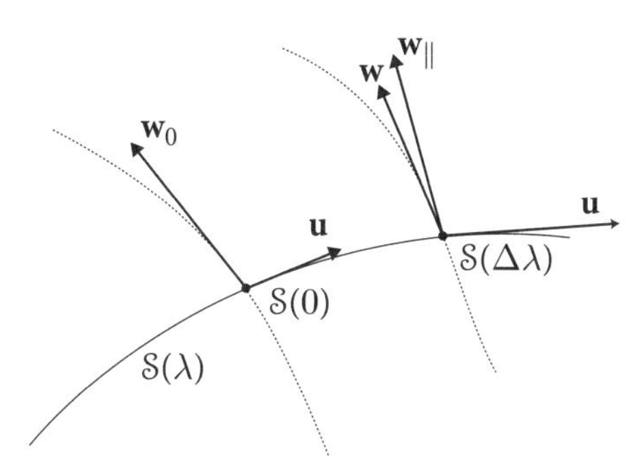

Figure 7.1: The covariant derivative is a measure of the variability of the vector field $\mathbf{w}$. In this two-dimensional illustration, the vector $\mathbf{w}_0 = \mathbf{w}(\mathcal{S}(0))$ is *parallel transported* along the curve $\mathcal{S}(\lambda)$, and denoted by $\mathbf{w}_{||}$ at point $\mathcal{S}(\Delta\lambda)$. Its orientation is kept fixed with respect to the vector $\mathbf{u}$, which is tangent to the curve at $\mathcal{S}(\lambda)$. Also, its magnitude is unchanged. The difference between $\mathbf{w}(\mathcal{S}(\Delta\lambda))$ and $\mathbf{w}_{||}$ is the directional derivative $\boldsymbol{\nabla}_{\mathbf{u}}\mathbf{w}$, as expressed in (7.46).

Thus, the change along the curve is used to characterize the variability of the vector field, as illustrated in figure 7.1.

Let us assume that parallel vectors at two neighboring points of the manifold differ infinitesimally from one another. Formally, we write the components of these vectors as w^i and $w^i + \delta w^i$. Furthermore, we assume that the infinitesimal difference, δw^i, is a linear function of displacement, δx^i, and that vector $\mathbf{u} + a\,\mathbf{w}$ at one point corresponds to $\mathbf{u}' + a\,\mathbf{w}'$, at the other, where a denotes a constant. Definition (7.43) may be written as

$$\frac{\mathrm{d}w^k}{\mathrm{d}\lambda} + \Gamma^k_{ji}\, w^i\, \frac{\mathrm{d}x^j}{\mathrm{d}\lambda} = 0. \tag{7.47}$$

which implies that

$$\delta w^k = -\,\Gamma^k_{ji}\, w^i\, \delta x^j. \tag{7.48}$$

There is no intrinsic meaning to parallelism on a manifold, and, consequently, there is no explicit formula for parallel transport: it requires a connection, which is not an intrinsic entity of a manifold. Given a connection, we *can* define the corresponding parallel transport. For any connection, if we transport a vector along a closed loop, the resulting vector does not, in general, coincide with the original one, an issue we discuss further in section 7.1.4.

Geodesics

A *geodesic* is a curve whose tangent vector $\mathbf{u}$ is parallel transported along itself, as illustrated in Figure 7.1. Analytically, a geodesic is defined by the *geodesic equation*:

$$\mathbf{u}\cdot(\boldsymbol{\nabla}\mathbf{u}) = \boldsymbol{\nabla}_{\mathbf{u}}\mathbf{u} = \mathbf{0}; \tag{7.49}$$

in components,

$$u^j(\partial_j u^k + \Gamma^k_{ji}\, u^i) = \frac{\mathrm{d}u^k}{\mathrm{d}\lambda} + \Gamma^k_{ji}\, u^j\, u^i = 0. \tag{7.50}$$

Invoking the equation for a curve, given by expression (4.14), $u^k = \mathrm{d}x^k/\mathrm{d}\lambda$, we may write the geodesic equation as

$$\frac{\mathrm{d}^2x^k}{\mathrm{d}\lambda^2} + \Gamma^k_{ji}\frac{\mathrm{d}x^j}{\mathrm{d}\lambda}\frac{\mathrm{d}x^i}{\mathrm{d}\lambda} = 0. \tag{7.51}$$

Solutions $x^k(\lambda)$ in a manifold are generalizations of straight lines in a plane.

Example 7.3 Geodesics in General Relativity

In general relativity, geodesics, $x^\mu(\lambda)$, generalize the notion of straight lines in spacetime. They are governed by the geodesic equation

$$\frac{\mathrm{d}^2x^\mu}{\mathrm{d}\lambda^2} + \Gamma^\mu_{\nu\tau}\frac{\mathrm{d}x^\nu}{\mathrm{d}\lambda}\frac{\mathrm{d}x^\tau}{\mathrm{d}\lambda} = 0. \tag{7.52}$$

Note that only the *torsion-free* connection coefficients $\Gamma^\mu_{\nu\tau} = \Gamma^\mu_{\tau\nu}$ govern geodesics.[a]

[a] See section 7.1.5 for a discussion of torsion.

7.1.5 Torsion and Curvature Tensors

The covariant derivative of a function f is a one-form, ∇f, and thus the covariant derivative of the covariant derivative of a function, $\nabla(\nabla \mathrm{f})$, is a $(0,2)$ tensor. Upon substituting $\omega = \nabla \mathrm{f}$ in expression (7.20) we find

$$\nabla(\nabla \mathrm{f}) = (\partial_i\partial_j f - \Gamma^k_{ij}\partial_k f)\,\mathbf{e}^i \otimes \mathbf{e}^j. \tag{7.53}$$

Allowing for anholonomicity, expressed by (4.72) and (4.73), this implies that

$$\begin{aligned}(\nabla_i\nabla_j - \nabla_j\nabla_i)f &= (\partial_i\partial_j - \partial_j\partial_i)f - (\Gamma^k_{ij} - \Gamma^k_{ji})\,\nabla_k f \\ &= [\mathbf{e}_i, \mathbf{e}_j]f - (\Gamma^k_{ij} - \Gamma^k_{ji})\,\nabla_k f \\ &= -\,(\Gamma^k_{ij} - \Gamma^k_{ji} - \tau^k_{ij})\,\nabla_k f.\end{aligned} \tag{7.54}$$

The left-hand side of the first equation is a $(0,2)$ tensor by construction. In the last equality, $\nabla_k f$ represents a $(0,1)$ tensor, so for these reasons,

$$t_{ij}{}^k \equiv \Gamma^k_{ij} - \Gamma^k_{ji} - \tau^k_{ij} \tag{7.55}$$

defines a $(1,2)$ tensor known as the *torsion tensor*. For a holonomic basis, we have

$$t_{ij}{}^k = \Gamma^k_{ij} - \Gamma^k_{ji}. \tag{7.56}$$

An alternative definition of torsion involves the linear map

$$\mathbf{t}(\mathbf{u},\mathbf{w}) \equiv \boldsymbol{\nabla}_{\mathbf{u}}\mathbf{w} - \boldsymbol{\nabla}_{\mathbf{w}}\mathbf{u} - [\mathbf{u},\mathbf{w}]. \tag{7.57}$$

In coordinates, accommodating anholonomicity, using the definition of the directional derivative (7.46) and the Lie bracket (4.73), we find that this implies

$$\begin{aligned} t_{jk}{}^{i}\, u^j\, w^k &= u^j(\partial_j w^k + \Gamma^k_{j\ell}\, w^\ell) - w^j(\partial_j u^k + \Gamma^k_{j\ell}\, u^\ell) \\ &\quad - (u^i\,\partial_i w^j - w^i\,\partial_i u^j - \tau^j_{ik}\, u^i\, w^k) \\ &= (\Gamma^i_{jk} - \Gamma^i_{kj} - \tau^k_{jk})\, u^j\, w^k = 0, \end{aligned} \tag{7.58}$$

in agreement with result (7.55).

We have seen that the covariant derivative of a vector $\mathbf{u}$ has components defined by (7.13). Inspired by the result (7.54), let us consider the result of applying the covariant-derivative operator twice to a vector, $\nabla_i\nabla_j u^k$, and subtracting the transpose of this $(1,2)$ tensor, $\nabla_j\nabla_i u^k$. We find the *Ricci identity*

$$(\nabla_i\nabla_j - \nabla_j\nabla_i)u^k = r_{ij\ell}{}^{k}\, u^\ell - t_{ij}{}^{\ell}\, \nabla_\ell\, u^k, \tag{7.59}$$

where $t_{ij}{}^{\ell}$ denote the components of the torsion tensor (7.55), and where $r_{ij\ell}{}^{k}$ denote the components of the *curvature tensor* or *Riemann tensor*,[5]

$$r_{ij\ell}{}^{k} \equiv \partial_i\Gamma^k_{j\ell} - \partial_j\Gamma^k_{i\ell} + \Gamma^k_{im}\,\Gamma^m_{j\ell} - \Gamma^k_{jm}\,\Gamma^m_{i\ell} - \tau^m_{ij}\,\Gamma^k_{m\ell}. \tag{7.60}$$

Since the left-hand side and the second term on the right-hand side of (7.59) define tensors, the elements $r_{ij\ell}{}^{k}$ must also define a tensor. For a holonomic basis, we have

$$r_{ij\ell}{}^{k} \equiv \partial_i\Gamma^k_{j\ell} - \partial_j\Gamma^k_{i\ell} + \Gamma^k_{im}\,\Gamma^m_{j\ell} - \Gamma^k_{jm}\,\Gamma^m_{i\ell}. \tag{7.61}$$

Example 7.4 Ricci Identity for Material Velocity

In continuum mechanics, Galilean spacetime is flat and torsion-free, which implies that the curvature and torsion tensors vanish. Thus, the Ricci identity (7.59) for the material velocity field $\mathbf{v}$ becomes

$$(\nabla_i\nabla_j - \nabla_j\nabla_i)v^k = 0. \tag{7.62}$$

We conclude that in continuum mechanics, covariant derivatives commute.

We note that the Riemann tensor is antisymmetric in its first two covariant indices:

$$r_{ij\ell}{}^{k} = -\, r_{ji\ell}{}^{k}. \tag{7.63}$$

[5] Our definition of the Riemann tensor has the same placement of the indices as Wald (1984) but the opposite sign.

To gain an insight into the geometrical meaning of the Riemann tensor, we note that for a two-dimensional manifold, there is only one independent component. This component is the Gaussian curvature of a surface (Penrose, 2004), whose value depends only on the metric and is, as expected given a tensorial formulation, an intrinsic property of a space independent of the coordinate system. A sphere, a cylinder, and a hyperboloid have positive, zero, and negative Gaussian curvatures, respectively. In higher dimensions, there are more components, and such a reduction is not possible.

More generally, curvature is defined in terms of the linear map

$$\mathbf{r}(\mathbf{u},\mathbf{w})\,\mathbf{z} \equiv \boldsymbol{\nabla}_{\mathbf{u}}\,\boldsymbol{\nabla}_{\mathbf{w}}\,\mathbf{z} - \boldsymbol{\nabla}_{\mathbf{w}}\boldsymbol{\nabla}_{\mathbf{u}}\,\mathbf{z} - \boldsymbol{\nabla}_{[\mathbf{u},\mathbf{w}]}\,\mathbf{z}. \tag{7.64}$$

Using the repeated application of the definition of the directional derivative and invoking the nonholonomic Lie bracket (4.74), it is a matter of tedious algebra to show that

$$\mathbf{r}(\mathbf{e}_i,\mathbf{e}_j)\,\mathbf{z} = \mathbf{e}_k\, r_{ij\ell}{}^{k}\, z^{\ell}, \tag{7.65}$$

thereby equating the curvature tensor with the Riemann tensor.

Geometrical Meaning of Torsion and Curvature Tensors

The torsion and curvature tensors capture the behavior of a vector that is parallel transported around a loop, as discussed in section 7.1.4. As a result of torsion, the vector may not return to the same location. As a result of curvature, the vector, even though returning to the same location, may exhibit a different orientation. Let us consider these two effects in more detail.

Consider a parallelogram spanned by two vectors $\mathbf{u}$ and $\mathbf{v}$ located in the tangent space. We may think of the edge of this parallelogram as defining a loop, with a sense of direction defined by $\mathbf{u}$. If we feed the torsion tensor, with components $t_{ij}{}^{k}$, two vectors whose components are u^i and v^j, it returns a vector whose components are $w^k = t_{ij}{}^{k}\, u^i\, v^j$; in the language of machines, $\mathbf{w}(\cdot) = \mathbf{t}(\mathbf{u},\mathbf{v},\cdot)$, leaving the one-form slot of the torsion tensor open. One may regard torsion as a *vector-valued two-form*, as discussed in section 8.9.4. The vector $\mathbf{w}$ indicates the direction in which $\mathcal{S}$ tends to move as a result of the torsion; it captures the distance or *dislocation* between the starting and ending points of the loop if it were cut at $\mathcal{S}$. We conclude that the torsion tensor induces *helicity*.

The curvature tensor captures the change of orientation of the vector $\mathbf{w}$ if parallel transported around the loop defined by vectors $\mathbf{u}$ and $\mathbf{v}$. Let $\mathbf{w}$ denote the vector at the start of the loop, and $\tilde{\mathbf{w}}$ the vector at the end of the loop. Tensor $r_{ijk}{}^{\ell}\, u^i\, v^j$ defines a machine, $\mathbf{r}(\mathbf{u},\mathbf{v},\cdot,\cdot)$, on the loop, which if fed vector $\mathbf{w}$ returns $\mathbf{z}(\cdot) = \tilde{\mathbf{w}}(\cdot) - \mathbf{w}(\cdot) = \mathbf{r}(\mathbf{u},\mathbf{v},\mathbf{w},\cdot)$, whose components are $z^{\ell} = r_{ijk}{}^{\ell}\, u^i\, v^j\, w^k$. One may regard curvature as a *tensor-valued two-form*, as discussed in section 8.9.7. The curvature tensor captures the *disclination*, $\mathbf{z}$, between the orientation of vectors $\tilde{\mathbf{w}}$ and $\mathbf{w}$, a property known in differential geometry as the *holonomy* of the manifold. The curvature tensor measures the extent to which the metric tensor is not locally isometric to Euclidean space.

We could justify the rank and type of the curvature tensor a priori using our familiarity with tensor algebra. The change, $\tilde{\mathbf{w}} - \mathbf{w} = \mathbf{z}$, which is a vector, is proportional to components u^i and v^j, which constitute the loop; this loop can be viewed as a parallelogram built from $\mathbf{u}$ and $\mathbf{v}$. This change is also proportional to $\mathbf{w}$ itself, whose components are w^k.

Hence, $z^\ell = r_{ijk}{}^\ell\, u^i\, v^j\, w^k$, where $r_{ijk}{}^\ell$ is a $(1,3)$ tensor that describes the proportionality and depends on curvature.

7.1.6 Bianchi Identities

It is a matter of tedious algebra (see, e.g., Lovelock & Rund, 1989) to show that the Riemann and torsion tensors satisfy the *Bianchi identities*

$$\nabla_i\, r_{jk\ell}{}^m + \nabla_k\, r_{ij\ell}{}^m + \nabla_j r_{ki\ell}{}^m = t_{ij}{}^n\, r_{kn\ell}{}^m + t_{ki}{}^n\, r_{jn\ell}{}^m + t_{jk}{}^n\, r_{in\ell}{}^m, \tag{7.66}$$

$$\begin{aligned}\nabla_i t_{jk}{}^\ell + \nabla_k t_{ij}{}^\ell + \nabla_j t_{ki}{}^\ell &= r_{ijk}{}^\ell + r_{kij}{}^\ell + r_{jki}{}^\ell \\ &\quad + t_{ij}{}^m\, t_{km}{}^\ell + t_{ki}{}^m\, t_{jm}{}^\ell + t_{jk}{}^m\, t_{im}{}^\ell.\end{aligned} \tag{7.67}$$

Note the cyclicity in the indices i, j, and k. These important equations link the Riemann and torsion tensors. Once we introduce the notion of a metric, equating the indices m and k in expression (7.66) and k and ℓ in expression (7.67) leads to contracted Bianchi identities, as discussed in section 7.1.11. The contracted Bianchi identities play a central role in the theory of general relativity.

7.1.7 Torsion-Free Connection

If we demand that the covariant derivative of the covariant derivative of a function, $\boldsymbol{\nabla}(\boldsymbol{\nabla}\mathrm{f})$, be symmetric, in the sense of definition (5.20), then

$$\boldsymbol{\nabla}(\boldsymbol{\nabla}\mathrm{f}) = [\boldsymbol{\nabla}(\boldsymbol{\nabla}\mathrm{f})]^t. \tag{7.68}$$

which, using the component expression (7.54), implies that the connection coefficients must be *torsion-free*,

$$t_{jk}{}^i = 0. \tag{7.69}$$

In that case

$$(\nabla_i\nabla_j - \nabla_j\nabla_i)\, u^k = r_{ij\ell}{}^k\, u^\ell. \tag{7.70}$$

As expected, the torsion, or curvature, is expressed in terms of second derivatives, akin to the notion of an inflection point in standard calculus. It is a remarkable result that, for a torsion-free connection, the difference between $\nabla_i\nabla_j u^k$ and $\nabla_j\nabla_i u^k$ is simply a linear combination of the components of the vector u^k, namely: $r_{ij\ell}{}^k\, u^\ell$.

In the absence of torsion, the Bianchi identities (7.66) and (7.67) reduce to

$$\nabla_i\, r_{jk\ell}{}^m + \nabla_k\, r_{ij\ell}{}^m + \nabla_j r_{ki\ell}{}^m = 0, \tag{7.71}$$

$$r_{ijk}{}^\ell + r_{kij}{}^\ell + r_{jki}{}^\ell = 0. \tag{7.72}$$

7.1.8 Covariant Derivative of the Metric Tensor

Let us investigate whether we can find a covariant-derivative operator that preserves the metric:

$$\nabla \mathbf{g} = \mathbf{0}. \tag{7.73}$$

If so, the process of raising and lowering indices commutes with covariant differentiation. In components, using expression (7.21), we have

$$\nabla_i\, g_{jk} = \partial_i g_{jk} - \Gamma^{\ell}_{ij}\, g_{\ell k} - \Gamma^{\ell}_{ik}\, g_{j\ell} = 0. \tag{7.74}$$

What are the implications of the last equation for the connection coefficients? Let us define

$$\Gamma^{\ell}_{ij}\, g_{\ell k} \equiv \Gamma_{ijk}, \tag{7.75}$$

and

$$t_{ijk} \equiv t_{ij}{}^{\ell}\, g_{\ell k}, \tag{7.76}$$

$$\tau_{ijk} \equiv \tau^{\ell}_{ij}\, g_{\ell k}. \tag{7.77}$$

Then, we have the three permutations

$$\partial_i g_{jk} = \Gamma_{ijk} + \Gamma_{ikj}, \tag{7.78}$$

$$\partial_j g_{ki} = \Gamma_{jki} + \Gamma_{jik}, \tag{7.79}$$

$$\partial_k g_{ij} = \Gamma_{kij} + \Gamma_{kji}. \tag{7.80}$$

Upon subtracting the second and third equations from the first equation, using relationship (7.55), we find that

$$\begin{aligned}\Gamma_{jki} = {} & \tfrac{1}{2}(\partial_k\, g_{ij} + \partial_j g_{ki} - \partial_i g_{jk}) \\ & + \tfrac{1}{2}(\tau_{ijk} + \tau_{ikj} + \tau_{jki}) \\ & + \tfrac{1}{2}(t_{ijk} + t_{ikj} + t_{jki}).\end{aligned} \tag{7.81}$$

Thus, if the metric and the torsion are given, also including the structure coefficients, we may compute the connection coefficients via

$$\begin{aligned}\Gamma^{i}_{jk} = {} & \tfrac{1}{2}\, g^{i\ell}(\partial_k\, g_{\ell j} + \partial_j g_{k\ell} - \partial_\ell g_{jk}) \\ & + \tfrac{1}{2}\,(\tau^{i}_{jk} + g^{i\ell}\, \tau^{m}_{\ell j}\, g_{mk} + g^{i\ell}\, \tau^{m}_{\ell k}\, g_{mj}) \\ & + K_{jk}{}^{i}.\end{aligned} \tag{7.82}$$

The tensor with elements

$$K_{jk}{}^{i} = \tfrac{1}{2}\,(t_{jk}{}^{i} + g^{i\ell}\, t_{\ell j}{}^{m}\, g_{mk} + g^{i\ell}\, t_{\ell k}{}^{m}\, g_{mj}), \tag{7.83}$$

involving only the elements of the torsion tensor (7.55), is known as the *contortion tensor*. Since the torsion tensor is proper, so is the contortion tensor.

We have

$$t_{i\ell}{}^{\ell} = \Gamma^{\ell}_{i\ell} - \Gamma^{\ell}_{\ell i} - \tau^{\ell}_{i\ell}, \tag{7.84}$$

and, using the antisymmetry $\tau^{k}_{ij} = -\tau^{k}_{ji}$ of the structure coefficients,

$$t_{\ell i}{}^{\ell} = \Gamma^{\ell}_{\ell i} - \Gamma^{\ell}_{i\ell} - \tau^{\ell}_{\ell i} = -(\Gamma^{\ell}_{i\ell} - \Gamma^{\ell}_{\ell i} - \tau^{\ell}_{i\ell}) = -t_{i\ell}{}^{\ell}. \tag{7.85}$$

Thus, by equating an upper and lower index of the connection coefficients (7.82),

$$\Gamma^{i}_{ji} = \tfrac{1}{2} g^{i\ell} \partial_j g_{i\ell}, \tag{7.86}$$

and

$$\Gamma^{i}_{ij} = \tfrac{1}{2} g^{i\ell} \partial_j g_{i\ell} + \tau^{i}_{ij} + t_{ij}{}^{i}. \tag{7.87}$$

It is important to note that (7.86) involves only the metric and not the torsion.

Let us determine what this means in terms of the square root of the determinant of the metric $\bar{g}$, given by expression (5.93). The inverse of the metric tensor may be written in terms of the determinant of the metric tensor, $\overline{\overline{\det \mathbf{g}}}$, and its matrix of cofactors, $\overline{\overline{\mathrm{Cof}}}^{ij}$, as

$$g^{ij} = (\overline{\overline{\det \mathbf{g}}})^{-1}\, \overline{\overline{\mathrm{Cof}}}^{ij}. \tag{7.88}$$

According to matrix algebra,[6]

$$\partial_i \overline{\overline{\det \mathbf{g}}} = \overline{\overline{\mathrm{Cof}}}^{jk}\, \partial_i g_{jk} = \overline{\overline{\det \mathbf{g}}}\, g^{jk}\, \partial_i g_{jk}. \tag{7.89}$$

Using definitions (5.93) and (5.95), we conclude that

$$\underline{g}\, \partial_i \bar{g} = \tfrac{1}{2} g^{jk}\, \partial_i g_{jk}. \tag{7.90}$$

Upon combining expressions (7.86) and (7.90), we find that

$$\partial_i \bar{g} = \Gamma^{j}_{ij}\, \bar{g}. \tag{7.91}$$

Equations (7.90) and (7.91) are useful expressions in the context of tensor algebra on Riemannian manifolds.

If the connection is holonomic and torsion-free, the covariant derivative preserves the metric tensor if the connection coefficients are given by

$$\Gamma^{i}_{jk} = \tfrac{1}{2} g^{i\ell} (\partial_k g_{\ell j} + \partial_j g_{\ell k} - \partial_\ell g_{jk}) \equiv \left\{ {i \atop j\,k} \right\}. \tag{7.92}$$

[6] Alternatively, to avoid invoking matrix algebra, this result may be obtained by differentiating the square of expression (5.102) and using Cramer's rule (5.164) for the inverse metric tensor.

Torsion-free connection coefficients (7.92) are called *Christoffel symbols of the second kind*. *Christoffel symbols of the first kind* are

$$\Gamma_{jki} = \tfrac{1}{2}(\partial_j g_{ik} + \partial_k g_{ij} - \partial_i g_{jk}) \equiv [jk, i]. \tag{7.93}$$

Example 7.5 Polar Coordinates

To illustrate the Christoffel symbol, we consider the transformation between polar coordinates, $x = \{r, \theta\}$, and Cartesian coordinates, $x' = \{x^{1'}, x^{2'}\}$:

$$x = x^{1'} = \Phi^{1'}(r, \theta) = r\cos\theta, \tag{7.94}$$

and

$$y = x^{2'} = \Phi^{2'}(r, \theta) = r\sin\theta. \tag{7.95}$$

The transformation matrix is

$$(\Lambda^{i'}{}_i) = \begin{pmatrix} \partial_1\Phi^{1'} & \partial_2\Phi^{1'} \\ \partial_1\Phi^{2'} & \partial_2\Phi^{2'} \end{pmatrix} = \begin{pmatrix} \cos\theta & -r\sin\theta \\ \sin\theta & r\cos\theta \end{pmatrix}. \tag{7.96}$$

The Cartesian components of the metric tensor are $g_{i'j'} = \delta_{i'j'}$. Thus, the metric tensor transforms from Cartesian to polar coordinates according to

$$(g_{ij}) = (g_{i'j'}\, \Lambda^{i'}{}_i\, \Lambda^{j'}{}_j) = (\delta_{i'j'}\, \Lambda^{i'}{}_i\, \Lambda^{j'}{}_j) = \begin{pmatrix} 1 & 0 \\ 0 & r^2 \end{pmatrix}, \tag{7.97}$$

with inverse

$$(g^{ij}) = \begin{pmatrix} 1 & 0 \\ 0 & 1/r^2 \end{pmatrix}. \tag{7.98}$$

The Christoffel symbols may now be determined from expression (7.92); we obtain $\Gamma^1_{11} = 0$, $\Gamma^1_{22} = -r$, and $\Gamma^2_{12} = 1/r$. Since the Christoffel symbol is symmetric in its lower indices, it follows from the last equality that $\Gamma^2_{21} = 1/r$. The remaining four components of the Christoffel symbol are zero.

Example 7.6 Lagrangian Connection Coefficients

In example 7.33, we introduced the covariant derivative of the material velocity $\mathbf{v}$. In continuum mechanics, it is convenient to choose Cartesian Eulerian coordinates, such that the Eulerian connection coefficients Γ^i_{jk} vanish. The Lagrangian connection coefficients may be expressed in terms of the Lagrangian components of the metric tensor and its inverse as

$$\Gamma^I_{JK} = \frac{1}{2} g^{IL} (\partial_K g_{LJ} + \partial_J g_{KL} - \partial_L g_{JK}). \tag{7.99}$$

These connection coefficients are time-dependent and evolve with the flow of matter. Thanks to the symmetric nature of the metric tensor (5.84), the Lagrangian connection coefficients are torsion-free:

$$\Gamma^I_{JK} = \Gamma^I_{KJ}. \tag{7.100}$$

As stated in section 7.1.2, the Christoffel symbol is not a tensor, since it does not transform according to tensorial rules. Notably, its components are zero in Cartesian coordinate systems, which, if it were a tensor, would imply that they would be zero in all systems.

7.1.9 Mixed Covariant Derivative in Tetrad Basis

Let us consider the covariant derivative of a vector and transform it to a coordinate-free tetrad basis, introduced in section 5.3.1. We have, using (5.46),

$$\begin{aligned}
\nabla_i u^j &= \partial_i u^j + \Gamma^j_{ik} u^k \\
&= \partial_i (e^j{}_{j'} u^{j'}) + \Gamma^j_{ik} e^k{}_{k'} u^{k'} \\
&= e^j{}_{j'} \partial_i u^{j'} + u^{k'} \partial_i e^j_{k'} + e^j{}_{j'} e^{j'}{}_\ell \Gamma^\ell_{ik} e^k{}_{k'} u^{k'} \\
&= e^j{}_{j'} e_i{}^{i'} [e_{i'}{}^m \partial_m u^{j'} + e_{i'}{}^m e^{j'}{}_\ell (\partial_m e^\ell{}_{k'} + \Gamma^\ell_{mk} e^k{}_{k'}) u^{k'}] \\
&= e^j{}_{j'} e_i{}^{i'} (e_{i'}{}^m \partial_m u^{j'} + \omega^{j'}_{i'k'} u^{k'}),
\end{aligned} \tag{7.101}$$

where we have defined the frame connection coefficients:

$$\omega^{k'}_{i'j'} \equiv e_{i'}{}^i e^{k'}{}_k (\partial_i e^k{}_{j'} + \Gamma^k_{ij} e^j{}_{j'}). \tag{7.102}$$

Thus, the covariant derivative in the frame basis is

$$\nabla_{i'} u^{j'} = e_{i'}{}^i \partial_i u^{j'} + \omega^{j'}_{i'k'} u^{k'}. \tag{7.103}$$

Sometimes this covariant derivative is expressed in the mixed representation (e.g., Carroll, 2004)

$$\tilde{\nabla}_i u^{j'} \equiv \partial_i u^{j'} + \omega^{j'}_{ik'} u^{k'}, \tag{7.104}$$

where

$$\omega^{j'}_{ik'} \equiv e_{i'}{}^i \omega^{j'}_{i'k'}. \tag{7.105}$$

Using this notation, we can rewrite expression (7.102) in the form

$$\partial_i e^k{}_{j'} = \Gamma^k_{ij} e^j{}_{j'} - e^k{}_{k'} \omega^{k'}_{ij'}. \tag{7.106}$$

This result motivates the introduction of the *mixed covariant derivative* of an object with mixed components $T^{ij\cdots}{}_{k\ell\cdots}{}^{i'j'\cdots}{}_{k'\ell'\cdots}$:

$$\begin{aligned}\tilde{\nabla}_n T^{ij\cdots}{}_{k\ell\cdots}{}^{i'j'\cdots}{}_{k'\ell'\cdots} &= \partial_n T^{ij\cdots}{}_{k\ell\cdots}{}^{i'j'\cdots}{}_{k'\ell'\cdots} \\ &+ \Gamma^i_{nm}\, T^{mj\cdots}{}_{k\ell\cdots}{}^{i'j'\cdots}{}_{k'\ell'\cdots} + \cdots \\ &+ \omega^{i'}_{nm'}\, T^{ij\cdots}{}_{k\ell\cdots}{}^{m'j'\cdots}{}_{k'\ell'\cdots} + \cdots \\ &- \Gamma^m_{nk}\, T^{ij\cdots}{}_{m\ell\cdots}{}^{i'j'\cdots}{}_{k'\ell'\cdots} - \cdots \\ &- \omega^{m'}_{nk'}\, T^{ij\cdots}{}_{k\ell\cdots}{}^{i'j'\cdots}{}_{m'\ell'\cdots} - \cdots ,\end{aligned} \tag{7.107}$$

which generalizes definition (7.21). Using this definition, we can rewrite expression (7.106) in the form

$$\tilde{\nabla}_i\, e_{i'}{}^j = 0, \tag{7.108}$$

which, thanks to the duality of frames and coframes, also implies that

$$\tilde{\nabla}_i\, e_j{}^{i'} = 0. \tag{7.109}$$

We conclude that the mixed covariant derivative of a frame or a coframe vanishes. This means they can be freely moved in and out of mixed covariant differentiation.

7.1.10 Spin Connection

In section 5.10.4, we considered the metric tensor in a tetrad coframe basis. In the tetrad basis, as discussed in section 7.1.9, the mixed covariant derivative (7.107) of the metric is determined by

$$\nabla_i\, g_{j'k'} = \partial_i\, g_{j'k'} - \omega^{\ell'}_{ij'}\, g_{\ell'k'} - \omega^{\ell'}_{ik'}\, g_{j'\ell'} = 0, \tag{7.110}$$

where in the last equality, we used the fact that, in the absence of nonmetricity, the covariant derivative of the metric vanishes. Suppose we select an orthogonal frame basis of the form expressed by (5.120). In that case, expression (7.110) implies

$$\omega_{ij'k'} = \omega_{ik'j'}, \tag{7.111}$$

where we have used the convention $\omega_{ij'k'} = \omega^{\ell'}_{ij'}\, g_{\ell'k'}$. We conclude that the connection coefficients of an orthogonal frame are antisymmetric in their two coframe indices. Such a connection is called a *spin connection* (e.g., Carroll, 2004).

7.1.11 Contracted Bianchi Identities

With the introduction of the metric tensor, we may lower the last index on the Riemann tensor (7.60):

$$r_{ijk\ell} = r_{ijk}{}^m\, g_{m\ell}. \tag{7.112}$$

What are the symmetries of this tensor? We have already seen in expression (7.63) that

$$r_{ijk\ell} = -\, r_{jik\ell}. \tag{7.113}$$

Using (7.74) and the definition of the covariant derivative (7.21), it is readily shown that

$$\nabla_i \nabla_j g_{k\ell} - \nabla_j \nabla_i g_{k\ell} = -\, r_{ijk}{}^{n} g_{n\ell} - r_{ij\ell}{}^{n} g_{kn} - t_{ij}{}^{n} \nabla_n g_{k\ell} = 0, \tag{7.114}$$

which implies the symmetry

$$r_{ijk\ell} = -\, r_{ij\ell k}. \tag{7.115}$$

As a consequence

$$r_{ijk}{}^{k} = 0. \tag{7.116}$$

In the absence of torsion, Wald (1984, Chapter 3.2) establishes four key properties of the Riemann tensor (7.60), which, using the order in which they are listed by Wald, are captured by equations (7.63), (7.72), (7.115), and (7.71). When the connection is not torsion-free, the generalized key properties become (7.63), (7.67), (7.115), and (7.66).

The components of the *Ricci tensor* are defined by the following contraction of the Riemann tensor (7.60)

$$r_{ij} \equiv r_{kij}{}^{k} = \partial_k \Gamma^k_{ij} - \partial_i \Gamma^k_{kj} + \Gamma^k_{km} \Gamma^m_{ij} - \Gamma^k_{im} \Gamma^m_{kj} - \tau^m_{ki} \Gamma^k_{mj} = -\, r_{ikj}{}^{k}. \tag{7.117}$$

In general, the Ricci tensor is asymmetric, unless the connection is torsion-free. Using (7.113) and (7.115), it may now be shown that

$$r_{ij} = r_{kij}{}^{k} = r_{kij\ell}\, g^{k\ell} = r_{ik\ell j}\, g^{k\ell}, \tag{7.118}$$

that is, the Ricci tensor is obtained by contracting either the first and fourth or the second and third indices of the Riemann tensor. Loosely speaking, the Ricci tensor measures the degree to which a unit ball in a manifold deviates from the Euclidean unit sphere.

We may define the *Ricci scalar* or *Ricci curvature* in terms of the Ricci tensor (7.117) as

$$r \equiv g^{ij}\, r_{ij}. \tag{7.119}$$

The Ricci scalar captures the difference in volume between a unit ball in the manifold and the Euclidean unit sphere. Note that the Ricci tensor does not require the introduction of a metric, but the Ricci scalar does. It is traditional to use the same symbol r for the Riemann tensor, Ricci tensor, and Ricci scalar, distinguishing them by their number of indices.

The *Einstein tensor* is defined in terms of the Ricci tensor (7.117) and Ricci scalar (7.119) as

$$E_i{}^{j} \equiv r_i{}^{j} - \tfrac{1}{2}\, r\, \delta_i{}^{j}. \tag{7.120}$$

Like the Ricci tensor, only when the connection is torsion-free is the Einstein tensor symmetric. Upon contracting the m and k indices in the Bianchi identity (7.66), using the properties of the Ricci tensor (7.118), we find the *contracted Bianchi identity*

$$\nabla_j E_i{}^j = t_{k\ell}{}^j \left(\tfrac{1}{2} r_{ij}{}^{k\ell} + r_j{}^\ell \delta^k{}_i\right), \tag{7.121}$$

where we have used (7.118). In the absence of torsion, we are left with

$$\nabla_j E_i{}^j = 0. \tag{7.122}$$

The *Cartan tensor* or *Palatini torsion tensor* or *modified torsion tensor* is defined in terms of the torsion tensor (7.55) as

$$C_{ij}{}^k \equiv t_{ij}{}^k + t_i \delta^k{}_j - t_j \delta^k{}_i, \tag{7.123}$$

where the components of the *torsion form* are defined by the contraction

$$t_i \equiv t_{ji}{}^j = -t_{ij}{}^j. \tag{7.124}$$

Upon contracting the k and ℓ indices in the Bianchi identity (7.67), it may be shown that the Cartan tensor satisfies the contracted Bianchi identity

$$\nabla_k C_{ij}{}^k = r_{ij} - r_{ji} + t_k t_{ij}{}^k, \tag{7.125}$$

where we have used (7.116).

It is important to note that the Einstein and Cartan tensors (7.120) and (7.123) arise naturally when contracting the Bianchi identities (7.66) and (7.67) to obtain (7.121) and (7.125). They play a central role in the Einstein–Cartan theory of general relativity (e.g., Hehl et al., 1976; Trautman, 2006; Hehl & Weinberg, 2007). In that case, the metric tensor describes the four-dimensional *spacetime manifold*.

Example 7.7 Field Equations

In general relativity without torsion, as originally envisioned by Einstein (Einstein, 1915, 1918), there is a balance between *geometry* in the form of the Einstein tensor

$$E_\mu{}^\nu = r_\mu{}^\nu - \tfrac{1}{2} r \delta_\mu{}^\nu, \tag{7.126}$$

and *physics* in the form of the *stress-energy tensor* $\sigma_\mu{}^\nu$,

$$E_\mu{}^\nu = \kappa \sigma_\mu{}^\nu. \tag{7.127}$$

The constant κ is determined in terms of Newton's gravitational constant, G, and the speed of light in vacuum, c, by $\kappa = 8\pi G/c^4$. The balance (7.127) defines the Einstein field equations of general relativity without torsion.

The theory of general relativity with spin and torsion was initially proposed by Cartan (1923, 1924, 1925). It was rediscovered in the 1960s by Kibble (1961) and Sciama (1962, 1964), summarized by Hehl et al. (1976) and Trautman (2006), and remains a viable theory of relativity (Trautman, 2006; Hehl & Weinberg, 2007). In this formulation, in addition to the field equation (7.127), there is also a balance between the Cartan tensor or modified torsion tensor

$$C_{\mu\nu}{}^{\tau} = t_{\mu\nu}{}^{\tau} + t_{\alpha\mu}{}^{\alpha}\,\delta_{\nu}{}^{\tau} - t_{\alpha\nu}{}^{\alpha}\,\delta_{\mu}{}^{\tau}, \tag{7.128}$$

and the *spin tensor* $M_{\mu\nu}{}^{\tau}$:

$$C_{\mu\nu}{}^{\tau} = \kappa\, M_{\mu\nu}{}^{\tau}. \tag{7.129}$$

The Einstein–Cartan field equations (7.127) and (7.129) may be obtained from a variational principle, as explored in example 8.29 It is noteworthy that, while the field equation (7.127) is a partial differential equation, the spin field equation (7.129) is an algebraic identity. This means that spin vanishes if and only if torsion does too.

Example 7.8 Cosmological Constant and Dark Energy

To counteract the force of gravity and maintain a stationary universe, which was the prevailing belief at the time, Einstein (1917) introduced the *cosmological constant* Λ as an adjustment to the field equations (7.127) in the following form

$$E_{\mu}{}^{\nu} + \Lambda\,\delta_{\mu}{}^{\nu} = \kappa\,\sigma_{\mu}{}^{\nu}. \tag{7.130}$$

Einstein regretted his "greatest blunder" and subsequently set the cosmological constant equal to zero, but the discovery of the universe's accelerating expansion in the 1990's implied that it may have a positive value. Since then, it has been discovered that around two-thirds of the mass-energy density of the universe can be attributed to so-called *dark energy*, and a nonzero cosmological constant is the simplest explanation for this.

Example 7.9 Dynamic Equations

In the context of general relativity, the contracted Bianchi identities (7.121) and (7.125) may be expressed in the form

$$\breve{\nabla}_{\nu} E_{\mu}{}^{\nu} = \tfrac{1}{2}\, C_{\sigma\alpha}{}^{\nu}\, r_{\mu\nu}{}^{\sigma\alpha} + E_{\nu}{}^{\alpha}\, t_{\mu\alpha}{}^{\nu}, \tag{7.131}$$

$$\breve{\nabla}_{\sigma} C_{\mu\nu}{}^{\sigma} = E_{\mu\nu} - E_{\nu\mu}. \tag{7.132}$$

where we have defined the *modified covariant derivative*

$$\breve{\nabla}_{\nu} \equiv \nabla_{\nu} + t_{\nu\beta}{}^{\beta}. \tag{7.133}$$

Upon using the field equations (7.127) and (7.129), we obtain the dynamic equations of Einstein–Cartan general relativity:

$$\breve{\nabla}_{\nu}\sigma_{\mu}{}^{\nu} = \tfrac{1}{2}\, M_{\sigma\alpha}{}^{\nu}\, r_{\mu\nu}{}^{\sigma\alpha} + \sigma_{\nu}{}^{\alpha}\, t_{\mu\alpha}{}^{\nu}, \tag{7.134}$$

$$\breve{\nabla}_{\sigma} M_{\mu\nu}{}^{\sigma} = \sigma_{\mu\nu} - \sigma_{\nu\mu}. \tag{7.135}$$

In the absence of torsion, expressions (7.134) and (7.135) reduce to the dynamic equations of Einstein general relativity:

$$\nabla_{\nu}\sigma_{\mu}{}^{\nu} = 0, \tag{7.136}$$

$$\sigma_{\mu\nu} = \sigma_{\nu\mu}. \tag{7.137}$$

Thus, in Einstein relativity the stress-energy tensor is symmetric, but in Einstein–Cartan relativity it is not.

In example 8.29, we derive the field and dynamical equations of Einstein, Einstein–Cartan, and metric-affine general relativity from variational principles.

In the absence of torsion, (7.125) implies the symmetry of the Ricci tensor:

$$r_{ij} = r_{ji}. \tag{7.138}$$

In that case, we deduce from the definition of the Ricci tensor in terms of the Riemann tensor, (7.117), that the Riemann tensor has an important additional symmetry:

$$r_{ijk\ell} = r_{ikj\ell} = r_{ki\ell j} = r_{k\ell ij}. \tag{7.139}$$

In the second equality, we have used the previously established symmetries of the Riemann tensor (7.63) and (7.115) simultaneously.

Example 7.10 Inertial Frames

In example 5.8, we introduced a coordinate-free tetrad basis, $\{e_a\}$, of tensors in general relativity. We have the freedom to choose these tetrads. One option is to choose them such that they define an *inertial frame*. To accomplish this, based on definition (7.43) of parallel transport, the time-like frame basis element, e_0, must define a geodetic:

$$\nabla_{e_0} e_0 = \omega_{00}^{a}\, e_a = 0, \tag{7.140}$$

thereby making its acceleration vanish, rendering the frame inert. Here ω_{ab}^{c} denote the tetrad connection coefficients discussed in section 7.1.9. A further constraint is to choose a *nonspinning inertial frame* in the sense

$$\nabla_{e_0} e_a = \omega^b_{0a}\, e_b = 0, \qquad a = 0, 1, 2, 3, \tag{7.141}$$

thereby parallel transporting the tetrad basis along the worldline of an observer.

7.1.12 Covariant Derivative of Tensor Densities and Capacities

In section 5.12, we discussed tensor densities and capacities whose components transform according to rules (5.60) and (5.61). Upon taking the covariant derivative of expressions (5.132) and (5.133), we find that

$$\nabla_m \overline{T}^{i_1\cdots i_p}{}_{j_1\cdots j_q} = \overline{g}^w \left(\nabla_m T^{i_1\cdots i_p}{}_{j_1\cdots j_q} + w\, T^{i_1\cdots i_p}{}_{j_1\cdots j_q}\, \Gamma^k_{mk}\right), \tag{7.142}$$

and

$$\nabla_m \underline{T}^{i_1\cdots i_p}{}_{j_1\cdots j_q} = \underline{g}^w \left(\nabla_m T^{i_1\cdots i_p}{}_{j_1\cdots j_q} - w\, T^{i_1\cdots i_p}{}_{j_1\cdots j_q}\, \Gamma^k_{mk}\right), \tag{7.143}$$

where we have used relationship (7.91).

7.1.13 Nonmetricity

In the theory of material defects, *nonmetricity* is used to capture *point defects*. Similarly, *metric-affine gravitation theory* allows for nonmetricity in addition to torsion and curvature (e.g., Hehl et al., 1995; Vitagliano et al., 2011), thereby extending the Einstein–Cartan theory of relativity.

A non-vanishing covariant derivative of the metric tensor is captured by the *nonmetricity tensor*:

$$\boldsymbol{\nabla}\mathbf{g} = \mathbf{Q}. \tag{7.144}$$

In components, using expression (7.21), we have

$$\nabla_i\, g_{jk} = \partial_i\, g_{jk} - \Gamma^\ell_{ij}\, g_{\ell k} - \Gamma^\ell_{ik}\, g_{j\ell} = Q_{ijk}. \tag{7.145}$$

The nonmetricity tensor has the symmetry

$$Q_{ijk} = Q_{ikj}. \tag{7.146}$$

In the presence of nonmetricity, the connection coefficients (7.82) are modified to become

$$\begin{aligned}\Gamma^i_{jk} &= \tfrac{1}{2}\, g^{i\ell}(\partial_k\, g_{\ell j} + \partial_j\, g_{k\ell} - \partial_\ell\, g_{jk}) \\ &\quad + \tfrac{1}{2}\,(\tau^i_{jk} + g^{i\ell}\, \tau^m_{\ell j}\, g_{mk} + g^{i\ell}\, \tau^m_{\ell k}\, g_{mj}) \\ &\quad + K_{jk}{}^i - N_{jk}{}^i,\end{aligned} \tag{7.147}$$

where ${K_{jk}}^i$ denote the elements of the contortion tensor (7.83), and where we have introduced the *disformation tensor* with elements

$$N^i{}_{jk} = \tfrac{1}{2}\, g^{i\ell}\,(Q_{j\ell k} + Q_{kj\ell} - Q_{\ell jk}), \tag{7.148}$$

which exhibits the symmetry

$$N^i{}_{jk} = N^i{}_{kj}. \tag{7.149}$$

In the presence of nonmetricity, expression (7.114) is modified to become

$$\nabla_i Q_{jk\ell} - \nabla_j Q_{ik\ell} = -\,{r_{ijk}}^n\, g_{n\ell} - {r_{ij\ell}}^n\, g_{kn} - {t_{ij}}^n\, Q_{nk\ell}, \tag{7.150}$$

and we no longer have the symmetry (7.115):

$$r_{ijk\ell} \neq -\,r_{ij\ell k}. \tag{7.151}$$

Example 7.11 Metric-Affine Gravitation Theory

In Einstein and Einstein–Cartan relativity, the nonmetricity tensor

$$Q_{\mu\nu\tau} = \nabla_\mu\, g_{\nu\tau} \tag{7.152}$$

vanishes, $Q_{\mu\nu\tau} = 0$, and the curvature tensor exhibits the symmetry

$$r_{\mu\nu\tau\sigma} = -\,r_{\mu\nu\sigma\tau}, \tag{7.153}$$

but in *metric-affine gravitation theory*, it does not.

Example 7.12 Nonmetricity in Continuum Mechanics

In continuum mechanics, nonmetricity is associated with *point defects*. There are three types, namely, (1) a vacancy defect, which is created when an atom is missing from an otherwise perfect lattice, (2) an interstitial defect, which is a constituent or non-constituent atom inserted in the inter-atom space of an otherwise perfect lattice, and (3) a substitutional defect, which is a non-constituent atom inserted in the lattice space of an otherwise perfect lattice, as illustrated in figure 7.2 (see, e.g., Roychowdhury & Gupta, 2013, for a review).

Example 7.13 Maxwell's Equations

In this box, we summarize Maxwell's equations (see, e.g., Jackson, 1998), which are tensorial equations that form the foundation of electrodynamics, optics, and

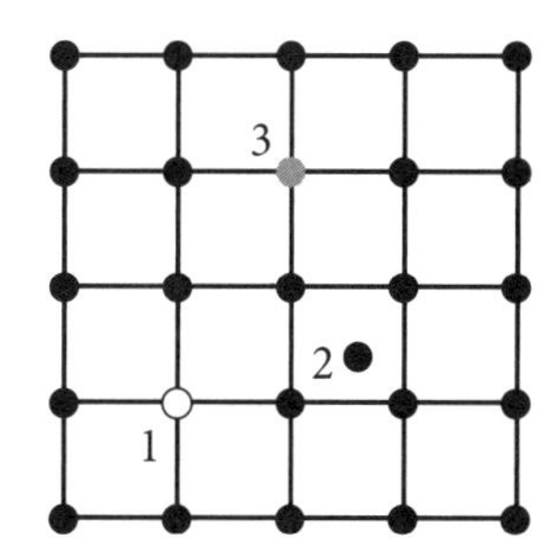

Figure 7.2: Illustration of the three types of crystallographic *point defects*, which are geometrically identified with *nonmetricity*. (1) A *vacancy defect*, which is created when an atom is missing from an otherwise perfect lattice. (2) An *interstitial defect*, which is a constituent or non-constituent atom inserted in the inter-atom space of an otherwise perfect lattice. (3) A *substitutional defect*, which is a non-constituent atom inserted in the lattice of an otherwise perfect lattice.

(geo)dynamo theory. These equations involve the Euler and covariant derivatives. In this Box, we present the classical form of Maxwell's equations, using operations from classical vector calculus, such as the divergence and curl. In Box 8.4, we present a modern view based on differential geometry.

Gauss's law for the electric field

Gauss's law describes the relationship between an electric field and the electric charges that cause it. Gauss's law for the *electric field* **E** is

$$\nabla \cdot \mathbf{E} = \rho/\epsilon_0, \tag{7.154}$$

where ρ denotes the *electric charge density* and ϵ_0 the *permittivity* of free space, or the *universal electric constant.*

Gauss's law for the magnetic field

Gauss's law for the *magnetic field* **B** is

$$\nabla \cdot \mathbf{B} = \mathbf{0}. \tag{7.155}$$

There are no sources for the magnetic field: no magnetic monopoles.

Faraday's law

Faraday's law describes how a time-varying magnetic field induces an electric field:

$$\nabla \times \mathbf{E} = -\, d_t \mathbf{B}. \tag{7.156}$$

Ampère's law

Ampère's law states that magnetic fields can be generated in two ways: by electrical current (the original "Ampère's law") and by changing electric fields ("Maxwell's correction"):

$$\nabla \times \mathbf{B} = \mu_0 \left(\mathbf{j} + \epsilon_0 \, \mathrm{d}_t \mathbf{E}\right), \tag{7.157}$$

where μ_0 denotes the *permeability* of free space, or the *universal magnetic constant.*

Continuity equation

Upon taking the divergence of (7.157) and using Gauss's law for the electric field (7.154) we obtain the *continuity equation* or current conservation law

$$\mathrm{d}_t \rho + \nabla \cdot \mathbf{j} = \mathbf{0}. \tag{7.158}$$

Maxwell's equations in vacuum

In the absence of charges and currents, Maxwell's equations take the form

$$\nabla \cdot \mathbf{E} = \mathbf{0}, \tag{7.159}$$

$$\nabla \cdot \mathbf{B} = \mathbf{0}, \tag{7.160}$$

$$\nabla \times \mathbf{E} = -\, \mathrm{d}_t \mathbf{B}, \tag{7.161}$$

$$\nabla \times \mathbf{B} = \mu_0 \, \epsilon_0 \, \mathrm{d}_t \mathbf{E}. \tag{7.162}$$

These equations lead to $\mathbf{E}$ and $\mathbf{B}$ satisfying the wave equations

$$\mathrm{d}_t^2 \, \mathbf{E} = c^2 \, \nabla^2 \mathbf{E}, \qquad \mathrm{d}_t^2 \, \mathbf{B} = c^2 \, \nabla^2 \mathbf{B}, \tag{7.163}$$

where c denotes the *speed of light*

$$c = \frac{1}{\sqrt{\mu_0 \, \epsilon_0}}. \tag{7.164}$$

Thus, Ampère's law (7.157) may be rewritten as

$$\nabla \times \mathbf{B} = \mu_0 \, \mathbf{j} + \frac{1}{c^2} \, \mathrm{d}_t \mathbf{E}. \tag{7.165}$$

Ohm's law

Ohm's law defines the current density $\mathbf{j}$ in terms of the electric and magnetic fields as

$$\mathbf{j} = \sigma \left(\mathbf{E} + \mathbf{v} \times \mathbf{B}\right), \tag{7.166}$$

where σ denotes the *electrical conductivity*. Relation (7.166) is a generalization of the simple "voltage = current × resistance" law to a moving conductor.

Induction equation

When the velocity $\mathbf{v}$ is much smaller than the speed of light c, Ampère's law (7.165) may be approximated by

$$\nabla \times \mathbf{B} = \mu_0 \, \mathbf{j}. \tag{7.167}$$

From Ohm's law (7.166) we then find that

$$\mathbf{E} = \frac{1}{\sigma} \mathbf{j} - \mathbf{v} \times \mathbf{B} = \frac{1}{\sigma \mu_0} \nabla \times \mathbf{B} - \mathbf{v} \times \mathbf{B}, \tag{7.168}$$

and Faraday's law (7.156) then reduces to the *induction equation*, namely

$$\mathrm{d}_t \mathbf{B} = \nabla \times (\mathbf{v} \times \mathbf{B}) + \eta \, \nabla^2 \mathbf{B}, \tag{7.169}$$

where

$$\eta = \frac{1}{\sigma \mu_0}, \tag{7.170}$$

is the *magnetic diffusivity*. The induction equation plays a central role in magnetohydrodynamics, specifically in the generation of the magnetic fields of the Earth and Sun. Since convective motions in the Earth's core and Sun are slow compared to the speed of light, the induction equation is an excellent representation of terrestrial and solar dynamo action.

Lorentz force

The *Lorentz force* is the electromagnetic force per unit volume:

$$\mathbf{f} = \rho \mathbf{E} + \mathbf{j} \times \mathbf{B}, \tag{7.171}$$

The electric force per unit volume is $\rho \, \mathbf{E}$, and the magnetic force is $\mathbf{j} \times \mathbf{B}$.

Maxwell's equations in matter

The *electric displacement* field, electric induction, or electric flux density is given by

$$\mathbf{D} = \epsilon_0 \, \mathbf{E} + \mathbf{P}, \tag{7.172}$$

and the *magnetizing field*, auxiliary magnetic field, or magnetic field intensity is

$$\mathbf{H} = \frac{1}{\mu_0} \mathbf{B} - \mathbf{M}, \tag{7.173}$$

where $\mathbf{P}$ is the *polarization field* and $\mathbf{M}$ the *magnetization field*. Often we can write

$$\mathbf{D} = \epsilon \, \mathbf{E}, \tag{7.174}$$

$$\mathbf{H} = \mathbf{B}/\mu, \tag{7.175}$$

where ϵ and μ are scalars. The *macroscopic Maxwell's equations* are then

$$\nabla \cdot \mathbf{D} = \rho_f, \tag{7.176}$$

$$\nabla \cdot \mathbf{B} = \mathbf{0}, \tag{7.177}$$

$$\nabla \times \mathbf{E} = -\, \mathrm{d}_t \mathbf{B}, \tag{7.178}$$

$$\nabla \times \mathbf{H} = \mathbf{j}_f + \mathrm{d}_t \mathbf{D}, \tag{7.179}$$

where ρ_f denotes the free charge density (excluding the bound charge), and $\mathbf{j}_f$ the free current density (excluding the bound current).

7.2 Euler Derivative

The *Euler derivative* of a vector, denoted as $\mathrm{d}_t\mathbf{u}$, is obtained by taking the rate of change of the components of the vector relative to a time-independent coordinate system,

$$\mathrm{d}_t \mathbf{u} = (\partial_t u^i)\, \mathbf{e}_i. \tag{7.180}$$

The Euler derivative of a general tensor $\mathbf{T}$ is given in terms of its components by

$$(\mathrm{d}_t \mathbf{T})^{i_1 \cdots i_p}{}_{j_1 \cdots j_q} = \partial_t T^{i_1 \cdots i_p}{}_{j_1 \cdots j_q}. \tag{7.181}$$

We compare the Euler derivative with the Lie derivative, discussed in section 7.3, below, in figure 7.3.

Example 7.14 Material Derivative

In continuum mechanics, the *material derivative* or *substantial derivative* of a general tensor field $\mathbf{T}$ is defined in terms of the Euler derivative, the material velocity $\mathbf{v}$, and the covariant derivative as

$$\mathrm{D}_t \mathbf{T} \equiv \mathrm{d}_t \mathbf{T} + \mathbf{v} \cdot \nabla \mathbf{T}. \tag{7.182}$$

This derivative combines the local change of a tensor field, $\mathrm{d}_t\mathbf{T}$, with a term due to advection, $\mathbf{v} \cdot \nabla\mathbf{T}$. Thus, it captures the rate of change of a tensor field experienced by an observer who "rides along" with an element of the continuum.

Example 7.15 Corotational Material Derivative

In continuum mechanics, *corotational material derivatives* capture the rate of change experienced by an observer who comoves and *corotates* with an element of the continuum. For a general tensor $\mathbf{T}$, the corotational material derivative due to material velocity $\mathbf{v}$ and a generic spin tensor $\boldsymbol{\Omega}$ is defined as

$$(\overset{\circ\Omega}{\mathrm{D}}_t\, \mathbf{T})^{ij\cdots}{}_{k\ell\cdots} \equiv (\mathrm{D}_t\, \mathbf{T})^{ij\cdots}{}_{k\ell\cdots} - \Omega^i{}_m\, T^{mj\cdots}{}_{k\ell\cdots} - \Omega^j{}_m\, T^{im\cdots}{}_{k\ell\cdots} - \cdots$$

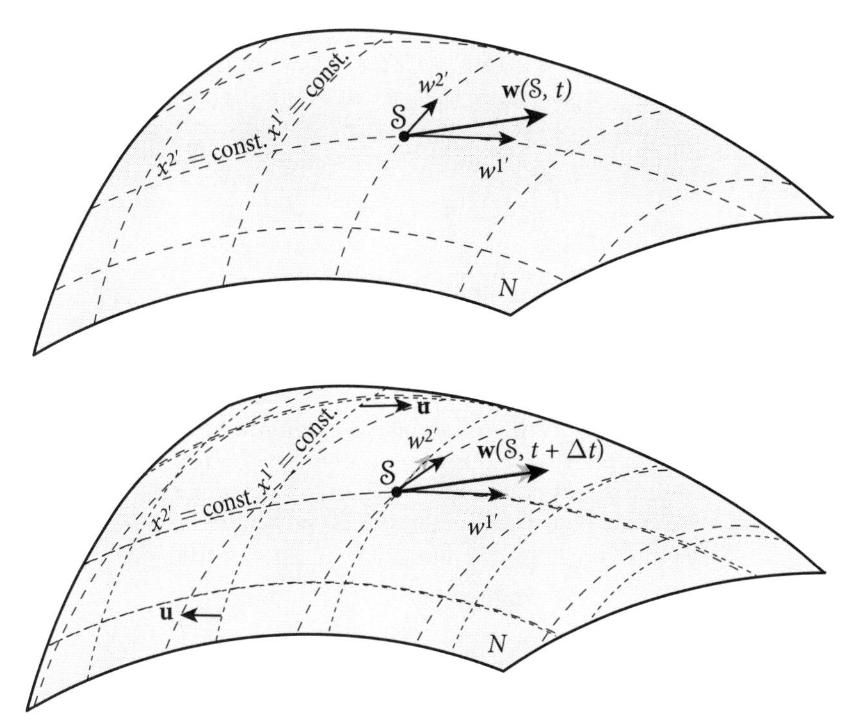

Figure 7.3: The coordinate system $x' = \{x^{1'}, x^{2'}\}$ moves with the flow defined by the vector field $\mathbf{u}$; in this case, it moves with a horizontal shear flow. Top: At time t, the convected coordinate lines $x^{1'} = \text{const.}$ and $x^{2'} = \text{const.}$ are shown by the dashed lines. The components of the vector $\mathbf{w}(\mathcal{S}, t)$ at location $\mathcal{S}$ and time t in manifold N relative to the convected coordinate system x' are $w^{1'}$ and $w^{2'}$. Bottom: At time $t + \Delta t$, the coordinate lines $x^{1'} = \text{const.}$ and $x^{2'} = \text{const.}$ have been convected by the flow of vector field $\mathbf{u}$. The long-dashed lines show the current coordinate lines at time $t + \Delta t$, and the short-dashed lines show the coordinate lines as they were at time t in the top panel. The comoving components of the vector $\mathbf{w}(\mathcal{S}, t + \Delta t)$ at location $\mathcal{S}$ and time $t + \Delta t$ are $w^{1'}$ and $w^{2'}$; the faint arrows show $\mathbf{w}(\mathcal{S}, t)$ and its components, as in the top panel. The convected components of the Lie derivative of the vector $\mathbf{w}$ relative to the flow $\mathbf{u}$ are $(\mathcal{L}_{\mathbf{u}}\mathbf{w})^{i'}(x', t') = \partial_{t'} w^{i'}(x', t')$, that is, the time rate of change of the convected components of vector $\mathbf{w}$. Thus, the Lie derivative of the vector $\mathbf{w}$ relative to the flow defined by vector field $\mathbf{u}$, $\mathcal{L}_{\mathbf{u}}\mathbf{w}$, provides a measure of change in the vector $\mathbf{w}$ by comparing it over time to a flow $\mathbf{u}$. In contrast, the spatial components of the Euler derivative of the vector, $\mathrm{d}_t\mathbf{w}$, are $\partial_t w^i(x, t)$: the rate of change is measured relative to the inertial frame x.

$$+T^{ij\cdots}{}_{m\ell\cdots}\,\Omega^m{}_k + T^{ij\cdots}{}_{km\cdots}\,\Omega^m{}_\ell + \cdots . \tag{7.183}$$

Note how each contravariant and covariant component is rotated by the spin tensor. Depending on the rate of rotation, these derivatives carry different names. For example, the *Zaremba–Jaumann rate* (Zaremba, 1903; Jaumann, 1911) of a general tensor is obtained by using the vorticity tensor (7.39) as the spin tensor, $\boldsymbol{\Omega} = \mathbf{W}$, in equation (7.183).

The Zaremba–Jaumann rate is an example of a so-called *objective rate*, which means that it is a measure of a time rate of change that is independent of the frame of reference. In other words, an objective rate is unaffected by rigid translations and rotations of the reference frame. The material time derivative is not objective, but the Lie derivative is.

7.3 Lie Derivative

Characterizing the spatial variability of a vector field requires the ability to compare its behavior at a given spatial location $\mathcal{S}$ with its behavior at a nearby location. As stated on page 85, the challenge is that one can compare tensors only *at the same location in the manifold*. The covariant derivative, discussed in section 7.1, provides one suitable measure of spatial variability. An alternative measure of the spatial variability of a tensor field involves evaluating the change in this field relative to the flow of a vector field (e.g., Spivak, 1965; Schutz, 1980).

7.3.1 Lie Derivative of Vectors

To introduce the Lie derivative, let us consider two vector fields, **u** and **w**, since such an introduction lends itself to a pictorial description: a vector from field **w** being transported along with the flow of field **u**. Transport along the congruence of the integral curves of **u**, as discussed in section 4.1.3, allows us to avoid the concept of *parallel transport*, which is required for the covariant derivative.

Suppose we seek to characterize the spatial variability of a vector field, **w**, relative to the time-dependent flow defined by another vector field, **u**. We allow both **u** and **w** to be time-dependent. To quantify this notion, we consider two local coordinate systems. We introduce an arbitrary time-independent coordinate system $x = \{x^i\}$ in the neighborhood of a spatial point of interest, $\mathcal{S}$, as illustrated in figure 7.3. As usual, the spatial location of a vector **w**, which is $\mathcal{S} = \mathcal{S}(x)$, may be expressed relative to the associated basis vectors $\mathbf{e}_i = \partial_i \mathcal{S}$ as

$$\mathbf{w} = w^i\,\mathbf{e}_i, \tag{7.184}$$

where $w^i(x,t)$ are the components of the vector **w** relative to the time-independent basis $\mathbf{e}_i(x)$.

To capture the change in **w** relative to the flow of field **u**, we introduce a time-dependent coordinate system $x' = \{x^{i'}\}$ that moves with the flow defined by vector field **u**. The diffeomorphic transformations between x and x' are

$$\begin{aligned} x^i &= \varphi^i(x', t'). \\ t &= t', \end{aligned} \tag{7.185}$$

with inverse

$$
\begin{aligned}
x^{i'} &= \Phi^{i'}(x,t). \\
t' &= t,
\end{aligned} \tag{7.186}
$$

such that, at a given spatial location $\mathcal{S}$,

$$
\partial_{t'}\varphi^i \equiv u^i; \tag{7.187}
$$

in other words, u^i are the components of the flow $\mathbf{u}$ with respect to the time-independent coordinate system. Note that transformations (7.185) are time-dependent versions of expressions for local coordinate changes in a manifold, discussed in section 3.3. We refer to the coordinate system x' as a *convected coordinate system*, that is, a coordinate system convected by the flow of the vector field, $\mathbf{u}$, as specified in definition (7.187). We introduce the parameter t' so that we can distinguish a partial derivative ∂_t, which implies holding the coordinates x^i fixed, from a partial derivative $\partial_{t'}$, which implies holding the coordinates $x^{i'}$ fixed.

We may express field $\mathbf{w}$ at location $\mathcal{S} = \mathcal{S}(x')$ relative to the associated convected basis vectors $\mathbf{e}_{i'} = \partial_{i'}\mathcal{S}$ as

$$
\mathbf{w} = w^{i'}\,\mathbf{e}_{i'}. \tag{7.188}
$$

In general, the Lie derivative associates the value of a tensor field, $\mathbf{T}$, at one point of the manifold with the value of this field at another point via the congruence provided by a vector field, $\mathbf{u}$. The congruence consists of the flow defined by nonintersecting integral curves of $\mathbf{u}$ that cover the manifold, as discussed in section 4.1.3. There is no intrinsic relation between $\mathbf{T}$ and $\mathbf{u}$.

We define the *Lie derivative* of the time-dependent vector $\mathbf{w}$ relative to the flow of vector field $\mathbf{u}$ as

$$
\mathcal{L}_{\mathbf{u}}\mathbf{w} \equiv \left(\partial_{t'} w^{i'}\right)\mathbf{e}_{i'}. \tag{7.189}
$$

In words: we determine the rate of change of the convected components of the vector $\mathbf{w}$, $\partial_{t'} w^{i'}$, and use this rate to define the convected components of the Lie derivative of $\mathbf{w}$ relative to the flow $\mathbf{u}$, $(\mathcal{L}_{\mathbf{u}}\mathbf{w})^{i'}$. If the convected components are time-independent, that is, $\partial_{t'} w^{i'} = 0$, then the Lie derivative of the vector $\mathbf{w}$ relative to the flow $\mathbf{u}$ equals zero. Note that time independence of the convected components does not imply time independence of $\mathbf{w}$; furthermore, in general, the Lie derivative is not zero—even for time-independent fields—unless the Lie bracket vanishes. Thus, we have determined a way of characterizing the change in a vector field relative to the flow of another vector field. An important implication of definition (7.189) is that the Lie derivative of a frame $\mathbf{e}_{i'}$ vanishes

$$
\mathcal{L}_{\mathbf{u}}\mathbf{e}_{i'} = \mathbf{0}. \tag{7.190}
$$

The challenge is finding the expression of the Lie derivative in the original coordinate system, x. To establish this, we need to use transformation rules (4.16),

$$
\mathbf{e}_{i'} = \mathbf{e}_i\,\lambda^i{}_{i'} \qquad \text{and} \qquad \mathbf{e}_i = \mathbf{e}_{i'}\,\Lambda^{i'}{}_i, \tag{7.191}
$$

and (4.22)

$$w^{i'} = \Lambda^{i'}{}_i \, w^i \qquad \text{and} \qquad w^i = \lambda^i{}_{i'} \, w^{i'}, \tag{7.192}$$

for bases and components, respectively. In the present case, these transformations are time-dependent:

$$\lambda^i{}_{i'}(x', t') = \partial_{i'} \varphi^i(x', t'), \qquad \Lambda^{i'}{}_i(x, t) = \partial_i \Phi^{i'}(x, t). \tag{7.193}$$

To find an expression for the Lie derivative (7.189) in coordinate system x, we may use expression (7.191) to transform the basis vectors $\mathbf{e}_{i'}$ to $\mathbf{e}_i$, but we need to be careful with the transformation of the components, $\partial_{t'} w^{i'}$, because they involve differentiation in time while holding the convected coordinates, x', fixed. With this caveat, upon taking the rate of change of

$$w^{i'}(x', t') = \Lambda^{i'}{}_i(\varphi(x', t'), t') \, w^i(\varphi(x', t'), t'), \tag{7.194}$$

we find

$$\partial_{t'} w^{i'} = [\partial_t \Lambda^{i'}{}_i + (\partial_j \Lambda^{i'}{}_i) \, \partial_{t'} \varphi^j] \, w^i + \Lambda^{i'}{}_i \, (\partial_t w^i + u^j \, \partial_j w^i). \tag{7.195}$$

Upon differentiating $x^{i'} = \Phi^{i'}(\varphi(x', t'), t')$, with respect to time t', holding the coordinates x' fixed, we find, using the chain rule,

$$0 = \partial_t \Phi^{i'} + (\partial_i \Phi^{i'}) \, \partial_{t'} \varphi^i = \partial_t \Phi^{i'} + \Lambda^{i'}{}_i \, u^i, \tag{7.196}$$

which, using the fact that $u^{i'} = \Lambda^{i'}{}_i \, u^i$, implies

$$u^{i'} = -\, \partial_t \Phi^{i'}. \tag{7.197}$$

The signs in expressions (7.187) and (7.197) reflect the relative motion of the coordinate systems, x and x'. Thus, expression (7.195) becomes

$$\begin{aligned}
(\mathcal{L}_{\mathbf{u}} \mathbf{w})^{i'} &= \partial_{t'} w^{i'} \\
&= [-\partial_i u^{i'} + (\partial_j \Lambda^{i'}{}_i) \, u^j] \, w^i + \Lambda^{i'}{}_i \, (\partial_t w^i + u^j \, \partial_j w^i) \\
&= [-\partial_i (\Lambda^{i'}_j \, u^j) + (\partial_j \Lambda^{i'}{}_i) \, u^j] \, w^i + \Lambda^{i'}{}_i \, (\partial_t w^i + u^j \, \partial_j w^i) \\
&= \Lambda^{i'}{}_i \, (\partial_t w^i + u^j \, \partial_j w^i - w^j \, \partial_j u^i) \\
&= \Lambda^{i'}{}_i \, (\mathcal{L}_{\mathbf{u}} \mathbf{w})^i.
\end{aligned} \tag{7.198}$$

Comparing the last line and the penultimate line of equation (7.198), we conclude that the components of the Lie derivative in coordinate system x are

$$(\mathcal{L}_{\mathbf{u}} \mathbf{w})^i = \partial_t w^i + u^j \, \partial_j w^i - w^j \, \partial_j u^i. \tag{7.199}$$

The first summand on the right-hand side describes the explicit dependence of $\mathbf{w}(\mathcal{S}, t)$ on t, while holding $\mathcal{S}$ fixed. The second and third summands describe the variability of $\mathbf{w}$ with respect to the flow of $\mathbf{u}$. If we choose $\mathbf{u} = \mathbf{w}$, expression (7.199) becomes

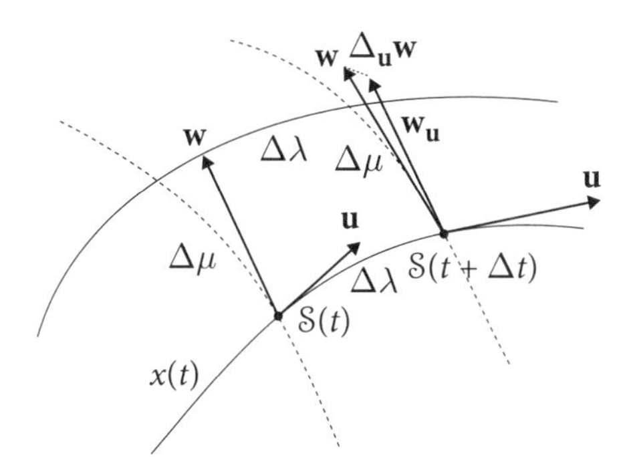

Figure 7.4: Illustration of the Lie derivative of a vector $\mathbf{w}$, which belongs to the vector field whose integral curves are dashed, with respect to the vector field $\mathbf{u}$, marked by the solid integral curves. The Lie derivative is the limit of $\Delta_\mathbf{u}\mathbf{w}/\Delta t$, as Δt tends to zero. Vector $\mathbf{w}$ is *Lie-dragged* along curve $x(t)$ by the flow of field $\mathbf{u}$, which results in $\mathbf{w_u}$. To picture it, imagine field $\mathbf{u}$ as a flow of water, and let vector $\mathbf{w}$ be drawn as an arrow on this flow at $\mathcal{S}(t)$. The distance between points $\mathcal{S}(t)$ and $\mathcal{S}(t+\Delta t)$ is exaggerated for this illustration. Since Δt tends to zero, the distance is infinitesimal. Consequently, it does not matter if we compare the Lie-dragged vector to the value of the field at $\mathcal{S}(t)$ or $\mathcal{S}(t+\Delta t)$.

$$(\mathcal{L}_\mathbf{w}\mathbf{w})^i = \partial_t w^i. \tag{7.200}$$

Based on (7.199), seting $w^i = 1$, we conclude that the Lie derivative of a basis vector is

$$\mathcal{L}_\mathbf{u}\mathbf{e}_i = -\,\partial_i u^j\,\mathbf{e}_j, \tag{7.201}$$

which should be contrasted with (7.190).

Equation (7.198) demonstrates that the components of the Lie derivative of a vector constitute a vector; they transform according to $(\mathcal{L}_\mathbf{u}\mathbf{w})^{i'} = \Lambda^{i'}{}_i\,(\mathcal{L}_\mathbf{u}\mathbf{w})^i$.

7.3.2 Geometrical Interpretation

To visualize the Lie derivative, consider the flow of vector field $\mathbf{u}$ defined by integral curves $x(t)$, as illustrated in figure 7.4.

Let the vector $\mathbf{w}$ be located at $\mathcal{S}(t)$.[7] After time Δt, an image of this vector is moved by the flow to a point $\mathcal{S}(t+\Delta t)$, in which another vector, $\mathbf{w}(\mathcal{S}(t+\Delta t))$, resides. The Lie derivative is the infinitesimal difference between $\mathbf{w}(\mathcal{S}(t))$ moved by the flow of $\mathbf{u}$ to $\mathcal{S}(t+\Delta t)$ and vector $\mathbf{w}(\mathcal{S}(t+\Delta t))$, which is field $\mathbf{w}$ at that point; we denote this difference by $\Delta_\mathbf{u}\mathbf{w}$, and hence,

$$\mathcal{L}_\mathbf{u}\mathbf{w} = \lim_{\Delta t\to 0}\frac{\Delta_\mathbf{u}\mathbf{w}}{\Delta t}, \tag{7.202}$$

which amounts to expression (7.189).

[7] Again, we abuse the notation: $\mathcal{S}(t) \equiv \mathcal{S}(x(t))$.

7.3.3 Autonomous Lie Derivative

Recalling expression (4.67), we can write expression (7.199) as

$$\mathcal{L}_{\mathbf{u}}\mathbf{w} = \mathrm{d}_t\mathbf{w} + [\mathbf{u}, \mathbf{w}] \equiv \mathrm{d}_t\mathbf{w} + \pounds_{\mathbf{u}}\mathbf{w}, \tag{7.203}$$

where we denote the *autonomous Lie derivative* by $\pounds_{\mathbf{u}}$. Expression (7.203) shows that the autonomous Lie derivative of a vector reduces to the Lie bracket:

$$\pounds_{\mathbf{u}}\mathbf{w} = [\mathbf{u}, \mathbf{w}]. \tag{7.204}$$

Thus, as a consequence of the anti-symmetry of the Lie bracket (4.68),

$$\pounds_{\mathbf{u}}\mathbf{w} = -\pounds_{\mathbf{w}}\mathbf{u}, \tag{7.205}$$

as illustrated in figure 4.9. It follows that, for three vector fields, $\mathbf{w}$, $\mathbf{u}$, and $\mathbf{w}$, the Jacobi identity holds:

$$\pounds_{\mathbf{w}}\pounds_{\mathbf{u}}\mathbf{w} - \pounds_{\mathbf{u}}\pounds_{\mathbf{w}}\mathbf{w} = \pounds_{[\mathbf{w},\mathbf{u}]}\mathbf{w}. \tag{7.206}$$

7.3.4 Lie Derivative of One-Forms

Next, let us consider the Lie derivative of a one-form, $\boldsymbol{\omega}$, which we define as

$$\mathcal{L}_{\mathbf{u}}\boldsymbol{\omega} \equiv (\partial_{t'}\omega_{i'})\,\mathbf{e}^{i'}, \tag{7.207}$$

with the implication that the Lie derivative of a coframe $\mathbf{e}^{i'}$ vanishes

$$\mathcal{L}_{\mathbf{u}}\mathbf{e}^{i'} = \mathbf{0}. \tag{7.208}$$

What are the components of this derivative in coordinates x? Invoking rules (4.45):

$$\omega_{i'} = \omega_i\,\lambda^i{}_{i'} \qquad \text{and} \qquad \omega_i = \omega_{i'}\,\Lambda^{i'}{}_i, \tag{7.209}$$

and considering the rate of change of

$$\omega_{i'}(x', t') = \omega_i(\varphi(x', t'), t')\,\lambda^i{}_{i'}(x', t'), \tag{7.210}$$

we find

$$\begin{aligned}
(\mathcal{L}_{\mathbf{u}}\boldsymbol{\omega})_{i'} &= \partial_{t'}\omega_{i'} \\
&= (\partial_t\omega_i + u^j\,\partial_j\omega_i)\,\lambda^i{}_{i'} + \omega_i\,\partial_{i'}\partial_{t'}\varphi^i \\
&= (\partial_t\omega_i + u^j\,\partial_j\omega_i)\,\lambda^i{}_{i'} + \omega_i\,\partial_j u^i\,\lambda^j{}_{i'}
\end{aligned}$$

$$= (\partial_t \omega_i + u^j\, \partial_j \omega_i + \omega_j\, \partial_i u^j)\, \lambda^i{}_{i'}$$
$$= (\mathcal{L}_{\mathbf{u}} \boldsymbol{\omega})_i\, \lambda^i{}_{i'}. \tag{7.211}$$

Hence, the components of the Lie derivative in x are

$$(\mathcal{L}_{\mathbf{u}} \boldsymbol{\omega})_i = \partial_t \omega_i + u^j\, \partial_j \omega_i + \omega_j\, \partial_i u^j. \tag{7.212}$$

Examining transformation (7.211), we see that the components of the Lie derivative of a one-form transform according to the tensorial rule. Based on (7.212), setting $\omega_i = 1$, we conclude that the Lie derivative of a basis one-form is

$$\mathcal{L}_{\mathbf{u}} \mathbf{e}^i = \partial_j u^i\, \mathbf{e}^j, \tag{7.213}$$

which should be contrasted with (7.208).

7.3.5 Lie Derivative of (p, q) Tensors

The Lie derivative of a (p, q) tensor, $\mathbf{T}$, is defined relative to the convected basis vectors and one-forms by

$$\mathcal{L}_{\mathbf{u}} \mathbf{T} \equiv (\partial_{t'} T^{i'_1 \cdots i'_p}{}_{j'_1 \cdots j'_q})\, \mathbf{e}_{i'_1} \otimes \cdots \otimes \mathbf{e}_{i'_p} \otimes \mathbf{e}^{j'_1} \otimes \cdots \otimes \mathbf{e}^{j'_q}. \tag{7.214}$$

Using rule (5.34), it may be shown that in the x coordinates the components of the Lie derivative are

$$\begin{aligned}(\mathcal{L}_{\mathbf{u}} \mathbf{T})^{i_1 \cdots i_p}{}_{j_1 \cdots j_q} &= \partial_t T^{i_1 \cdots i_p}{}_{j_1 \cdots j_q} + u^m\, \partial_m T^{i_1 \cdots i_p}{}_{j_1 \cdots j_q} \\ &+ T^{i_1 \cdots i_p}{}_{m \cdots j_q}\, \partial_{j_1} u^m + \cdots + T^{i_1 \cdots i_p}{}_{j_1 \cdots m}\, \partial_{j_q} u^m \\ &- T^{m \cdots i_p}{}_{j_1 \cdots j_q}\, \partial_m u^{i_1} - \cdots - T^{i_1 \cdots m}{}_{j_1 \cdots j_q} \partial_m u^{i_p}.\end{aligned} \tag{7.215}$$

which transform as a (p, q) tensor.

Example 7.16 Definition of the Lie Derivative in Continuum Mechanics

In continuum mechanics, the Lie derivative of a (p, q) tensor $\mathbf{T}$ relative to the flow of matter $\mathbf{v}$ may be defined in terms of a pullback from the spatial manifold $\mathcal{S}$ to the referential manifold $\mathcal{B}$ and a pushforward in the opposite direction with the motion φ as

$$\mathcal{L}_{\mathbf{v}} \mathbf{T} \equiv \varphi_*\, \partial_T \varphi^* \mathbf{T}. \tag{7.216}$$

In Lagrangian coordinates, this derivative takes the simple form

$$\mathcal{L}_{\mathbf{v}}\mathbf{T} = (\partial_T T^{I_1\cdots I_p}{}_{J_1\cdots J_q})\, \mathbf{e}_{I_1} \otimes \cdots \otimes \mathbf{e}_{I_p} \otimes \mathbf{e}^{J_1} \otimes \cdots \otimes \mathbf{e}^{J_q}. \tag{7.217}$$

For example, the Lie derivative of the metric tensor relative to the flow of matter may be expressed in Lagrangian components as

$$\mathcal{L}_{\mathbf{v}}\mathbf{g} = (\partial_T g_{IJ})\, \mathbf{e}^I \otimes \mathbf{e}^J. \tag{7.218}$$

If we consider a torsion-free connection for a covariant derivative—which means that equation (7.69) holds—we may replace the partial derivatives in expression (7.216) with covariant derivatives:

$$\begin{aligned}(\mathcal{L}_{\mathbf{u}}\mathbf{T})^{i_1\cdots i_p}{}_{j_1\cdots j_q} &= (\mathrm{D}_t\mathbf{T})^{i_1\cdots i_p}{}_{j_1\cdots j_q} \\ &\quad + T^{i_1\cdots i_p}{}_{m\cdots j_q}\, \nabla_{j_1} u^m + \cdots + T^{i_1\cdots i_p}{}_{j_1\cdots m}\, \nabla_{j_q} u^m \\ &\quad - T^{m\cdots i_p}{}_{j_1\cdots j_q}\, \nabla_m u^{i_1} - \cdots - T^{i_1\cdots m}{}_{j_1\cdots j_q}\, \nabla_m u^{i_p}.\end{aligned} \tag{7.219}$$

Here we encounter the *material derivative* $\mathrm{D}_t\mathbf{T}$, introduced in example 7.14, which has the x components

$$(\mathrm{D}_t\mathbf{T})^{i_1\cdots i_p}{}_{j_1\cdots j_q} \equiv \partial_t T^{i_1\cdots i_p}{}_{j_1\cdots j_q} + u^m\, \nabla_m T^{i_1\cdots i_p}{}_{j_1\cdots j_q}. \tag{7.220}$$

Expressions (7.216) and (7.219) are equivalent to one another, since the symmetry of the connection coefficients results in canceling these coefficients in expression (7.219), which becomes expression (7.216). In general, unlike the covariant derivative, the Lie derivative does not require a connection; it uses the flow of vector field **u**. We emphasize that expression (7.219) is not a generalization of the Lie derivative. Formally, there is an irreducible distinction between the Lie derivative and the covariant derivative, since the former produces a (p, q) tensor and the latter a $(p, q+1)$ tensor

7.3.6 Lie Derivative of Functions

A simple example of the Lie derivative of a tensor is the Lie derivative of a function, which is a $(0, 0)$ tensor. Following expression (7.219), we find that

$$\mathcal{L}_{\mathbf{u}}\mathrm{f} = \partial_t f + u^i\, \partial_i f = \mathrm{d}_t\mathrm{f} + \mathbf{u} \cdot \boldsymbol{\nabla}\mathrm{f}. \tag{7.221}$$

This result may be obtained also by taking the time rate of change of a time-dependent version of expression (5.39): $f'(x', t') = f(\varphi(x', t'), t')$. Thus, the Lie derivative of a function measures the rate of change of that function relative to a flow.

7.3.7 Lie Derivative of Metric Tensors

Let us consider the action of the Lie derivative on the metric tensor introduced in section 7.1.8. By definition, the components of the Lie derivative in convected coordinates,

x', are

$$(\mathcal{L}_{\mathbf{u}}\mathbf{g})_{i'j'} = \partial_{t'} g_{i'j'}. \tag{7.222}$$

As shown in section 7.1.8, the covariant derivative of the metric tensor vanishes, $\boldsymbol{\nabla}\mathbf{g} = \mathbf{0}$; since the spatial components of the metric tensor, g_{ij}, are time-independent, $\partial_t g_{ij} = 0$, using expression (7.219), we find that, for a torsion-free connection, the x components of the Lie derivative of the metric tensor are

$$(\mathcal{L}_{\mathbf{u}}\mathbf{g})_{ij} = g_{jk}\,\nabla_i u^k + g_{ik}\,\nabla_j u^k, \tag{7.223}$$

where we used the fact that lowering indices and covariant differentiation commute. Written in invariant notation we have

$$\mathcal{L}_{\mathbf{u}}\mathbf{g} = (\boldsymbol{\nabla}\mathbf{u})\cdot\mathbf{g} + \mathbf{g}\cdot(\boldsymbol{\nabla}\mathbf{u})^t. \tag{7.224}$$

The *deformation-rate tensor* associated with a flow of material $\mathbf{u}$ is defined as the $(1,1)$ tensor

$$\mathbf{d} \equiv \tfrac{1}{2}[\boldsymbol{\nabla}\mathbf{u} + (\boldsymbol{\nabla}\mathbf{u})^t], \tag{7.225}$$

and thus we deduce that the Lie derivative of the metric tensor equals twice the strain rate:

$$\mathcal{L}_{\mathbf{u}}\mathbf{g} = 2\,\mathbf{g}\cdot\mathbf{d}. \tag{7.226}$$

The Lie derivative of the metric tensor captures how norms of vectors and angles between vectors are affected by the flow $\mathbf{u}$.

Example 7.17 Lie Derivative of the Metric in Continuum Mechanics

In continuum mechanics, the Lie derivative of the metric with respect to the material velocity $\mathbf{v}$ is given by

$$\mathcal{L}_{\mathbf{v}}\mathbf{g} = 2\,\mathbf{g}\cdot\mathbf{D}, \tag{7.227}$$

where $\mathbf{D}$ denotes the deformation-rate tensor (7.37). Using expression (7.218) for the Lie derivative of the metric tensor relative to the flow of matter, in Lagrangian coordinates equation (7.227) takes the form

$$\partial_T g_{IJ} = 2\,D_{IJ}. \tag{7.228}$$

A summary of the properties of the metric tensor in continuum mechanics is given in table 7.1.

A *Killing vector field* is a field for which the Lie derivative of the metric is zero:

$$\mathcal{L}_{\mathbf{u}}\mathbf{g} = \mathbf{0}, \tag{7.229}$$

	EULERIAN	LAGRANGIAN
Metric	$\mathbf{g} = g_{ij}\,\mathbf{e}^i \otimes \mathbf{e}^j$	$\mathbf{g} = g_{IJ}\,\mathbf{e}^I \otimes \mathbf{e}^J$
Components	$g_{ij}(x)$	$g_{IJ}(X,T)$
Inverse	$\mathbf{g}^{-1} = g^{ij}\,\mathbf{e}_i \otimes \mathbf{e}_j$	$\mathbf{g}^{-1} = g^{IJ}\,\mathbf{e}_I \otimes \mathbf{e}_J$
Components	$g^{ij}(x)$	$g^{IJ}(X,T)$
Property	$g^{ik}\,g_{kj} = g_{jk}\,g^{ki} = \delta^i{}_j$	$g^{IK}\,g_{KJ} = g_{JK}\,g^{KI} = \delta^I{}_J$
Transformation	$g_{ij} = g_{IJ}\,(F^{-1})^I{}_i\,(F^{-1})^J{}_j$	$g_{IJ} = g_{ij}\,F^i{}_I\,F^j{}_J$
Covariant derivative	$\nabla_k\,g_{ij} = 0$	$\nabla_K\,g_{IJ} = 0$
Covariant derivative	$\nabla_k\,g^{ij} = 0$	$\nabla_K\,g^{IJ} = 0$
Spatial derivative	$\partial_k\,g_{ij} = g_{\ell j}\,\Gamma^\ell_{ki} + g_{i\ell}\,\Gamma^\ell_{kj}$	$\partial_K\,g_{IJ} = g_{LJ}\,\Gamma^L_{KI} + g_{IL}\,\Gamma^L_{KJ}$
Temporal derivative	$\partial_t g_{ij} = 0$	$\partial_T g_{IJ} = \nabla_I V_J + \nabla_J V_I$
Connection	$\Gamma^i_{jk} =$ $\frac{1}{2}\,g^{i\ell}(\partial_k\,g_{\ell j} + \partial_j\,g_{\ell k} - \partial_\ell\,g_{jk})$	$\Gamma^I_{JK} =$ $\frac{1}{2}\,g^{IL}\,(\partial_K\,g_{LJ} + \partial_J\,g_{LK} - \partial_L\,g_{JK})$
Temporal derivative	$\partial_t \Gamma^i_{jk} = 0$	$\partial_T \Gamma^I_{JK} = \nabla_J \nabla_K V^I$
Lie derivative	$(\mathcal{L}_{\mathbf{v}}\,\mathbf{g})_{ij} = \nabla_i v_j + \nabla_j v_i$	$(\mathcal{L}_{\mathbf{v}}\,\mathbf{g})_{IJ} = \nabla_I V_J + \nabla_J V_I$
Determinant	$\overline{g} \equiv \left[\overline{\overline{\det(g_{ij})}}\right]^{1/2}$	$\overline{G} = \left[\overline{\overline{\det(g_{IJ})}}\right]^{1/2}$
Inverse	$\underline{g} = 1/\overline{g}$	$\underline{G} = 1/\overline{G}$
Transformation	$\overline{g} = F^{-1}\,\overline{G}$	$\overline{G} = F\,\overline{g}$
Spatial derivative	$\underline{g}\,\partial_i \overline{g} = \Gamma^j_{ij}$	$\underline{G}\,\partial_I \overline{G} = \Gamma^J_{IJ}$
Temporal derivative	$\underline{g}\,\partial_t \overline{g} = 0$	$\underline{G}\,\partial_T \overline{G} = \frac{1}{2}\,g^{IJ}\,\partial_T g_{IJ} = \nabla_I V^I$

Table 7.1: Properties of the metric tensor $\mathbf{g}$ and its inverse $\mathbf{g}^{-1}$ in continuum mechanics. Its spatial or Eulerian components, g_{ij}, are time-independent and typically chosen once and for all. For a Cartesian Eulerian representation, $g_{ij} = \delta_{ij}$. Its comoving or Lagrangian components, g_{IJ}, are time-dependent and evolve according to the motion (3.12). Transformations between the two sets of components are accomplished based on the deformation gradient (4.26) and its inverse (4.27).

which we can state as a skewsymmetric covariant derivative: $\nabla\mathbf{u} = -(\nabla\mathbf{u})^t$. Such flows preserve inner products and, hence, norms and angles.

We make an important observation. Because the Lie derivative of the metric is nonzero, unlike the covariant derivative, *raising and lowering indices do not commute with the Lie derivative.* Thus, one must specify the kind of tensor to which one is applying the Lie derivative, as we explore in example 7.18.

Example 7.18 Distinguishing Rank-2 Tensors

A rank-2 tensor, such as the deformation-rate tensor (7.37), may be expressed as a $(2,0)$, $(1,1)$, or $(0,2)$ tensor, with components D^{IJ}, $D^I{}_J = D_J{}^I$, or D_{IJ}, respectively. Most of the time, these distinctions are immaterial because in a Riemannian manifold we can raise and lower indices with the metric tensor and its inverse. However, we noted the problem of raising and lowering indices not commuting with the Lie derivative. This leads to a desire to be cognizant of the nature of the tensor of which we are taking the Lie derivative. As discussed by Marsden & Hughes (1983), this may be accomplished by borrowing notation from music in the form of the *accidentals* "sharp" ($\sharp$), "natural" ($\natural$), and "flat" ($\flat$) to label $(2,0)$, $(1,1)$, and $(0,2)$ versions of a rank-2 tensor. Thus, the deformation-rate tensor $\mathbf{D}$ has three versions, a sharp version $\mathbf{D}^\sharp$ with elements D^{IJ}, a natural version $\mathbf{D}^\natural$ with elements $D^I{}_J = D_J{}^I$, and a flat version $\mathbf{D}^\flat$ with elements D_{IJ}. In this notation, equation (7.227) for the Lie derivative of the metric relative to the flow of matter becomes

$$\mathcal{L}_{\mathbf{v}}\mathbf{g} = 2\,\mathbf{g}\cdot\mathbf{D}^\natural = 2\,\mathbf{D}^\flat. \tag{7.230}$$

To reduce clutter, we continue to use $\mathbf{D}$ to denote the natural form $\mathbf{D}^\natural$.

7.3.8 Lie Derivative of Levi-Civita Tensor

Let us consider the action of the Lie derivative on the Levi-Civita tensor ϵ, introduced in section 5.13. By definition, the components of its Lie derivative with respect to flow $\mathbf{u}$ in convected coordinates, x', are

$$(\mathcal{L}_{\mathbf{u}}\epsilon)_{i'j'k'\dots} = \partial_{t'}\epsilon_{i'j'k'\dots} = \partial_{t'}(\bar{g}'\,\underline{\epsilon}_{i'j'k'\dots}) = \underline{\epsilon}_{i'j'k'\dots}\,\partial_{t'}\bar{g}', \tag{7.231}$$

where we used the fact that the Levi-Civita symbols $\underline{\epsilon}_{i'j'k'\dots}$ are independent of time. The square roots of the convected and inertial determinants of the metric, $\bar{g}'$ and $\bar{g}$, respectively, are related by rule (5.92):

$$\bar{g}'(x',t') = \bar{g}(\varphi(x',t'))\,\lambda(x',t'). \tag{7.232}$$

Upon differentiating this relation with respect to time t', holding the convected coordinates, x', fixed, we find

$$\partial_{t'}\bar{g}' = \bar{g}\,\partial_{t'}\lambda + \lambda\, u^i\, \partial_i \bar{g}, \tag{7.233}$$

where we used definition (7.187). Upon differentiating definition (5.56) with respect to time t' and using expression (7.187), we find

$$\partial_{t'}\lambda = \lambda\, \partial_i u^i, \tag{7.234}$$

and thus, using expression (7.91), we find that

$$\partial_{t'}\bar{g}' = \bar{g}\,\lambda\,(\partial_i u^i + u^i\,\Gamma^j_{ij}) = \bar{g}'\,\nabla_i u^i = \bar{g}'\,\mathrm{div}\,\mathbf{u}. \tag{7.235}$$

Thus, we obtain the desired Lie derivative:

$$\mathcal{L}_{\mathbf{u}}\boldsymbol{\epsilon} = \boldsymbol{\epsilon}\,\mathrm{div}\,\mathbf{u}. \tag{7.236}$$

The Lie derivative of the Levi-Civita tensor, $\boldsymbol{\epsilon}$, relative to flow $\mathbf{u}$ equals the divergence of this flow times the original volume form.

Example 7.19 Lie Derivative of the Volume Form

In continuum mechanics, the Lie derivative of the volume form, $\boldsymbol{\epsilon}$, introduced in example 5.13, with respect to the material velocity $\mathbf{v}$ is given by

$$\mathcal{L}_{\mathbf{v}}\boldsymbol{\epsilon} = \boldsymbol{\epsilon}\,\mathrm{div}\,\mathbf{v}. \tag{7.237}$$

We conclude that the divergence of the flow of matter controls variations in volume. A summary of the properties of the volume form in continuum mechanics is given in table 7.2.

Example 7.20 Lie Derivative of the Jacobian of the Motion

To further investigate the rate of change in volume, we differentiate equation (5.101) with respect to convected time T to find

$$\begin{aligned} 2\,\overline{G}\,\partial_T\overline{G} &= \tfrac{1}{2}\,\bar{\epsilon}^{IJK}\,(\partial_T g_{IL})\,g_{JM}\,g_{KN}\,\bar{\epsilon}^{LMN} \\ &= \bar{\epsilon}^{IJK}\,g_{IP}\,D^P{}_L\,g_{JM}\,g_{KN}\,\bar{\epsilon}^{LMN} \\ &= 2\,\overline{G}^2\,\mathrm{tr}\,(\mathbf{D}), \end{aligned} \tag{7.238}$$

where we used equation (7.228) in the second equality and equation (5.165) in the last equality. We conclude that

	EULERIAN	LAGRANGIAN
Volume form	$\boldsymbol{\epsilon} = \epsilon_{ijk}\, \mathbf{e}^i \otimes \mathbf{e}^j \otimes \mathbf{e}^k$	$\boldsymbol{\epsilon} = \epsilon_{IJK}\, \mathbf{e}^I \otimes \mathbf{e}^J \otimes \mathbf{e}^K$
Components	$\epsilon_{ijk} = \overline{g}\, \underline{\epsilon}_{\,ijk}$ $\epsilon_{ijk}(x)$	$\epsilon_{IJK} = \overline{G}\, \underline{\epsilon}_{\,IJK}$ $\epsilon_{IJK}(X, T)$
Transformation	$\epsilon_{ijk} =$ $\epsilon_{IJK}\, (F^{-1})^I{}_i\, (F^{-1})^J{}_j\, (F^{-1})^K{}_k$	$\epsilon_{IJK} = \epsilon_{ijk}\, F^i{}_I\, F^j{}_J\, F^k{}_K$
Covariant derivative	$\nabla_i \epsilon_{jk\ell} = 0$	$\nabla_I \epsilon_{JKL} = 0$
Spatial derivative	$\partial_i \epsilon_{jk\ell} = \Gamma^m_{im}\, \epsilon_{jk\ell}$	$\partial_I \epsilon_{JKL} = \Gamma^M_{IM}\, \epsilon_{JKL}$
Temporal derivative	$\partial_t \epsilon_{ijk} = 0$	$\partial_T \epsilon_{IJK} = \epsilon_{IJK}\, \nabla_L V^L$
Lie derivative	$(\mathcal{L}_{\mathbf{v}}\, \boldsymbol{\epsilon})_{ijk} = \epsilon_{ijk}\, \nabla_\ell v^\ell$	$(\mathcal{L}_{\mathbf{v}}\, \boldsymbol{\epsilon})_{IJK} = \epsilon_{IJK}\, \nabla_L V^L$

Table 7.2: Properties of volume form $\boldsymbol{\epsilon}$ in continuum mechanics. Its spatial or Eulerian components, $\epsilon_{ijk} = \overline{g}\, \underline{\epsilon}_{\,ijk}$, are time-independent and typically chosen once and for all. For a Cartesian Eulerian representation, $\overline{g} = 1$. Its comoving or Lagrangian components, $\epsilon_{IJK} = \overline{G}\, \underline{\epsilon}_{\,IJK}$, are time-dependent because of the time-dependence of the square root of the determinant of the Lagrangian metric $\overline{G}$. Transformations between the two sets of components are accomplished based on the deformation gradient (4.26) and its inverse (4.27).

$$\underline{G}\, \partial_T \overline{G} = \mathrm{tr}\,(\mathbf{D}). \tag{7.239}$$

This expression may be integrated to obtain an explicit expression for volume changes, namely,

$$\overline{G} = \overline{G}_0\, \exp\left[\int_0^T \mathrm{tr}(\mathbf{D})\, dT'\right]. \tag{7.240}$$

In terms of the Jacobian of the motion (6.28), we have

$$J = \exp\left[\int_0^T \mathrm{tr}(\mathbf{D})\, dT'\right], \tag{7.241}$$

such that

$$\partial_T J = J\, \mathrm{tr}(\mathbf{D}) = J\, \boldsymbol{\nabla} \cdot \mathbf{v}, \tag{7.242}$$

and $J(X, 0) = 1$, as required.

We note an important relationship. In view of expressions (7.224) and (7.225), equation (7.235) implies that

$$\underline{g}'\,\partial_{t'}\overline{g}' = \operatorname{div}\mathbf{u} = \nabla_{i'}u^{i'} = g^{i'j'}\,d_{i'j'} = \tfrac{1}{2}\,g^{i'j'}\,(\mathcal{L}_{\mathbf{u}}\mathbf{g})_{i'j'}. \tag{7.243}$$

Thus, using expression (7.222), we find

$$\underline{g}'\,\partial_{t'}\overline{g}' = \tfrac{1}{2}\,g^{i'j'}\,\partial_{t'}g_{i'j'}, \tag{7.244}$$

which complements expression (7.90); in convected coordinates, we write

$$\underline{g}'\,\partial_{i'}\overline{g}' = \tfrac{1}{2}\,g^{j'k'}\,\partial_{i'}g_{j'k'}. \tag{7.245}$$

Chapter 8

DIFFERENTIAL FORMS

Life need not be easy, provided only that it is not empty.
—LISE MEITNER

In section 4.2, the term "one-form" was introduced, implying the existence of entities called "two-forms" or "three-forms" and so on. Such forms, also known as *exterior differential forms*, are a type of tensor with practical physical applications. While we only provide an introduction to differential forms in this chapter, readers seeking a more detailed discussion are encouraged to consult books such as Flanders (1989) and Do Carmo (1994).

8.1 Definition

A k-form is a *completely antisymmetric*[1] $(0, k)$ tensor $\boldsymbol{\omega}$.[2] Specifically, its components satisfy

$$\boldsymbol{\omega}_{\pi(i_1,\ldots,i_k)} = (\operatorname{sgn}\pi)\, \omega_{i_1\cdots i_k}, \tag{8.1}$$

for all permutations π of the indices, $i_1, \ldots, i_k$. The sign of the permutation, $\operatorname{sgn}\pi$, is positive for even permutations and negative for odd permutations.

To gain insight into the formulation of k-forms, consider a $(0, 2)$ tensor,

$$\mathbf{T} = T_{ij}\, \mathrm{d}x^i \otimes \mathrm{d}x^j = T_{ij}\, \mathbf{e}^i \otimes \mathbf{e}^j. \tag{8.2}$$

It can be expressed in terms of its symmetric and antisymmetric parts, as defined by (5.26). We have

$$\begin{aligned} \mathbf{T} &= \tfrac{1}{2}\,(T_{ij} + T_{ji})\, \tfrac{1}{2}\,(\mathbf{e}^i \otimes \mathbf{e}^j + \mathbf{e}^j \otimes \mathbf{e}^i) + \tfrac{1}{2}\,(T_{ij} - T_{ji})\, \tfrac{1}{2}\,(\mathbf{e}^i \otimes \mathbf{e}^j - \mathbf{e}^j \otimes \mathbf{e}^i) \\ &= \tfrac{1}{2}\, \widehat{T}_{ij}\, (\mathbf{e}^i \otimes \mathbf{e}^j + \mathbf{e}^j \otimes \mathbf{e}^i) + \tfrac{1}{2}\, \widetilde{T}_{ij}\, (\mathbf{e}^i \otimes \mathbf{e}^j - \mathbf{e}^j \otimes \mathbf{e}^i). \end{aligned} \tag{8.3}$$

[1] Antisymmetric in all its components.
[2] Throughout this book, we use bold Greek letters to denote forms.

By analogy with identification (4.40), this result prompts us to define the basis elements of completely antisymmetric $(0,2)$ tensors, which are *two-forms*,[3] as

$$\mathrm{d}x^i \wedge \mathrm{d}x^j \equiv \mathbf{e}^i \otimes \mathbf{e}^j - \mathbf{e}^j \otimes \mathbf{e}^i, \tag{8.4}$$

where we introduce the symbol $\wedge$ to define the bases of forms.[4] The two-form basis is antisymmetric,

$$\mathrm{d}x^i \wedge \mathrm{d}x^j = -\,\mathrm{d}x^j \wedge \mathrm{d}x^i, \tag{8.5}$$

which implies that

$$\mathrm{d}x^i \wedge \mathrm{d}x^i = 0. \tag{8.6}$$

In a three-dimensional manifold, there are three linearly independent two-form basis elements, namely, $\mathrm{d}x^1 \wedge \mathrm{d}x^2$, $\mathrm{d}x^1 \wedge \mathrm{d}x^3$, and $\mathrm{d}x^2 \wedge \mathrm{d}x^3$. Geometrically, $\mathrm{d}x^i \wedge \mathrm{d}x^j$ measures the signed area of a parallelepiped in $\mathbb{R}^3$ projected onto the $x^i x^j$-plane. Similarly, $\mathrm{d}x^i$ measures the length of a projection on the x^i-axis. In general, the k-form bases measure the volume of a k-dimensional parallelepiped projected onto the coordinates. As discussed further in section 8.11, we do not view $\mathrm{d}x^i$ as an infinitesimal quantity but as a density, which can be integrated to give a length along a curve and—in general—a volume. In two dimensions, we see that the anticommutation of the wedge product indicates that $\mathrm{d}x^i \wedge \mathrm{d}x^j$ is a density with respect to an oriented area measure.

Using components, a general two-form in an n-dimensional manifold may be expressed as

$$\boldsymbol{\omega} = \omega_{ij}\, \mathbf{e}^i \otimes \mathbf{e}^j = \tfrac{1}{2}\, \omega_{ij}\, \mathrm{d}x^i \wedge \mathrm{d}x^j. \tag{8.7}$$

The factor $1/2$ ensures that $\mathrm{d}x^i \wedge \mathrm{d}x^j$ and $\mathrm{d}x^j \wedge \mathrm{d}x^i$ are not counted as two independent summands, since they are linearly dependent due to property (8.5). Alternatively, to avoid this factor, $\boldsymbol{\omega}$ may be written as

$$\boldsymbol{\omega} = \sum_{i<j} \omega_{ij}\, \mathrm{d}x^i \wedge \mathrm{d}x^j, \tag{8.8}$$

where $i<j$ means that—for this double summation—one sums over $i = \{1, \ldots, n-1\}$, and—for a given i—one sums over $j = \{i+1, \ldots, n\}$. To exemplify summation (8.8), consider a two-form in $\mathbb{R}^3$. For $i=1$, we let $j=2$ and 3 to get $\omega_{12}\, \mathrm{d}x^1 \wedge \mathrm{d}x^2 + \omega_{13}\, \mathrm{d}x^1 \wedge \mathrm{d}x^3$; for $i=2$, we let $j=3$ to get $\omega_{23}\, \mathrm{d}x^2 \wedge \mathrm{d}x^3$. Thus, we obtain

$$\boldsymbol{\omega} = \omega_{12}\, \mathrm{d}x^1 \wedge \mathrm{d}x^2 + \omega_{13}\, \mathrm{d}x^1 \wedge \mathrm{d}x^3 + \omega_{23}\, \mathrm{d}x^2 \wedge \mathrm{d}x^3, \tag{8.9}$$

where we recognize the three two-form basis elements. Such an *increasing-order summation* may be used for an arbitrary k-form.

In an n-dimensional manifold, there are $n(n-1)/2$ independent basis elements $\mathrm{d}x^i \wedge \mathrm{d}x^j$, which we can write as $\binom{n}{2}$: the number of two-element sets that can be chosen among

[3] These basis elements are tensor capacities introduced in section 5.6.

[4] In section 8.2.2, we formalize the use of this symbol in the context of the *wedge product*.

n elements. The number of linearly independent two-element bases is the dimension of the two-form space in $\mathbb{R}^3$. This is a notable exception; in general, the dimensions are not the same.

A three-form in an n-dimensional manifold may be expressed as

$$\boldsymbol{\omega} = \omega_{ijk}\, \mathbf{e}^i \otimes \mathbf{e}^j \otimes \mathbf{e}^k = \tfrac{1}{3!}\, \omega_{ijk}\, \mathrm{d}x^i \wedge \mathrm{d}x^j \wedge \mathrm{d}x^k, \tag{8.10}$$

or, to avoid the factor 1/3!, as

$$\boldsymbol{\omega} = \sum_{i<j<k} \omega_{ijk}\, \mathrm{d}x^i \wedge \mathrm{d}x^j \wedge \mathrm{d}x^k, \tag{8.11}$$

where the basis elements of three-forms are defined by

$$\begin{aligned} \mathrm{d}x^i \wedge \mathrm{d}x^j \wedge \mathrm{d}x^k \equiv\, & \mathbf{e}^i \otimes \mathbf{e}^j \otimes \mathbf{e}^k + \mathbf{e}^j \otimes \mathbf{e}^k \otimes \mathbf{e}^i + \mathbf{e}^k \otimes \mathbf{e}^i \otimes \mathbf{e}^j \\ & - \mathbf{e}^j \otimes \mathbf{e}^i \otimes \mathbf{e}^k - \mathbf{e}^k \otimes \mathbf{e}^j \otimes \mathbf{e}^i - \mathbf{e}^i \otimes \mathbf{e}^k \otimes \mathbf{e}^j. \end{aligned} \tag{8.12}$$

Any odd permutation of the indices, i, j, k, changes the sign of $\mathrm{d}x^i \wedge \mathrm{d}x^j \wedge \mathrm{d}x^k$. The dimension of the space spanned by three-forms in an n-dimensional manifold is $\binom{n}{3} = n(n-1)(n-2)/6$. Thus, in a three-dimensional manifold, there is just a single three-form:

$$\boldsymbol{\omega} = \omega_{123}\, \mathrm{d}x^1 \wedge \mathrm{d}x^2 \wedge \mathrm{d}x^3, \tag{8.13}$$

and the single basis element of a three-form in a three-dimensional manifold is

$$\begin{aligned} \mathrm{d}x^1 \wedge \mathrm{d}x^2 \wedge \mathrm{d}x^3 \equiv\, & \mathbf{e}^1 \otimes \mathbf{e}^2 \otimes \mathbf{e}^3 + \mathbf{e}^2 \otimes \mathbf{e}^3 \otimes \mathbf{e}^1 + \mathbf{e}^3 \otimes \mathbf{e}^1 \otimes \mathbf{e}^2 \\ & - \mathbf{e}^2 \otimes \mathbf{e}^1 \otimes \mathbf{e}^3 - \mathbf{e}^3 \otimes \mathbf{e}^2 \otimes \mathbf{e}^1 - \mathbf{e}^1 \otimes \mathbf{e}^3 \otimes \mathbf{e}^2. \end{aligned} \tag{8.14}$$

In terms of the Levi-Civita capacity defined in section 5.7, we can rewrite expression (8.14) as

$$\mathrm{d}x^1 \wedge \mathrm{d}x^2 \wedge \mathrm{d}x^3 = \underline{\epsilon}_{ijk}\, \mathbf{e}^i \otimes \mathbf{e}^j \otimes \mathbf{e}^k = \underline{\epsilon}_{ijk}\, \mathrm{d}x^i \otimes \mathrm{d}x^j \otimes \mathrm{d}x^k, \tag{8.15}$$

establishing a three-form basis as a tensor capacity.

Generally, a k-form may be expressed as

$$\boldsymbol{\omega} = \omega_{i_1 \cdots i_k}\, \mathbf{e}^{i_1} \otimes \cdots \otimes \mathbf{e}^{i_k} = \frac{1}{k!}\, \omega_{i_1 \cdots i_k}\, \mathrm{d}x^{i_1} \wedge \cdots \wedge \mathrm{d}x^{i_k}, \tag{8.16}$$

or as

$$\boldsymbol{\omega} = \sum_{i_1 < \cdots < i_k} \omega_{i_1 \cdots i_k}\, \mathrm{d}x^{i_1} \wedge \cdots \wedge \mathrm{d}x^{i_k}. \tag{8.17}$$

The corresponding k-form-space is $\binom{n}{k}$-dimensional, where the binomial coefficient for nonnegative n and k is

$$\binom{n}{k} = \frac{n!}{k!\,(n-k)!}, \tag{8.18}$$

and 0 for $n<k$. This value is the number of elements $\mathrm{d}x^1 \wedge \cdots \wedge \mathrm{d}x^k$ that can be composed of n one-form basis, $\mathrm{d}x^1, \ldots, \mathrm{d}x^n$, with no distinction being made between $\mathrm{d}x^1 \wedge \cdots \mathrm{d}x^i \wedge \mathrm{d}x^j \cdots \wedge \mathrm{d}x^k$ and $\mathrm{d}x^1 \wedge \cdots \mathrm{d}x^j \wedge \mathrm{d}x^i \cdots \wedge \mathrm{d}x^k$.

Examining this expression, we can infer properties of k-forms in an n-dimensional manifold. All zero-form spaces are one-dimensional. All one-form spaces are n-dimensional; notably, the cotangent space, which is a one-form space, has the same dimensions as the manifold. If $k=n$, the k-form space is also one-dimensional. For all k-forms, $k \leq n$; in other words, including a zero-form, there are $n+1$ forms in an n-dimensional manifold.

More generally, in terms of coframes, introduced in example 5.8, a k-form is expressed as

$$\boldsymbol{\omega} = \sum_{i_1<\cdots<i_k} \omega_{i_1\cdots i_k} \mathbf{e}^{i_1} \wedge \cdots \wedge \mathbf{e}^{i_k}. \tag{8.19}$$

8.2 Operations on Forms

8.2.1 Addition

We may add a k-form and an ℓ-form if and only if $k=\ell$, which results in a form of the same type. We can illustrate properties of addition by considering the sum of two two-forms in a three-dimensional manifold:

$$\begin{aligned}\boldsymbol{\omega}+\boldsymbol{\eta} = {} & (\omega_{12}+\eta_{12})\, \mathrm{d}x^1 \wedge \mathrm{d}x^2 + (\omega_{13}+\eta_{13})\, \mathrm{d}x^1 \wedge \mathrm{d}x^3 \\ & + (\omega_{23}+\eta_{23})\, \mathrm{d}x^2 \wedge \mathrm{d}x^3,\end{aligned} \tag{8.20}$$

resulting in a third two-form in the manifold. In general, for two k-forms, we may write their sum as

$$\boldsymbol{\omega}+\boldsymbol{\eta} = \frac{1}{k!}\left(\omega_{i_1\cdots i_k}+\eta_{i_1\cdots i_k}\right) \mathrm{d}x^{i_1} \wedge \cdots \wedge \mathrm{d}x^{i_k}. \tag{8.21}$$

Since multiplication of a k-form by a scalar function results in another k-form, the k-forms at a point of an n-dimensional manifold constitute an $\binom{n}{k}$-dimensional linear space.[5]

8.2.2 Exterior Product

In an n-dimensional manifold, if $\boldsymbol{\omega}$ is a k-form,

$$\boldsymbol{\omega} = \frac{1}{k!}\, \omega_{i_1\cdots i_k}\, \mathrm{d}x^{i_1} \wedge \cdots \wedge \mathrm{d}x^{i_k}, \tag{8.22}$$

and $\boldsymbol{\eta}$ is an ℓ-form,

$$\boldsymbol{\eta} = \frac{1}{\ell!}\, \eta_{j_1\cdots j_\ell}\, \mathrm{d}x^{j_1} \wedge \cdots \wedge \mathrm{d}x^{j_\ell}, \tag{8.23}$$

[5] See chapter 2.

then their *exterior product*, denoted by the symbol $\wedge$ and known also as the *wedge product*, $\omega \wedge \eta$, is a $(k+\ell)$-form given by

$$\omega \wedge \eta = \frac{1}{k!\,\ell!}\,\omega_{i_1 \cdots i_k}\,\eta_{j_1 \cdots j_\ell}\,dx^{i_1} \wedge \cdots \wedge dx^{i_k} \wedge dx^{j_1} \wedge \cdots \wedge dx^{j_\ell}. \tag{8.24}$$

The operation of multiplication is anticommutative. If ω is a k-form and η is an ℓ-form, then $\omega \wedge \eta = (-1)^{k\ell}\,\eta \wedge \omega$; it is also distributive and associative.[6] There is a close association between exterior multiplication and Lie algebras, where we require multiplication to be anticommutative.

Based on these definitions, the wedge product of two-form α and one-form β produces three-form $\alpha \wedge \beta$ This three-form has three slots that accept vectors. It may be readily shown that if we feed it the three vectors, $\mathbf{u}_1$, $\mathbf{u}_2$, $\mathbf{u}_3$, it returns

$$\begin{aligned}(\alpha \wedge \beta)(\mathbf{u}_1, \mathbf{u}_2, \mathbf{u}_3) = \alpha(\mathbf{u}_1, \mathbf{u}_2)\,\omega(\mathbf{u}_3) - \alpha(\mathbf{u}_1, \mathbf{u}_3)\,\omega(\mathbf{u}_2)\\ + \alpha(\mathbf{u}_2, \mathbf{u}_3)\,\omega(\mathbf{u}_1).\end{aligned} \tag{8.25}$$

More generally, if α is a k-form, β an ℓ-form, and ω an m-form, then the wedge product is associative:

$$\alpha \wedge (\beta \wedge \omega) = (\alpha \wedge \beta) \wedge \omega, \tag{8.26}$$

and satisfies the commutation relation

$$\alpha \wedge \beta = (-1)^{k\ell} \beta \wedge \alpha. \tag{8.27}$$

So if the product $k\,\ell$ is odd, say the exterior product of a two-form with a one-form, the wedge product is commutative, whereas if $k\,\ell$ is even, say the exterior product of two one-forms, the wedge product is anticommutative.

In general, we denote the standard basis element of a k-form as

$$dx^i \wedge dx^j \wedge \cdots \wedge dx^k, \qquad i < j < \cdots < k, \tag{8.28}$$

where

$$dx^i \wedge dx^j \wedge \cdots \wedge dx^k = \sum_\pi (\operatorname{sgn} \pi)\,\mathbf{e}^{\pi(i)} \otimes \mathbf{e}^{\pi(j)} \otimes \cdots \otimes \mathbf{e}^{\pi(k)}. \tag{8.29}$$

The sum is over all permutations of the indices $i, j, \ldots, k$. Thus, if α is a k-form,

[6] See chapter 2.

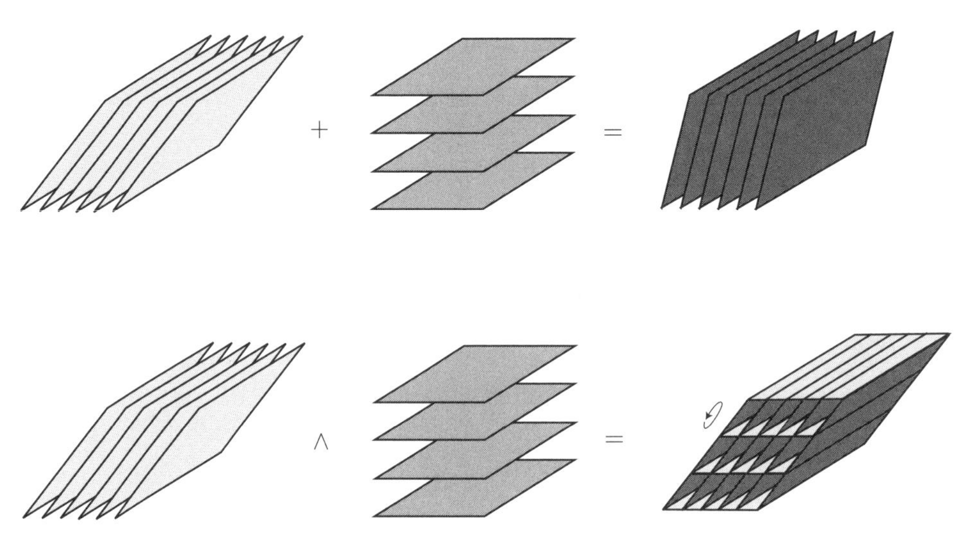

Figure 8.1: Top row: the sum of two one-forms results in another one-form. This is a three-dimensional version of the two-dimensional illustration of the addition of one-forms shown in figure 4.8. Bottom row: the wedge product of the two one-forms produces a two-form, which has a honeycomb structure consisting of "circulation tubes," with a sense of circulation as indicated by the small loop.

$$\begin{aligned}\boldsymbol{\alpha} &= \alpha_{i\cdots j}\, \mathrm{d}x^i \otimes \cdots \otimes \mathrm{d}x^j \\ &= \frac{1}{k!}\, \alpha_{i\cdots j}\, \mathrm{d}x^i \wedge \cdots \wedge \mathrm{d}x^j \\ &= \sum_{i<\cdots<j} \alpha_{i\cdots j}\, \mathrm{d}x^i \wedge \cdots \wedge \mathrm{d}x^j. \end{aligned} \tag{8.30}$$

Note the difference between the sum of two one-forms, $\mathrm{d}x^1 + \mathrm{d}x^2$, which is a one-form, and the wedge product of two one-forms, for example $\mathrm{d}x^1 \wedge \mathrm{d}x^2$, which is a two-form, as illustrated in figure 8.1. The wedge product $\boldsymbol{\omega} \wedge \boldsymbol{\alpha}$ may be represented by intersecting surfaces of $\boldsymbol{\omega}$ and $\boldsymbol{\alpha}$, plus a sense of circulation: a collection of "circulation tubes."

Figure 8.2 shows the "egg-crate" structure of the three-form in a three-dimensional manifold in terms of three one-forms. If you "feed" a three-form three vectors, it returns the number of egg-crate cells contained in the volume spanned by the vectors. In particular, if you feed the Cartesian three-form basis element three vectors **u**, **v**, and **w**, it returns the volume spanned by the three vectors, that is, $\epsilon_{ijk}\, u^i\, v^j\, w^k$.

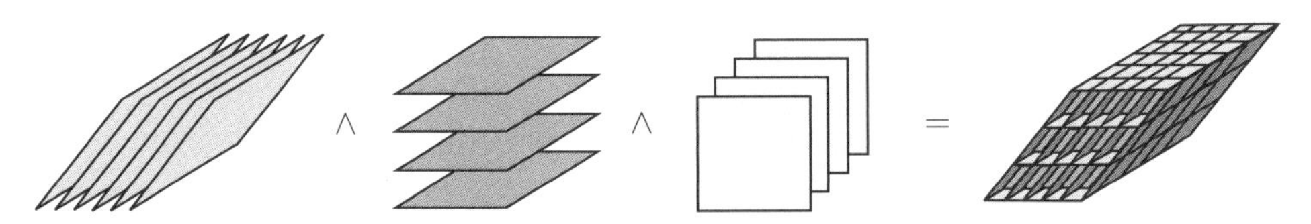

Figure 8.2: The wedge product of three one-forms results in a three-form. The three-form has an "egg-crate" structure. If "fed" three vectors, it returns the number of egg-crate cells contained in the volume spanned by the vectors.

8.2.3 Interior Product

The *interior product* or *pull down* of a vector $\mathbf{u}$ and a k-form ω is a $(k-1)$-form defined by

$$\begin{aligned}\mathbf{u} \rfloor \omega = \mathbf{i_u}\omega = \mathbf{u}\cdot\omega &\equiv u^{i_1}\, \omega_{i_1 i_2 \cdots i_k}\, \mathrm{d}x^{i_2} \otimes \cdots \otimes \mathrm{d}x^{i_k} \\ &= \frac{1}{(k-1)!}\, u^{i_1}\, \omega_{i_1 i_2 \cdots i_k}\, \mathrm{d}x^{i_2} \wedge \cdots \wedge \mathrm{d}x^{i_k},\end{aligned} \tag{8.31}$$

where $\rfloor$ denotes the *pull down* operator. Using the increasing-order summation, we write

$$\mathbf{u}\cdot\omega = \sum_{i_2<\cdots<i_k} u^{i_1}\, \omega_{i_1 i_2 \cdots i_k}\, \mathrm{d}x^{i_2} \wedge \cdots \wedge \mathrm{d}x^{i_k}, \tag{8.32}$$

thereby avoiding the factorial. We identify the interior product with the notion of a *flux* or *quantity*, which is an example of a pseudoform (e.g., Frankel, 2004).

A coordinate-free expression for the interior product between a vector $\mathbf{u}$ and a k-form ω is

$$\mathbf{i_u}\omega(\mathbf{u}_1, \mathbf{u}_2, \ldots, \mathbf{u}_k) \equiv \omega(\mathbf{u}, \mathbf{u}_2, \ldots, \mathbf{u}_k), \tag{8.33}$$

effectively filling the first vector slot of ω with the vector $\mathbf{u}$. For a k-form ω and another form α we have

$$\mathbf{i_u}(\omega \wedge \alpha) = (\mathbf{i_u}\omega) \wedge \alpha + (-1)^k\, \omega \wedge (\mathbf{i_u}\alpha). \tag{8.34}$$

8.3 *k*-Vectors

With every k-form ω, that is, a skewsymmetric $(0,k)$ tensor, we may associate a completely antisymmetric $(k,0)$ tensor called a *k-vector*, which takes the form

$$\omega = \omega^{i_1\cdots i_k}\, \mathbf{e}_{i_1} \otimes \cdots \otimes \mathbf{e}_{i_k} = \frac{1}{k!}\, \omega^{i_1\cdots i_k}\, \partial i_1 \wedge \cdots \wedge \partial_{i_k}, \tag{8.35}$$

or as

$$\omega = \sum_{i_1<\cdots<i_k} \omega^{i_1\cdots i_k}\, \partial_{i_1} \wedge \cdots \wedge \partial_{i_k}. \tag{8.36}$$

The components of this k-vector may be obtained from the components of a k-form by raising all the indices:

$$\omega^{j_1\cdots j_k} = g^{j_1\ell_1} \cdots g^{j_k\ell_k}\, \omega_{\ell_1\cdots\ell_k}. \tag{8.37}$$

The *norm of a k-form*, consistent with the norm of a one-form (5.115), is defined by

$$\begin{aligned}\|\omega\|^2 &\equiv \frac{1}{k!}\, g^{j_1\ell_1} \cdots g^{j_k\ell_k}\, \omega_{j_1\cdots j_k}\, \omega_{\ell_1\cdots\ell_k} \\ &= \frac{1}{k!}\, \omega^{j_1\cdots j_k}\, \omega_{j_1\cdots j_k} \\ &= \sum_{j_1<\cdots<j_k} \omega^{j_1\cdots j_k}\, \omega_{j_1\cdots j_k}.\end{aligned} \tag{8.38}$$

As an example, for two vectors $\mathbf{u} = u^i\,\mathbf{e}_i$ and $\mathbf{v} = v^i\,\mathbf{e}_i$, the two-vector $\mathbf{u} \wedge \mathbf{v}$ in $\mathbb{R}^3$ has Cartesian components

$$\begin{aligned} \mathbf{u} \wedge \mathbf{v} = &(u^1\,v^2 - v^1\,u^2)\,\mathbf{e}_1 \wedge \mathbf{e}_2 \\ &+ (u^1\,v^3 - v^1\,u^3)\,\mathbf{e}_1 \wedge \mathbf{e}_3 \\ &+ (u^2\,v^3 - v^2\,u^3)\,\mathbf{e}_2 \wedge \mathbf{e}_3. \end{aligned} \tag{8.39}$$

We recognize the components of the vector *cross product* $\mathbf{u} \times \mathbf{v}$: $u^2\,v^3 - v^2\,u^3, u^3\,v^1 - v^3\,u^1, u^1\,v^2 - v^1\,u^2$. In higher dimensional spaces, the exterior product may be viewed as a generalization of the cross product.

To give a geometrical interpretation to the three-vector in $\mathbb{R}^3$, consider the wedge product of three vectors $\mathbf{w}$, $\mathbf{u}$, and $\mathbf{v}$ in Cartesian coordinates:

$$\begin{aligned} \mathbf{w} \wedge \mathbf{u} \wedge \mathbf{v} = &\left[w^3(u^1\,v^2 - v^1\,u^2) - w^2(u^1\,v^3 - v^1\,u^3)\right. \\ &\left. + w^1(u^2\,v^3 - v^2\,u^3)\right]\mathbf{e}_1 \wedge \mathbf{e}_2 \wedge \mathbf{e}_3. \end{aligned} \tag{8.40}$$

We recognize the volume spanned by the three vectors:

$$\mathbf{w} \cdot (\mathbf{u} \times \mathbf{v}) = w^3(u^1\,v^2 - v^1\,u^2) - w^2(u^1\,v^3 - v^1\,u^3) + w^1(u^2\,v^3 - v^2\,u^3). \tag{8.41}$$

8.4 Hodge Dual

In n dimensions, the *Hodge dual* or *Hodge star* of a k-vector $\boldsymbol{\omega}$ is an $(n-k)$-form $*\,\boldsymbol{\omega}$ with components

$$(*\,\boldsymbol{\omega})_{i_1 \cdots i_{n-k}} = \sum_{j_1 < \cdots < j_k} \epsilon_{j_1 \cdots j_k i_1 \cdots i_{n-k}}\,\omega^{j_1 \cdots j_k}, \tag{8.42}$$

where $\epsilon_{j_1 \cdots j_k i_1 \cdots i_{n-k}}$ is the Levi-Civita pseudotensor defined by (5.134).

A modification of the definition of the dual, adopted by, for example, Appel (2005) and Carroll (2004), uses the relation (8.37) to take a k-form—rather than a k-vector—to an $(n-k)$-form by using mixed components for the Levi-Civita pseudotensor:

$$(*\,\boldsymbol{\omega})_{i_1 \cdots i_{n-k}} = \sum_{j_1 < \cdots < j_k} \epsilon^{j_1 \cdots j_k}{}_{i_1 \cdots i_{n-k}}\,\omega_{j_1 \cdots j_k}. \tag{8.43}$$

Since the application of the dual involves the Levi-Civita tensor (5.134), which in turn involves the metric through its determinant, definitions (8.42) and (8.43) are equivalent in the sense that they both produce an $(n-k)$-form. What is of importance is that $*\,\boldsymbol{\omega}$ *always* denotes a form, whose components are equivalently given by (8.42) and (8.43).

We note that

$$*\,*\,\boldsymbol{\omega} = (-1)^{k(n-k)}\,\boldsymbol{\omega}; \tag{8.44}$$

in other words, a dual of a dual is the original form (up to a sign). Finally, we also note that, for a general k-form,

$$\frac{1}{k!}\,\omega_{j_1\ldots j_k}\,\delta^{j_1\ldots j_k}{}_{i_1\ldots i_k} = \omega_{i_1\ldots i_k}, \tag{8.45}$$

where $\delta^{j_1\ldots j_k}{}_{i_1\ldots i_k}$ denotes the generalized Kronecker delta tensor defined in equation (5.145).

As introduced in section 8.3, in $\mathbb{R}^3$ in Cartesian coordinates $\{x, y, z\}$, we have the following relationship between two-forms and one-forms

$$*\,\mathrm{d}x = \mathrm{d}y \wedge \mathrm{d}z, \qquad *\,\mathrm{d}y = \mathrm{d}z \wedge \mathrm{d}x, \qquad *\,\mathrm{d}z = \mathrm{d}x \wedge \mathrm{d}y. \tag{8.46}$$

We conclude that one-forms and two-forms in $\mathbb{R}^3$ both span a three-dimensional linear space.

In $\mathbb{R}^3$ in Cartesian components, we may use expressions (8.39) and (8.40) in combination with the definition of the Hodge dual to write

$$*(\mathbf{u} \wedge \mathbf{v}) = \mathbf{u} \times \mathbf{v}, \tag{8.47}$$

$$*(\mathbf{w} \wedge \mathbf{u} \wedge \mathbf{v}) = \mathbf{w} \cdot (\mathbf{u} \times \mathbf{v}), \tag{8.48}$$

further defining the relationship between the cross and wedge products in $\mathbb{R}^3$.

8.5 Volumes

Under a coordinate change from $\{x^i\}$ to $\{x^{i'}\}$ in a three-dimensional manifold, using the expression for the determinant (5.56), the three-form basis $\mathrm{d}x^1 \wedge \mathrm{d}x^2 \wedge \mathrm{d}x^3$ changes as

$$\mathrm{d}x^{1'} \wedge \mathrm{d}x^{2'} \wedge \mathrm{d}x^{3'} = \lambda^{-1}\,\mathrm{d}x^1 \wedge \mathrm{d}x^2 \wedge \mathrm{d}x^3, \tag{8.49}$$

which, following expression (8.15), we rewrite as

$$\underline{\epsilon}_{\,i'j'k'}\,\mathrm{d}x^{i'} \otimes \mathrm{d}x^{j'} \otimes \mathrm{d}x^{k'} = \lambda^{-1}\underline{\epsilon}_{\,ijk}\,\mathrm{d}x^{i} \otimes \mathrm{d}x^{j} \otimes \mathrm{d}x^{k}. \tag{8.50}$$

The result (5.61) identifies the basis three-form, like the Levi-Civita capacity, as a tensor capacity of weight 1. We saw in section 5.10 that the square-root of the determinant of the metric $\bar{g}$ transforms according to $\bar{g}' = \lambda\,\bar{g}$, where λ is the determinant (5.56). Thus, we conclude that the *volume element*,

$$\begin{aligned}
\epsilon &= \bar{g}\,\underline{\epsilon}_{\,123}\,\mathrm{d}x^1 \wedge \mathrm{d}x^2 \wedge \mathrm{d}x^3 \\
&= \epsilon_{ijk}\,\mathrm{d}x^i \otimes \mathrm{d}x^j \otimes \mathrm{d}x^k \\
&= \bar{g}'\,\underline{\epsilon}_{\,1'2'3'}\,\mathrm{d}x^{1'} \wedge \mathrm{d}x^{2'} \wedge \mathrm{d}x^{3'} \\
&= \epsilon_{i'j'k'}\,\mathrm{d}x^{i'} \otimes \mathrm{d}x^{j'} \otimes \mathrm{d}x^{k'},
\end{aligned} \tag{8.51}$$

transforms like a $(0, 3)$ tensor. The volume form ϵ is an example of a *pseudoform*, because it changes sign if we change the orientation of the manifold, as discussed in section 3.5 (see also, e.g., Frankel, 2004).

To make the connection between this definition of a volume and the more traditional view, imagine feeding the volume form $\boldsymbol{\epsilon}$ three little vectors $\Delta\mathbf{u} = \Delta u^i\,\partial_i$, $\Delta\mathbf{v} = \Delta v^j\,\partial_j$, and $\Delta\mathbf{w} = \Delta w^k\,\partial_k$. The result would be a scalar:

$$\boldsymbol{\epsilon}(\Delta\mathbf{u}, \Delta\mathbf{v}, \Delta\mathbf{w}) = \epsilon_{ijk}\,\Delta u^i\,\Delta v^j\,\Delta w^k = \Delta\mathbf{u}\cdot(\Delta\mathbf{v}\times\Delta\mathbf{w}). \tag{8.52}$$

This is precisely the volume of the parallelepiped spanned by the three vectors $\Delta\mathbf{u}$, $\Delta\mathbf{v}$, and $\Delta\mathbf{w}$.

The results in this section are readily generalized for n-dimensional volumes based upon the alternating tensor in an n-dimensional manifold and the corresponding volume form, which is a (pseudo) n-form:

$$\begin{aligned}
\boldsymbol{\epsilon} &= \bar{g}\,\underline{\epsilon}_{123\ldots}\,\mathrm{d}x^1\wedge\cdots\wedge\mathrm{d}x^n \\
&= \frac{1}{n!}\,\bar{g}\,\underline{\epsilon}_{i_1\cdots i_n}\,\mathrm{d}x^{i_1}\wedge\cdots\wedge\mathrm{d}x^{i_n} \\
&= \bar{g}\,\underline{\epsilon}_{i_1\cdots i_n}\,\mathrm{d}x^{i_1}\otimes\cdots\otimes\mathrm{d}x^{i_n} \\
&= \epsilon_{i_1\cdots i_n}\,\mathrm{d}x^{i_1}\otimes\cdots\otimes\mathrm{d}x^{i_n}.
\end{aligned} \tag{8.53}$$

Based upon the definition (8.43), we conclude that the Hodge dual of the volume form equals 1:

$$*\,\boldsymbol{\epsilon} = 1. \tag{8.54}$$

This duality reflects the fact that there is one unique volume element, $\epsilon_{123\ldots}$, which has a one-to-one correspondence to unity.

8.5.1 Properties

In this section, we consider definitions and properties related to the three-dimensional volume form (for four-dimensional versions, see Trautman, 1973, and example 8.1). We start with the representation

$$\epsilon_{ijk} = \bar{g}\,\underline{\epsilon}_{ijk}. \tag{8.55}$$

Next, expressed in a coframe basis, we define the one-forms

$$\boldsymbol{\epsilon}_{ij} \equiv \epsilon_{ijk}\,\mathbf{e}^k, \tag{8.56}$$

the two-forms

$$\boldsymbol{\epsilon}_i \equiv \tfrac{1}{2}\,\mathbf{e}^j\wedge\boldsymbol{\epsilon}_{ij} = \tfrac{1}{2!}\,\epsilon_{ijk}\,\mathbf{e}^j\wedge\mathbf{e}^k, \tag{8.57}$$

and the volume three-forms

$$\boldsymbol{\epsilon} \equiv \tfrac{1}{3}\,\mathbf{e}^i\wedge\boldsymbol{\epsilon}_i = \tfrac{1}{3!}\,\epsilon_{ijk}\,\mathbf{e}^i\wedge\mathbf{e}^j\wedge\mathbf{e}^k. \tag{8.58}$$

This cascade of forms has the following properties:

$$\mathbf{e}^k\wedge\boldsymbol{\epsilon}_{ij} = \delta_j{}^k\,\boldsymbol{\epsilon}_i - \delta_i{}^k\,\boldsymbol{\epsilon}_j, \tag{8.59}$$

$$\mathbf{e}^i\wedge\boldsymbol{\epsilon}_j = \delta_j{}^i\,\boldsymbol{\epsilon}. \tag{8.60}$$

A final useful equality is

$$\mathbf{e}^i \wedge \mathbf{e}^j \wedge \mathbf{e}^k = \epsilon^{ijk}\, \boldsymbol{\epsilon}, \tag{8.61}$$

which may be verified by selecting random values of i, j, and k.

Example 8.1 Properties of the Four-Dimensional Volume Form

In this example, we consider definitions and properties related to the four-dimensional volume form (Trautman, 1973), which is relevant in general relativity. We start with the representation

$$\epsilon_{\mu\nu\sigma\tau} = \bar{g}\, \underline{\epsilon}_{\mu\nu\sigma\tau}, \tag{8.62}$$

where $\bar{g} = \sqrt{-\det(g_{\mu\nu})}$ denotes the square root of the negative determinant of the spacetime metric tensor, which is a scalar capacity of weight one, and $\underline{\epsilon}_{\mu\nu\sigma\tau}$ the alternating symbol, which is a tensor density of weight one. Next, expressed in a coframe basis, we define the one-forms

$$\boldsymbol{\epsilon}_{\mu\nu\sigma} \equiv \epsilon_{\mu\nu\sigma\tau}\, \mathbf{e}^{\tau}, \tag{8.63}$$

the two-forms

$$\boldsymbol{\epsilon}_{\mu\nu} \equiv \tfrac{1}{2}\, \mathbf{e}^{\sigma} \wedge \boldsymbol{\epsilon}_{\mu\nu\sigma} = \tfrac{1}{2!}\, \epsilon_{\mu\nu\sigma\tau}\, \mathbf{e}^{\sigma} \wedge \mathbf{e}^{\tau}, \tag{8.64}$$

the three-forms

$$\boldsymbol{\epsilon}_{\mu} \equiv \tfrac{1}{3}\, \mathbf{e}^{\nu} \wedge \boldsymbol{\epsilon}_{\mu\nu} = \tfrac{1}{3!}\, \epsilon_{\mu\nu\sigma\tau}\, \mathbf{e}^{\nu} \wedge \mathbf{e}^{\sigma} \wedge \mathbf{e}^{\tau}, \tag{8.65}$$

and the volume four-form

$$\boldsymbol{\epsilon} \equiv \tfrac{1}{4}\, \mathbf{e}^{\mu} \wedge \boldsymbol{\epsilon}_{\mu} = \tfrac{1}{4!}\, \epsilon_{\mu\nu\sigma\tau}\, \mathbf{e}^{\mu} \wedge \mathbf{e}^{\nu} \wedge \mathbf{e}^{\sigma} \wedge \mathbf{e}^{\tau}. \tag{8.66}$$

This cascade of forms has the following properties:

$$\mathbf{e}^{\tau} \wedge \boldsymbol{\epsilon}_{\mu\nu\sigma} = \delta_{\sigma}{}^{\tau}\, \boldsymbol{\epsilon}_{\mu\nu} + \delta_{\nu}{}^{\tau}\, \boldsymbol{\epsilon}_{\sigma\mu} + \delta_{\mu}{}^{\tau}\, \boldsymbol{\epsilon}_{\nu\sigma}, \tag{8.67}$$

$$\mathbf{e}^{\sigma} \wedge \boldsymbol{\epsilon}_{\mu\nu} = \delta_{\nu}{}^{\sigma}\, \boldsymbol{\epsilon}_{\mu} - \delta_{\mu}{}^{\sigma}\, \boldsymbol{\epsilon}_{\nu}, \tag{8.68}$$

$$\mathbf{e}^{\nu} \wedge \boldsymbol{\epsilon}_{\mu} = \delta_{\mu}{}^{\nu}\, \boldsymbol{\epsilon}. \tag{8.69}$$

8.6 Surfaces

In a three-dimensional manifold, we regard a surface-area element, $\boldsymbol{\Sigma}$, as a two-form that is defined implicitly in terms of a unit vector called the *normal* $\hat{\mathbf{n}}$ by (see, e.g., Frankel, 2004)

$$\boldsymbol{\Sigma} \equiv \mathbf{i}_{\hat{\mathbf{n}}}\boldsymbol{\epsilon} = \hat{\mathbf{n}} \cdot \boldsymbol{\epsilon}; \tag{8.70}$$

in components, we write

$$\Sigma_{jk} = \hat{n}^i \, \epsilon_{ijk}. \tag{8.71}$$

Using the fact that $\hat{\mathbf{n}}$ is a unit vector, $\|\hat{\mathbf{n}}\|^2 = \hat{n}^i \, \hat{n}_i = 1$, we find that

$$\begin{aligned}
\boldsymbol{\Sigma} \wedge \hat{\mathbf{n}} &= (\hat{\mathbf{n}} \cdot \epsilon) \wedge \hat{\mathbf{n}} \\
&= \epsilon_{123} \, (\hat{n}^1 \, dx^2 \wedge dx^3 - \hat{n}^2 \, dx^1 \wedge dx^3 \\
&\qquad + \hat{n}^3 \, dx^1 \wedge dx^2) \wedge (\hat{n}_1 \, dx^1 + \hat{n}_2 \, dx^2 + \hat{n}_3 \, dx^3) \\
&= \epsilon_{123} \, (\hat{n}^1 \, \hat{n}_1 + \hat{n}^2 \, \hat{n}_2 + \hat{n}^3 \, \hat{n}_3) \, dx^1 \wedge dx^2 \wedge dx^3 \\
&= \boldsymbol{\epsilon}.
\end{aligned} \tag{8.72}$$

In an n-dimensional manifold, we have

$$\Sigma_{i_2 \cdots i_{n-1}} = \hat{n}^{i_1} \, \epsilon_{\, i_1 \cdots i_{n-1} i_n}, \tag{8.73}$$

and we may write in general

$$\boldsymbol{\Sigma} = \hat{\mathbf{n}} \cdot \boldsymbol{\epsilon}, \qquad \boldsymbol{\epsilon} = \boldsymbol{\Sigma} \wedge \hat{\mathbf{n}}, \tag{8.74}$$

that is, a surface $(n-1)$-form is obtained from the volume n-form by contracting it with the corresponding unit vector, and the volume n-form is the wedge product of a surface $(n-1)$-form with a one-form.

In equation (5.136), we expressed a surface element spanned by two non-parallel vectors $\mathbf{u}$ and $\mathbf{w}$ as the one-form

$$\hat{\mathbf{n}} \, dS \equiv \boldsymbol{\epsilon}(\,\cdot, \mathbf{u}, \mathbf{w}), \tag{8.75}$$

such that in components

$$\hat{n}_i \, dS = \epsilon_{ijk} \, u^j \, w^k. \tag{8.76}$$

If we feed the surface two-form $\boldsymbol{\Sigma}$ the vectors $\mathbf{u}$ and $\mathbf{w}$, it returns

$$\boldsymbol{\Sigma}(\mathbf{u}, \mathbf{w}) = \hat{n}^i \, \epsilon_{ijk} \, u^j \, w^k. \tag{8.77}$$

Upon contraction of the form description of a surface (8.76) with the unit vector with contravariant components $\hat{n}^i$, we see that the surface two-form returns the area spanned by the two vectors $\mathbf{u}$ and $\mathbf{w}$:

$$dS = \hat{n}^i \, \epsilon_{ijk} \, u^j \, w^k = \boldsymbol{\Sigma}(\mathbf{u}, \mathbf{w}). \tag{8.78}$$

The metric in a surface may be determined by pulling back the metric of the corresponding volume, as follows. Let $x = \{x^1, \ldots, x^n\}$ denote a set of n coordinates in an n-dimensional manifold N, and let $x' = \{x^{1'}, \ldots, x^{(n-1)'}\}$ denote a set of $(n-1)$ coordinates in an $(n-1)$-dimensional surface S. Suppose there is a map $x = \varphi(x')$ relating points in the surface S to points in N, with gradient $\lambda^i{}_{i'} = \partial_{i'} \varphi^i$. As discussed in example 6.1 in section 6.2.1, we may pull back the metric on the manifold N to the surface S as follows

$$(\varphi_* \mathbf{g})_{i'j'} = g_{ij} \, \lambda^i{}_{i'} \, \lambda^j{}_{j'}. \tag{8.79}$$

Note that $\varphi_* \mathbf{g}$ is a $(0,2)$ tensor in the $(n-1)$-dimensional surface S, whereas $\mathbf{g}$ is a $(0,2)$ tensor in the n-dimensional manifold N. The square root of the metric in N is $\overline{g}$, whereas the square root of the metric in S is $\overline{g}'$. The volume element in N, a skewsymmetric $(0,n)$ tensor, has elements

$$\epsilon_{i_1\cdots i_n} = \overline{g}\, \underline{\epsilon}_{\,i_1\cdots i_n}, \tag{8.80}$$

whereas the volume element in S, a skewsymmetric $(0,n-1)$ tensor, has elements

$$\epsilon_{i'_1\cdots i'_{n-1}} = \overline{g}'\, \underline{\epsilon}_{\,i'_1\cdots i'_{n-1}}. \tag{8.81}$$

8.7 Exterior Derivative

The *exterior derivative* of a k-form,

$$\omega = \frac{1}{k!}\, \omega_{i_1\cdots i_k}\, \mathrm{d}x^{i_1} \wedge \cdots \wedge \mathrm{d}x^{i_k}, \tag{8.82}$$

is a $(k+1)$-form defined by

$$\mathrm{d}\omega \equiv \frac{1}{k!}\, \frac{\partial \omega_{i_1\cdots i_k}}{\partial x^{i_0}}\, \mathrm{d}x^{i_0} \wedge \mathrm{d}x^{i_1} \wedge \cdots \wedge \mathrm{d}x^{i_k}. \tag{8.83}$$

To exemplify the derivative in a three-dimensional manifold, for a scalar function f, which is a zero-form, we write

$$\mathrm{df} = \frac{\partial f}{\partial x^i}\, \mathrm{d}x^i, \tag{8.84}$$

which we recognize as a one-form and which is the standard differential.

Two successive applications of the exterior derivative result in the zero tensor,

$$\mathrm{d}(\mathrm{d}\omega) = \frac{1}{k!}\, \frac{\partial^2 \omega_{i_1\cdots i_k}}{\partial x^{j_0} \partial x^{i_0}}\, \mathrm{d}x^{j_0} \wedge \mathrm{d}x^{i_0} \wedge \mathrm{d}x^{i_1} \wedge \cdots \wedge \mathrm{d}x^{i_k} = \mathbf{0}, \tag{8.85}$$

since, by the equality of mixed partial derivatives, $\partial^2/\partial x^{j_0} \partial x^{i_0}$ is symmetric in i_0 and j_0, and $\mathrm{d}x^{j_0} \wedge \mathrm{d}x^{i_0} \wedge \mathrm{d}x^{i_1} \wedge \cdots \wedge \mathrm{d}x^{i_k}$ is, by definition (8.27), antisymmetric in these indices. We can concisely write

$$\mathrm{d}(\mathrm{d}\omega) = \mathrm{d}^2\omega = \mathbf{0}. \tag{8.86}$$

A *closed form* is a differential form ω whose exterior derivative is zero:

$$\mathrm{d}\omega = \mathbf{0}. \tag{8.87}$$

An *exact form* is a differential n-form α that is the exterior derivative of another differential $(n-1)$-form $\omega : \alpha = \mathrm{d}\omega$. Because $\mathrm{d}^2 = \mathbf{0}$, every exact form is closed. According to *Poincaré's lemma*, locally, every closed n-form α is exact; in other words, if $\mathrm{d}\alpha = \mathbf{0}$, then there exists an $(n-1)$-form ω such that $\alpha = \mathrm{d}\omega$.

Exterior derivatives are subject to a modified Leibniz rule of differentiation. If ω is a k-form and η a p-form,

$$\mathrm{d}\,(\omega \wedge \eta) = \mathrm{d}\omega \wedge \eta + (-1)^k\,(\omega \wedge \mathrm{d}\eta). \tag{8.88}$$

As a special case, if η is a scalar, f, that is a zero-form, then

$$\mathrm{d}\,(\mathrm{f}\,\omega) = \mathrm{f}\,\mathrm{d}\omega - \omega \wedge \mathrm{df} = \mathrm{df} \wedge \omega + \mathrm{f}\,\mathrm{d}\omega. \tag{8.89}$$

8.7.1 Coordinate-Free Definition

Definition (8.83) relies on the introduction of local coordinates x^i. Let us determine a definition of the exterior derivative that does not invoke a local coordinate system. First, we return to the definition of the exterior derivative of a function f, which is the one-form

$$\mathrm{df}(\mathbf{u}) \equiv \frac{\mathrm{df}}{\mathrm{d}\lambda}, \tag{8.90}$$

the derivative along the curve defined by the vector field $\mathbf{u}$, as discussed in section 4.1.2. This is a component-free definition of the exterior derivative of a function. An invariant definition of the exterior derivative of a k-form ω in terms of $k+1$ vector fields $\mathbf{u}_1, \ldots, \mathbf{u}_{k+1}$ is as follows:

$$\begin{aligned} \mathrm{d}\omega(\mathbf{u}_1, \ldots, \mathbf{u}_{k+1}) &= \sum_{i=1}^{k+1} (-1)^{i-1}\, \mathrm{d}(\omega(\mathbf{u}_1, \ldots, \widehat{\mathbf{u}}_i, \ldots, \mathbf{u}_{k+1}))(\mathbf{u}_i) \\ &+ \sum_{i<j}^{k} (-1)^{i+j}\, \omega([\mathbf{u}_i, \mathbf{u}_j], \mathbf{u}_1, \ldots, \widehat{\mathbf{u}}_i, \ldots, \widehat{\mathbf{u}}_j, \ldots, \mathbf{u}_{k+1}), \end{aligned} \tag{8.91}$$

where $[\mathbf{u}_i, \mathbf{u}_j]$ denotes the Lie bracket introduced in section 4.4, and $\widehat{\mathbf{u}}_i$ means "remove slot i." Note how we feed the exterior derivative of the k-form ω—the $k+1$-form $\mathrm{d}\omega$—$k+1$ vectors, whereas we feed the k-form ω just k vectors, one for each of their slots, thereby producing numbers. The quantity $\omega(\mathbf{u}_1, \ldots, \mathbf{u}_{i-1}, \mathbf{u}_{i+1}, \ldots, \mathbf{u}_{k+1})$ is such a number (a zero-form), and thus its exterior derivative $\mathrm{d}(\omega(\mathbf{u}_1, \ldots, \mathbf{u}_{i-1}, \mathbf{u}_{i+1}, \ldots, \mathbf{u}_{k+1}))$ is a one-form like (8.90), with one slot which we feed the vector $\mathbf{u}_i$.

As in definitions (7.57) and (7.64) of the torsion and curvature maps, respectively, the Lie bracket accommodates anholonomic bases, such as the orthonormal spherical unit vectors $\hat{\mathbf{r}}$, $\hat{\boldsymbol{\theta}}$, and $\hat{\boldsymbol{\kappa}}$, discussed in example 4.5. As a concrete example, for a one-form ω we have

$$\mathrm{d}\omega(\mathbf{u}, \mathbf{v}) = \mathrm{d}(\omega(\mathbf{v}))(\mathbf{u}) - \mathrm{d}(\omega(\mathbf{u}))(\mathbf{v}) - \omega([\mathbf{u}, \mathbf{v}]). \tag{8.92}$$

Note how the term $\omega([\mathbf{u}, \mathbf{v}])$ accommodates anholonomicity. Suppose the one-form ω is the basis coframe $\mathbf{e}^j$ as discussed in section 5.3.1. Then, using $\mathbf{e}^j(\mathbf{u}) = u^j$ and $\mathrm{d}(\mathbf{e}^j(\mathbf{v}))(\mathbf{u}) = u^i \partial_i v^j$, we have

$$\begin{aligned} \mathrm{d}\mathbf{e}^j(\mathbf{u}, \mathbf{v}) &= u^i\, \partial_i v^j - v^i\, \partial_i u^j - (u^i\, \partial_i v^j - v^i\, \partial_i u^j + u^i\, v^k\, \tau_{ik}^{\;j}) \\ &= -\,u^i\, v^k\, \tau_{ik}^{\;j}, \end{aligned} \tag{8.93}$$

where we have used the anholonomic Lie bracket (4.74). Thus we find the coordinate-free expression

$$\mathrm{d}\mathbf{e}^j = -\,\tau_{ik}^{\;j}\, \mathbf{e}^i \otimes \mathbf{e}^k = -\,\tfrac{1}{2}\, \tau_{ik}^{\;j}\, \mathbf{e}^i \wedge \mathbf{e}^k. \tag{8.94}$$

Only for a holonomic basis is $\mathrm{d}\mathbf{e}^j = \mathbf{0}$.

If the one-form $\boldsymbol{\omega}$ is the gradient of a function, df, then

$$\mathrm{d}^2\mathrm{f}(\mathbf{u},\mathbf{v}) = \mathrm{d}(\mathrm{df}(\mathbf{v}))(\mathbf{u}) - \mathrm{d}(\mathrm{df}(\mathbf{u}))(\mathbf{v}) - \mathrm{df}([\mathbf{u},\mathbf{v}]) = 0. \tag{8.95}$$

In components, this expresses an inequality of mixed partials:

$$\mathrm{d}^2\mathrm{f}(\mathbf{e}_i,\mathbf{e}_j) = (\partial_i\partial_j - \partial_j\partial_i - \tau_{ij}^{k}\,\partial_k)f = 0. \tag{8.96}$$

Example 8.2 Grad, Div, and All That[a]

In the context of Cartesian coordinates $\{x, y, z\}$ in $\mathbb{R}^3$, let us consider operations such as the exterior derivative and Hodge dual. Specifically, we examine the vector-calculus operations gradient, curl, and divergence.

Gradient

Consider a scalar-valued function, which is a zero-form. Upon taking its exterior derivative, we obtain a one-form, namely,

$$\mathrm{df} = \partial_x f\,\mathrm{d}x + \partial_y f\,\mathrm{d}y + \partial_z f\,\mathrm{d}z. \tag{8.97}$$

We recognize that the components of this one-form are the Cartesian components of the gradient of f, $\boldsymbol{\nabla}$f.

Divergence

Consider the Hodge dual of a vector:

$$*\,\mathbf{u} = u^x\,\mathrm{d}y\wedge\mathrm{d}z + u^y\,\mathrm{d}x\wedge\mathrm{d}z + u^z\,\mathrm{d}x\wedge\mathrm{d}y. \tag{8.98}$$

Upon taking its exterior derivative we obtain a three-form, namely,

$$\mathrm{d}*\mathbf{u} = \left(\partial_x u^x + \partial_y u^y + \partial_z u^z\right)\,\mathrm{d}x\wedge\mathrm{d}y\wedge\mathrm{d}z, \tag{8.99}$$

and upon taking its dual we find

$$*\,\mathrm{d}(*\mathbf{u}) = \partial_x u^x + \partial_y u^y + \partial_z u^z \equiv \mathrm{div}\,\mathbf{u}, \tag{8.100}$$

which is the divergence of $\mathbf{u}$.

Curl

Consider a one-form,

$$\boldsymbol{\omega} = \omega_x\,\mathrm{d}x + \omega_y\,\mathrm{d}y + \omega_z\,\mathrm{d}z. \tag{8.101}$$

Upon taking the exterior derivative of this one-form we obtain a two-form, namely,

$$\begin{aligned} \mathrm{d}\boldsymbol{\omega} &= \mathrm{d}\omega_x \wedge \mathrm{d}x + \mathrm{d}\omega_y \wedge \mathrm{d}y + \mathrm{d}\omega_z \wedge \mathrm{d}z \\ &= \left(\partial_y \omega_z - \partial_z \omega_y\right) \mathrm{d}y \wedge \mathrm{d}z + \left(\partial_x \omega_z - \partial_z \omega_x\right) \mathrm{d}x \wedge \mathrm{d}z \\ &\quad + \left(\partial_x \omega_y - \partial_y \omega_x\right) \mathrm{d}x \wedge \mathrm{d}y, \end{aligned} \tag{8.102}$$

which we write as $\operatorname{curl} \boldsymbol{\omega} \cdot \boldsymbol{\epsilon}$, since the vector, $\operatorname{curl} \boldsymbol{\omega}$, has components $(\epsilon^{ijk}\, \partial_j\, \omega_k) = (\partial_y\, \omega_z - \partial_z\, \omega_y, \partial_z\, \omega_x - \partial_x\, \omega_z, \partial_x\, \omega_y - \partial_y\, \omega_z)$, which we recognize to be the components of the curl.

In terms of the Hodge dual, we may write

$$* \,\mathrm{d}\boldsymbol{\omega} = \operatorname{curl} \boldsymbol{\omega}. \tag{8.103}$$

[a] A nod to Schey (2004).

8.7.2 Exact Forms

A k-form ω is said to be *exact* if there exists a $(k-1)$-form $\boldsymbol{\alpha}$ such that

$$\omega = \mathrm{d}\boldsymbol{\alpha}. \tag{8.104}$$

Thus, using the result (8.85), we deduce that the exterior derivative of an exact form is zero:

$$\mathrm{d}\omega = \mathrm{d}\mathrm{d}\boldsymbol{\alpha} = \mathbf{0}. \tag{8.105}$$

Example 8.3 Classical Vector Calculus

In this example, we draw connections between exterior calculus and classical vector calculus. As an example, consider the exterior derivative of the one-form $\boldsymbol{\omega} = \omega_j\, \mathrm{d}x^j$ in $\mathbb{R}^3$:

$$\begin{aligned} \mathrm{d}\boldsymbol{\omega} &= \partial_i \omega_j\, \mathrm{d}x^i \wedge \mathrm{d}x^j = -\,\partial_i \omega_j\, \mathrm{d}x^j \wedge \mathrm{d}x^i \\ &= \sum_{i<j} (\partial_i \omega_j - \partial_j \omega_i)\, \mathrm{d}x^i \wedge \mathrm{d}x^j. \end{aligned} \tag{8.106}$$

The components of this two-form are the components of the curl.

Now, let us suppose that the one-form $\boldsymbol{\omega}$ is exact. Then there exists a function, that is, a zero-form, ψ such that $\boldsymbol{\omega} = \mathrm{d}\psi$, and thus in components $\omega_i = \partial_i \psi$. The fact that $\mathrm{d}\boldsymbol{\omega} = \mathbf{0}$ implies, using (8.106), that

$$\operatorname{curl}(\operatorname{grad} \psi) = \mathbf{0}. \tag{8.107}$$

Next, let us assume that the exterior derivative of a one-form is nonzero, $\mathrm{d}\boldsymbol{\omega} \neq \mathbf{0}$. Then $\mathrm{dd}\boldsymbol{\omega} = \mathbf{0}$, which implies

$$\begin{aligned} \mathrm{dd}\boldsymbol{\omega} &= \sum_k \sum_{i<j} \partial_k(\partial_i\omega_j - \partial_j\omega_i)\, \mathrm{d}x^k \wedge \mathrm{d}x^i \wedge \mathrm{d}x^j \\ &= [\partial_1(\partial_2\omega_3 - \partial_3\omega_2) + \partial_2(\partial_3\omega_1 - \partial_1\omega_3) \\ &\quad + \partial_3(\partial_1\omega_2 - \partial_3\omega_2)]\, \mathrm{d}x^1 \wedge \mathrm{d}x^2 \wedge \mathrm{d}x^3 \\ &= \mathbf{0}; \end{aligned} \tag{8.108}$$

hence,

$$\mathrm{div}(\mathrm{curl}\,\boldsymbol{\omega}) = 0. \tag{8.109}$$

Example 8.4 Form-versions of Maxwell's Equations

In Box 7.13, we summarized the classical Maxwell's equations. In this box, we present a form-version of Maxwell's equations. We define the two-form *Faraday tensor*

$$\mathbf{F} = \tfrac{1}{2}\, F_{\alpha\beta}\, \mathbf{e}^{\alpha} \wedge \mathbf{e}^{\beta}, \tag{8.110}$$

combining the electric field $\mathbf{E}$ and the magnetic field $\mathbf{B}$, with components

$$(F_{\alpha\beta}) = \begin{pmatrix} 0 & -E_1/c & -E_2/c & -E_3/c \\ E_1/c & 0 & B_3 & -B_2 \\ E_2/c & -B_3 & 0 & B_1 \\ E_3/c & B_2 & -B_1 & 0 \end{pmatrix}. \tag{8.111}$$

Raising its indices with the metric of Minkowski spacetime yields

$$(F^{\mu\nu}) = \eta^{\mu\alpha}\, F_{\alpha\beta}\, \eta^{\beta\nu} = \begin{pmatrix} 0 & E^1/c & E^2/c & E^3/c \\ -E^1/c & 0 & B^3 & -B^2 \\ -E^2/c & -B^3 & 0 & B^1 \\ -E^3/c & B^2 & -B^1 & 0 \end{pmatrix}. \tag{8.112}$$

The Hodge dual of this two-form is another two-form with components

$$(*F_{\alpha\beta}) = \left(\tfrac{1}{2}\, \epsilon_{\alpha\beta\mu\nu}\, F^{\mu\nu}\right) = \begin{pmatrix} 0 & B_1/c & B_2/c & B_3/c \\ -B_1/c & 0 & E_3 & -E_2 \\ -B_2/c & -E_3 & 0 & E_1 \\ -B_3/c & E_2 & -E_1 & 0 \end{pmatrix}. \tag{8.113}$$

If we define the four-current vector

$$\mathbf{j} = j^{\alpha}\, \mathbf{e}_{\alpha} \tag{8.114}$$

with elements

$$(j^{\alpha}) = \begin{pmatrix} c\rho \\ j^1 \\ j^2 \\ j^3 \end{pmatrix}, \tag{8.115}$$

then the form-versions of Maxwell's equations are

$$\mathrm{d}\mathbf{F} = \mathbf{0}, \tag{8.116}$$

$$\mathrm{d} * \mathbf{F} = \mu_0 * \mathbf{j}, \tag{8.117}$$

and, using $\mathrm{dd} = \mathrm{d}^2 = \mathbf{0}$, the last equation implies that the four-current three-form $*\mathbf{j}$ is conserved

$$\mathrm{d} * \mathbf{j} = \mathbf{0}. \tag{8.118}$$

8.7.3 Commutativity with Pullback and Pushforward

In chapter 6, we discussed pullback and pushforward operations between manifolds. In this section, we investigate whether pullback and exterior differentiation commute. To do so, consider an n-dimensional manifold N and an m-dimensional manifold M, with corresponding coordinate systems $x = \{x^1, \ldots, x^n\}$ and $x' = \{x^{1'}, \ldots, x^{m'}\}$. We assume that there exists a map $\varphi : M \to N$ between points $\mathcal{S}'(x') \in M$ and points $\mathcal{S}(x) \in N$, $\mathcal{S} = \varphi(\mathcal{S}')$. Let $\boldsymbol{\omega}$ denote a one-form in the cotangent bundle of N

$$\boldsymbol{\omega} = \omega_i\, \mathbf{e}^i. \tag{8.119}$$

The pullback of this one-form to the cotangent bundle of M is the one-form

$$\varphi^* \boldsymbol{\omega} = \omega_i\, \lambda^i{}_{i'}\, \mathbf{e}^{i'}. \tag{8.120}$$

The exterior derivative of this one-form is a two-form in the cotangent bundle of M:

$$\mathrm{d}\varphi^* \boldsymbol{\omega} = \tfrac{1}{2}\, \partial_{i'} (\omega_i\, \lambda^i{}_{j'})\, \mathbf{e}^{i'} \wedge \mathbf{e}^{j'}. \tag{8.121}$$

On the other hand, the exterior derivative of $\boldsymbol{\omega}$ is a two-form in the cotangent bundle of N:

$$\mathrm{d}\boldsymbol{\omega} = \tfrac{1}{2}\, \partial_i \omega_j\, \mathbf{e}^i \wedge \mathbf{e}^j. \tag{8.122}$$

Its pullback is a two-form in the cotangent bundle of M:

$$\varphi^* \mathrm{d}\boldsymbol{\omega} = \tfrac{1}{2}\, (\partial_i \omega_j)\, \lambda^i{}_{i'}\, \lambda^j{}_{j'}\, \mathbf{e}^{i'} \wedge \mathbf{e}^{j'}$$

$$= \tfrac{1}{2}\,(\partial_{i'}\omega_j)\,\lambda^j{}_{j'}\,\mathbf{e}^{i'}\wedge\mathbf{e}^{j'}$$

$$= \tfrac{1}{2}\,\partial_{i'}(\omega_j\,\lambda^j{}_{j'})\,\mathbf{e}^{i'}\wedge\mathbf{e}^{j'} - \tfrac{1}{2}\,\omega_j\,(\partial_{i'}\lambda^j{}_{j'})\,\mathbf{e}^{i'}\wedge\mathbf{e}^{j'}. \tag{8.123}$$

Upon comparing expressions (8.121) and (8.123), we conclude that pullback and exterior differentiation commute,

$$\mathrm{d}\varphi^*\omega = \varphi^*\mathrm{d}\omega, \tag{8.124}$$

provided

$$\partial_{i'}\lambda^j{}_{j'} = \partial_{j'}\lambda^j{}_{i'}. \tag{8.125}$$

Since $\lambda^j{}_{i'} = \partial_{i'}\varphi^i$, and assuming that partial derivatives commute, this is the case.

In example 8.5, we consider a continuum with *defects* in the form of *dislocations* and *disclinations*. This requires the introduction of an *incompatible motion*, for which pullback and exterior differentiation do *not* commute. The concept of incompatibility in the context of defects has been extensively explored in the literature, notably by Kondo and his followers (e.g., Kondo, 1952, 1955a,b,c; Bilby et al., 1956; Bilby & Smith, 1956; Amari et al., 1961; Amari, 1962a,b; Holländer, 1960a,b,c; Nye, 1953; Mura, 1963, 1982; Kosevich, 1965; Nabarro, 1967; Noll, 1974; Sakata, 1971; Shimbo, 1971, 1981). Specifically, the concept of "no-more continuum" was introduced by Kondo (1963, 1964a,b) to describe the physics of a continuum with defects geometrically.

Example 8.5 Defects

In example 6.2, we introduced the notion of a referential manifold $\mathcal{B}$, capturing a state of mechanical equilibrium. As a result of the motion (3.12), a referential basis one-form $\mathring{\mathbf{e}}^I$ in the cotangent bundle of the referential manifold $\mathcal{B}$ at time $t=0$ evolves into a Lagrangian basis one-form $\mathbf{e}^I$ in the cotangent bundle of the spatial manifold $\mathcal{S}$ at time $t>0$. In terms of the pullback with the motion φ, we have the relationship

$$\mathring{\mathbf{e}}^I = \varphi^*\mathbf{e}^I. \tag{8.126}$$

The referential one-forms $\mathring{\mathbf{e}}^I$ in the cotangent bundle of the referential manifold $\mathcal{B}$ are suitably chosen and well-behaved, whereas the Lagrangian one-forms $\mathbf{e}^I$ in the cotangent bundle of the spatial manifold $\mathcal{S}$ may be torn and twisted as a result of an *incompatible motion*, causing *defects*, as illustrated in figure 8.3.

Similarly, let

$$\boldsymbol{\Gamma}^I_J \equiv \Gamma^I_{KJ}\,\mathbf{e}^K \tag{8.127}$$

denote Lagrangian *connection one-forms* in the spatial manifold $\mathcal{S}$.[a] These Lagrangian connection one-forms are also affected by the incompatible motion. Their pullback is

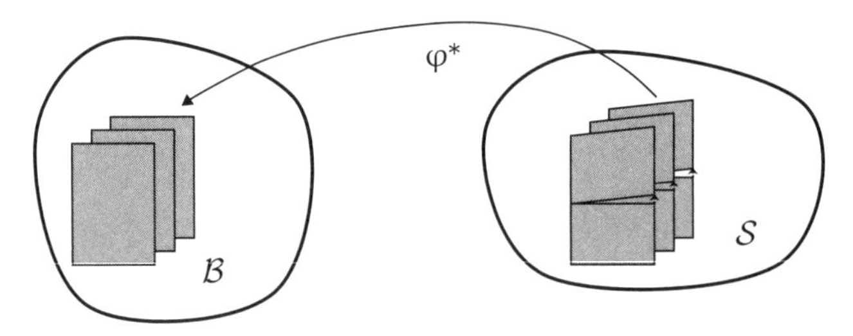

Figure 8.3: A Lagrangian one-form $\mathbf{e}^I$ in the cotangent bundle of the spatial manifold $\mathcal{S}$ twisted and torn by defects is pulled back by the motion φ to the cotangent bundle of the referential manifold $\mathcal{B}$ to yield the well-behaved referential one-forms $\mathring{\mathbf{e}}^I = \varphi^* \mathbf{e}^I$.

$$\varphi^* \mathbf{\Gamma}^I_J = \Gamma^I_{KJ}\, \mathring{\mathbf{e}}^K. \tag{8.128}$$

For a smooth motion, pullback and exterior differentiation commute, as expressed in (8.124). For an *incompatible motion* (Kleinert, 1989, 2008), partial derivatives with respect to Lagrangian coordinates X^I are noncommutable,

$$\partial_I \partial_J \varphi^i \neq \partial_J \partial_I \varphi^i, \tag{8.129}$$

or, in terms of the deformation gradient (4.26),

$$\partial_I F^i{}_J \neq \partial_J F^i{}_I. \tag{8.130}$$

This expression is a violation of requirement (8.125). For such an incompatible motion, taking the exterior derivative does *not* commute with the pullback. Instead, we write

$$\mathrm{d}\varphi^* \mathbf{e}^I = \varphi^* \mathrm{d}\mathbf{e}^I + \mathring{\boldsymbol{\alpha}}^I, \tag{8.131}$$

and

$$\mathrm{d}\varphi^* \mathbf{\Gamma}^I_J = \varphi^* \mathrm{d}\mathbf{\Gamma}^I_J + \mathring{\mathbf{\Xi}}^I_J, \tag{8.132}$$

where the two-forms

$$\mathring{\boldsymbol{\alpha}}^I = \tfrac{1}{2}\, \mathring{\alpha}_{KJ}{}^I\, \mathring{\mathbf{e}}^K \wedge \mathring{\mathbf{e}}^J \tag{8.133}$$

and

$$\mathring{\mathbf{\Xi}}^I_J = \tfrac{1}{2}\, \mathring{\Xi}_{KLJ}{}^I\, \mathring{\mathbf{e}}^K \wedge \mathring{\mathbf{e}}^L \tag{8.134}$$

capture, respectively, *dislocations* and *disclinations* in the referential manifold (see, e.g., Edelen & Lagoudas, 1988). As illustrated in figure 6.3, we may push these two-forms to the spatial manifold by "hanging" the referential components onto the Lagrangian coframe:

$$\boldsymbol{\alpha}^I = \varphi_* \mathring{\boldsymbol{\alpha}}^I = \tfrac{1}{2}\, \mathring{\alpha}_{KJ}{}^I\, \mathbf{e}^K \wedge \mathbf{e}^J, \tag{8.135}$$

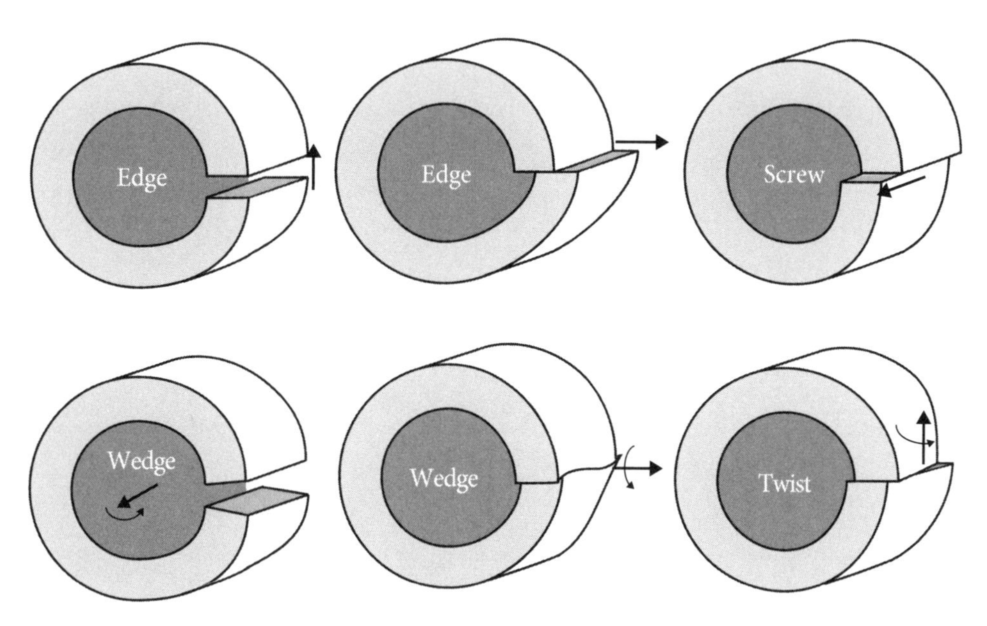

Figure 8.4: Illustrations of basic defects based on *Volterra's cut-and-weld protocols. Dislocations* correspond to an intrinsic torsion, whereas *disclinations* correspond to an intrinsic curvature. The cut parallel to the axis of the cylinder defines the *dislocation line. Top left*: Edge dislocation. The slip or *Burgers vector* (black arrow) is perpendicular to the dislocation line. The dislocation is of the form $\mathring{\alpha}_{IJ}{}^K$ where $I \neq K$ and $J \neq K$. *Top middle*: Edge dislocation. The slip/Burgers vector is perpendicular to the dislocation line. The dislocation is of the form $\mathring{\alpha}_{IJ}{}^K$ where $I \neq K$ and $J \neq K$. *Top right*: Screw dislocation. The slip/Burgers vector is tangential to the dislocation line. The dislocation is of the form $\mathring{\alpha}_{IJ}{}^K$ where $I = K$ or $J = K$. *Bottom left*: Wedge disclination. The rotation or *Frank vector* (black arrow) is tangential to the slip plane. The disclination is of the form $\mathring{\Xi}_{IJ}{}^{KL}$ where $I = K$ or $J = K$. *Bottom middle*: Wedge disclination. The rotation/Frank vector is tangential to the slip plane. The disclination is of the form $\mathring{\Xi}_{IJ}{}^{KL}$ where $I = K$ or $J = K$. *Bottom right*: Twist disclination. The rotation/Frank vector (black arrow) is perpendicular to the slip plane. The disclination is of the form $\mathring{\Xi}_{IJ}{}^{KL}$ where $I \neq K$ and $J \neq K$.

$$\Xi^I_J = \varphi_* \mathring{\Xi}^I_J = \tfrac{1}{2}\, \mathring{\Xi}_{KLJ}{}^I\, \mathbf{e}^K \wedge \mathbf{e}^L. \tag{8.136}$$

The *dislocation two-form* (8.135) is an example of a *vector-valued form*, and the *disclination two-form* (8.136) is an example of a *tensor-valued form*, as discussed in detail in section 8.9. The various types of dislocations and disclinations are illustrated in figure 8.4 based on the Volterra cut-and-weld protocols. Dislocations and disclinations are bound by a set of Bianchi identities, just like torsion and curvature in Einstein–Cartan general relativity, discussed in example 8.13 (see, e.g., Ruggiero & Tartaglia, 2003; Hehl & Obukhov, 2007; Tromp, 2025, for reviews).

[a] Connection one-forms are discussed in detail in section 8.9.3.

8.8 Lie Derivative of a Form

Using expression (7.216) on a general k-form $\boldsymbol{\alpha}$, we find that its Lie derivative is given by

$$
\begin{aligned}
(\mathcal{L}_{\mathbf{u}}\boldsymbol{\alpha})_{j_1\cdots j_k} = \partial_t\alpha_{j_1\cdots j_k} + u^m\,\partial_m\alpha_{j_1\cdots j_k} \\
+ \alpha_{m\cdots j_k}\,\partial_{j_1}u^m + \cdots + \alpha_{j_1\cdots m}\,\partial_{j_k}u^m,
\end{aligned}
\tag{8.137}
$$

which may be written in coordinate-free notation as (e.g., Marsden & Hughes, 1983; Frankel, 2004)

$$
\begin{aligned}
\mathcal{L}_{\mathbf{u}}\boldsymbol{\alpha} &= \mathrm{d}_t\boldsymbol{\alpha} + \mathbf{i}_{\mathbf{u}}\mathrm{d}\boldsymbol{\alpha} + \mathrm{d}(\mathbf{i}_{\mathbf{u}}\boldsymbol{\alpha}) \\
&= \mathrm{d}_t\boldsymbol{\alpha} + \mathbf{u}\cdot\,\mathrm{d}\boldsymbol{\alpha} + \mathrm{d}(\mathbf{u}\cdot\boldsymbol{\alpha}),
\end{aligned}
\tag{8.138}
$$

where d_t denotes the Euler time derivative (7.180).

Let us confirm formula (8.138) for the two-form

$$
\boldsymbol{\alpha} = \alpha_{ij}\,\mathbf{e}^i\otimes\mathbf{e}^j = \tfrac{1}{2}\,\alpha_{ij}\,\mathbf{e}^i\wedge\mathbf{e}^j. \tag{8.139}
$$

We have

$$
\mathbf{u}\cdot\boldsymbol{\alpha} = u^i\,\alpha_{ij}\,\mathbf{e}^j, \tag{8.140}
$$

and

$$
\begin{aligned}
\mathrm{d}(\mathbf{u}\cdot\boldsymbol{\alpha}) &= \partial_k(u^i\,\alpha_{ij})\,\mathbf{e}^k\wedge\mathbf{e}^j \\
&= \tfrac{1}{2}[\partial_i(u^k\,\alpha_{kj}) - \partial_j(u^k\,\alpha_{ki})]\,\mathbf{e}^i\wedge\mathbf{e}^j.
\end{aligned}
\tag{8.141}
$$

We also have

$$
\begin{aligned}
\mathrm{d}\boldsymbol{\alpha} &= \tfrac{1}{2}\,\partial_k\alpha_{ij}\,\mathbf{e}^k\wedge\mathbf{e}^i\wedge\mathbf{e}^j \\
&= \tfrac{1}{6}\,(\partial_k\alpha_{ij} - \partial_i\alpha_{kj} - \partial_j\alpha_{ik})\,\mathbf{e}^k\wedge\mathbf{e}^i\wedge\mathbf{e}^j,
\end{aligned}
\tag{8.142}
$$

and

$$
\mathbf{u}\cdot\mathrm{d}\boldsymbol{\alpha} = \tfrac{1}{2}\,u^k(\partial_k\alpha_{ij} - \partial_i\alpha_{kj} - \partial_j\alpha_{ik})\,\mathbf{e}^i\wedge\mathbf{e}^j. \tag{8.143}
$$

Now we find that

$$
\begin{aligned}
\mathrm{d}_t\boldsymbol{\alpha} + \mathbf{u}\cdot\,\mathrm{d}\boldsymbol{\alpha} + \mathrm{d}(\mathbf{u}\cdot\boldsymbol{\alpha}) &= \tfrac{1}{2}\,(\partial_t\alpha_{ij} + u^k\,\partial_k\alpha_{ij} - u^k\,\partial_i\alpha_{kj} - u^k\,\partial_j\alpha_{ik} \\
&\quad + \alpha_{kj}\,\partial_i u^k + u^k\,\partial_i\alpha_{kj} - \alpha_{ki}\,\partial_j u^k - u^k\,\partial_j\alpha_{ki})\,\mathbf{e}^i\wedge\mathbf{e}^j \\
&= \tfrac{1}{2}\,(\partial_t\alpha_{ij} + u^k\,\partial_k\alpha_{ij} + \alpha_{kj}\,\partial_i u^k + \alpha_{ik}\,\partial_j u^k)\,\mathbf{e}^i\wedge\mathbf{e}^j \\
&= \mathcal{L}_{\mathbf{u}}\boldsymbol{\alpha},
\end{aligned}
\tag{8.144}
$$

in agreement with (8.138).

In a spacetime formulation, combining space and time in a single spacetime manifold as discussed in example 3.3, equation (8.138) is known as the *Cartan magic formula* or *Cartan homotopy formula*

$$\mathcal{L}_{\mathbf{u}}\alpha = \mathbf{i}_{\mathbf{u}} \mathrm{d}\alpha + \mathrm{d}(\mathbf{i}_{\mathbf{u}}\alpha). \tag{8.145}$$

In this beautiful formula, the interior product, $\mathbf{i}$, relates the exterior derivative, d, and Lie derivative, $\mathcal{L}$, of differential forms.

As special cases, for an n-form α in an n-dimensional manifold we have $\mathrm{d}\alpha = 0$ and thus

$$\mathfrak{L}_{\mathbf{u}}\alpha = \mathrm{d}(\mathbf{u} \cdot \alpha), \tag{8.146}$$

whereas for a scalar function f we have $\mathbf{i}_{\mathbf{u}}\mathrm{f} = 0$ and thus

$$\mathfrak{L}_{\mathbf{u}}\mathrm{f} = \mathbf{u} \cdot \mathrm{df}. \tag{8.147}$$

Returning to expression (8.138) and letting $\alpha = \mathrm{d}\omega$ and using $\mathrm{dd}\omega = \mathbf{0}$, we find

$$\mathcal{L}_{\mathbf{u}}(\mathrm{d}\omega) = \mathrm{d}_t \mathrm{d}\omega + \mathrm{d}(\mathbf{u} \cdot \mathrm{d}\omega). \tag{8.148}$$

Using expression (8.138), we find

$$\mathbf{u} \cdot \mathrm{d}\omega = \mathcal{L}_{\mathbf{u}}\omega - \mathrm{d}_t\omega - \mathrm{d}(\mathbf{u} \cdot \omega), \tag{8.149}$$

which, upon substitution into expression (8.148) and using $\mathrm{dd}(\mathbf{u} \cdot \omega) = \mathbf{0}$, yields

$$\mathcal{L}_{\mathbf{u}}(\mathrm{d}\omega) = \mathrm{d}(\mathcal{L}_{\mathbf{u}}\omega) + \mathrm{d}_t(\mathrm{d}\omega) - \mathrm{d}(\mathrm{d}_t\omega). \tag{8.150}$$

Using the relationship between the autonomous Lie derivative and the Lie derivative expressed by equation (7.203), this implies that

$$\pounds_{\mathbf{u}}(\mathrm{d}\omega) = \mathrm{d}(\pounds_{\mathbf{u}}\omega). \tag{8.151}$$

The autonomous Lie derivative and the exterior derivative commute. The Euler derivative d_t and the exterior derivative d commute: $\mathrm{d}_t(\mathrm{d}\omega) = \mathrm{d}(\mathrm{d}_t\omega)$, as we can conclude from definition (8.83). Thus, examining expression (8.150), we establish that the Lie derivative and the exterior derivative also commute:

$$\mathcal{L}_{\mathbf{u}}(\mathrm{d}\omega) = \mathrm{d}(\mathcal{L}_{\mathbf{u}}\omega). \tag{8.152}$$

It may also be demonstrated that for two k-forms α and β and two scalars a and b

$$\mathcal{L}_{\mathbf{u}}(a\,\alpha + b\,\beta) = a\,\mathcal{L}_{\mathbf{u}}\alpha + b\,\mathcal{L}_{\mathbf{u}}\beta, \tag{8.153}$$

and

$$\mathcal{L}_{\mathbf{u}}(\alpha \wedge \beta) = (\mathcal{L}_{\mathbf{u}}\alpha) \wedge \beta + \alpha \wedge \mathcal{L}_{\mathbf{u}}\beta. \tag{8.154}$$

Finally, for a function f and a form α, we have

$$\mathcal{L}_{\mathbf{u}}(\mathrm{f}\,\alpha) = (\mathcal{L}_{\mathbf{u}}\mathrm{f})\,\alpha + \mathrm{f}\,\mathcal{L}_{\mathbf{u}}\alpha. \tag{8.155}$$

Let us consider the Lie derivative of volume forms. We use result (8.138) on a three-dimensional volume element (8.51) to write

$$\begin{aligned}\mathcal{L}_{\mathbf{u}}\epsilon &= \mathrm{d}(\mathbf{u}\cdot\epsilon)\\ &= \mathrm{d}(\bar{g}\,u^i \underline{\epsilon}_{ijk}\,\mathrm{d}x^j\otimes\mathrm{d}x^k) = \partial_\ell(\bar{g}\,u^\ell)\,\underline{\epsilon}_{ijk}\,\mathrm{d}x^i\otimes\mathrm{d}x^j\otimes\mathrm{d}x^k\\ &= \underline{g}\,\partial_\ell(\bar{g}\,u^\ell)\,\bar{g}\,\underline{\epsilon}_{ijk}\,\mathrm{d}x^i\otimes\mathrm{d}x^j\otimes\mathrm{d}x^k = \underline{g}\,\partial_\ell(\bar{g}\,u^\ell)\,\epsilon\\ &= \epsilon\,\mathrm{div}\,\mathbf{u}.\end{aligned} \tag{8.156}$$

Examining this expression, we see that the exterior derivative of the *flux*, $\mathbf{u}\cdot\epsilon$, which is the interior product (8.31) of the vector and the volume form, is related to the divergence of the vector field via

$$\mathrm{d}(\mathbf{u}\cdot\epsilon) = \epsilon\,\mathrm{div}\,\mathbf{u}. \tag{8.157}$$

8.9 Vector- and Tensor-Valued Forms

Consider a tensor with one contravariant index and k skew symmetric covariant indices:

$$\begin{aligned}\boldsymbol{\alpha} &= \alpha^j{}_{i_1\cdots i_k}\,\mathbf{e}_j\otimes\mathbf{e}^{i_1}\otimes\cdots\otimes\mathbf{e}^{i_k}\\ &= \tfrac{1}{k!}\,\alpha^j{}_{i_1\cdots i_k}\,\mathbf{e}_j\otimes\mathrm{d}x^{i_1}\wedge\ldots\wedge\mathrm{d}x^{i_k}\\ &\equiv \mathbf{e}_j\otimes\boldsymbol{\alpha}^j.\end{aligned} \tag{8.158}$$

The entity

$$\boldsymbol{\alpha}^j \equiv \tfrac{1}{k!}\,\alpha^j{}_{i_1\cdots i_k}\,\mathrm{d}x^{i_1}\wedge\ldots\wedge\mathrm{d}x^{i_k} \tag{8.159}$$

is a *vector-valued k-form* in the sense that

$$\boldsymbol{\alpha}^j(\mathbf{u}_1,\ldots,\mathbf{u}_k) = \alpha^j{}_{i_1\cdots i_k}\,u^{i_1}\;\cdots\;u^{i_k}, \tag{8.160}$$

which means that if fed k vectors, $\mathbf{u}_1$, . . ., $\mathbf{u}_k$, thereby filling all the vector slots, $\boldsymbol{\alpha}^j$ returns a vector—its component j—not a number. Trivially, a vector-valued zero-form, scalar α^i, defines a vector $\boldsymbol{\alpha}$:

$$\boldsymbol{\alpha} = \mathbf{e}_i\,\alpha^i. \tag{8.161}$$

As a three-dimensional example, consider the $(1,2)$ tensor $\mathbf{S}$ derived from the volume form ϵ:

$$\mathbf{S} = \epsilon^k{}_{ij}\,\mathbf{e}_k\otimes\mathbf{e}^i\otimes\mathbf{e}^j = \tfrac{1}{2}\,\epsilon^k{}_{ij}\,\mathbf{e}_k\otimes\mathrm{d}x^i\wedge\mathrm{d}x^j \equiv \mathbf{e}_k\otimes\mathbf{S}^k, \tag{8.162}$$

thereby defining a *vector-valued two-form*, $\mathbf{S}$. If fed two vectors, $\mathbf{u}$ and $\mathbf{w}$, the (pseudo)vector-valued two-form $\mathbf{S}^k$ returns the k-component of the cross product between the two vectors:

$$\mathbf{S}^k(\mathbf{u},\mathbf{w}) = (\mathbf{u}\times\mathbf{w})^k. \tag{8.163}$$

One can readily generalize the notion of vector-valued forms to *tensor-valued forms*. For example, the tensor

$$\boldsymbol{\alpha} \equiv \mathbf{e}_i \otimes \mathbf{e}^j \otimes \boldsymbol{\alpha}^i_j, \tag{8.164}$$

defines a $(1, 1)$ tensor-valued k-form,

$$\boldsymbol{\alpha}^i_j \equiv \frac{1}{k!} \alpha^i{}_{ji_1 \cdots i_k} \, \mathrm{d}x^{i_1} \wedge \ldots \wedge \mathrm{d}x^{i_k}, \tag{8.165}$$

in the sense

$$\boldsymbol{\alpha}^i_j(\mathbf{u}_1, \ldots, \mathbf{u}_k) = \alpha^i{}_{ji_1 \cdots i_k} \, u^{i_1} \cdots u^{i_k}, \tag{8.166}$$

that is, if fed k vectors it returns the components of a $(1, 1)$ tensor.

8.9.1 Transformations of Tensor-Valued Forms

Consider a tensor-valued form expressed in a coframe basis:

$$\boldsymbol{\alpha}^i_j \equiv \frac{1}{k!} \alpha^i{}_{ji_1 \cdots i_k} \, \mathbf{e}^{i_1} \wedge \ldots \wedge \mathbf{e}^{i_k}. \tag{8.167}$$

If we change the coframe basis, we may express it on the new basis as

$$\boldsymbol{\alpha}^i_j = \frac{1}{k!} \alpha^i{}_{ji'_1 \cdots i'_k} \, \mathbf{e}^{i'_1} \wedge \ldots \wedge \mathbf{e}^{i'_k}, \tag{8.168}$$

where the mixed components of this tensor-valued form are defined as

$$\alpha^i{}_{ji'_1 \cdots i'_k} \equiv \lambda^{i_1}{}_{i'_1} \cdots \lambda^{i_k}{}_{i'_k} \, \alpha^i{}_{ji_1 \cdots i_k}. \tag{8.169}$$

Equations (8.167) and (8.168) represent the *same* tensor-valued form, just expressed in different coframes.

Example 8.6 Stress as a Vector-Valued Two-Form

In examples 6.5 and 6.6, we discussed various descriptions of stress in continuum mechanics based on maps between the spatial and referential manifolds. In this example, we show that all these descriptions of stress may be captured in terms of asingle vector-valued two-form. It is important to represent stress as a *form* because only forms can be integrated. This is due to the generalization of Stokes's theorem, explained in detail in section 8.11. Integrations are necessary to consider the balance of linear and angular momentum, as illustrated in examples 8.20 and 8.21. With this objective, we introduce the Cauchy stress as a vector-valued (pseudo) two-form (e.g., Hehl & McCrea, 1986):

$$\boldsymbol{\sigma}^i \equiv \frac{1}{2} \sigma^{ij} \epsilon_{jk\ell} \, \mathbf{e}^k \wedge \mathbf{e}^\ell \qquad = \frac{1}{2} \sigma^i_{jk} \, \mathbf{e}^j \wedge \mathbf{e}^k, \tag{8.170}$$

where we have defined the components of the Eulerian vector-valued *Cauchy stress two-forms* as

$$\sigma^i_{jk} \equiv \sigma^{i\ell}\, \epsilon_{\ell jk}. \tag{8.171}$$

Note that the elements σ^i_{jk} of these vector-valued two-forms have three vector components i, and, because of the antisymmetry in their lower two indices, there are three independent combinations of j and k, for a total of nine independent components, which reflect the nine components of the Cauchy stress σ^{ij}. Using property (5.155), the inverse of relationship (8.171) is

$$\sigma^{i\ell} = \tfrac{1}{2}\, \epsilon^{\ell jk}\, \sigma^i_{jk}. \tag{8.172}$$

The Eulerian "vector-value" of the two-form (8.170) is signaled by the contravariant index i, and it is a pseudoform because the volume form is.

If we take the wedge product of the Cauchy stress two-form (8.170) with a coframe $\mathbf{e}^j$, we obtain

$$\begin{aligned} \boldsymbol{\sigma}^i \wedge \mathbf{e}^j &= \tfrac{1}{2}\, \sigma^{im}\, \epsilon_{mk\ell}\, \mathbf{e}^k \wedge \mathbf{e}^\ell \wedge \mathbf{e}^j \\ &= \sigma^{ik}\, \delta_k{}^j\, \boldsymbol{\epsilon} \\ &= \sigma^{ij}\, \boldsymbol{\epsilon}, \end{aligned} \tag{8.173}$$

where, in the second equality, we used one of the properties of the three-dimensional volume form, summarized (in four dimensions) in example 8.1. Symmetry of the Cauchy stress, $\sigma^{ij} = \sigma^{ji}$, is thus captured by the requirement

$$\boldsymbol{\sigma}^i \wedge \mathbf{e}^j = \boldsymbol{\sigma}^j \wedge \mathbf{e}^i. \tag{8.174}$$

If we change the underlying Eulerian form basis to a Lagrangian basis, we find

$$\boldsymbol{\sigma}^i = \tfrac{1}{2}\, \sigma^i_{JK}\, \mathbf{e}^J \wedge \mathbf{e}^K, \tag{8.175}$$

where we have defined the components of the Eulerian vector-valued Lagrangian two-forms as

$$\begin{aligned} \sigma^i_{JK} &\equiv F^j{}_J\, F^k{}_K\, \sigma^i_{jk} \\ &= \sigma^{ij}\, (F^{-1})^I{}_j\, \epsilon_{IJK} \\ &= J^{-1}\, P^{iI}\, \epsilon_{IJK}. \end{aligned} \tag{8.176}$$

In the last equality, we used relationship (6.50) between the Cauchy and first Piola–Kirchhoff stresses. Expression (8.176) is an example of a mixed component tensor (8.168).

As discussed in example 5.14, an oriented surface in the spatial manifold spanned by two vectors **u** and **w** is given in terms of a one-form with elements

$$\hat{n}_i\, dS = \epsilon_{ijk}\, u^j\, w^k. \tag{8.177}$$

When "fed" a surface spanned by two vectors **u** and **w** in the spatial manifold, the two-form $\boldsymbol{\sigma}^i$ returns the ith component of the *traction* on this spatial surface:

$$\boldsymbol{\sigma}^i(\mathbf{u},\mathbf{w}) = \sigma^i_{jk}\, u^j\, w^k = \sigma^{ij}\, \epsilon_{jk\ell}\, u^k\, w^\ell = \sigma^{ij}\, \hat{n}_j\, dS. \tag{8.178}$$

Therefore, we also refer to $\boldsymbol{\sigma}^i$ as the *traction two-forms*, which live in the spatial manifold. The three traction two-forms consist of one *normal traction two-form*, $\boldsymbol{\sigma}^i\, \hat{n}_i$, which captures the normal stress on the oriented surface, and two shear traction two-forms, $\boldsymbol{\sigma}^j\, (\delta^i{}_j - \hat{n}^i\, \hat{n}_j)$, which capture the shear stresses on the oriented surface.

If we define the coordinate-free $(1,2)$ *traction tensor* as

$$\boldsymbol{\sigma} \equiv \boldsymbol{\sigma}^i \otimes \mathbf{e}_i, \tag{8.179}$$

then, when fed two vectors **u** and **w**, this tensor returns the traction vector

$$\boldsymbol{\sigma}(\mathbf{u},\mathbf{w},\,\cdot\,) \equiv \boldsymbol{\sigma}^i(\mathbf{u},\mathbf{w}) \otimes \mathbf{e}_i = \sigma^{ij}\, \hat{n}_j\, dS\, \mathbf{e}_i, \tag{8.180}$$

as illustrated in figure 8.5.

The corresponding oriented surface spanned by two vectors $\mathring{\mathbf{u}} = \varphi^*\mathbf{u}$ and $\mathring{\mathbf{w}} = \varphi^*\mathbf{w}$ in the referential manifold is given in terms of the one-form

$$\mathring{\hat{n}}_I\, d\mathring{S} = \mathring{\epsilon}_{IJK}\, u^J\, w^k, \tag{8.181}$$

where we have used the fact that $\mathring{u}^J = u^J$ and $\mathring{w}^K = w^k$, as discussed in the caption of figure 6.3.

Nanson's relation (6.36) connects oriented surfaces in the spatial and referential manifolds, and this relationship motivates the definition of the first Piola–Kirchhoff stress as an Eulerian vector-valued (pseudo) two-form in the referential manifold:

$$\mathring{\mathbf{P}}^i \equiv \tfrac{1}{2}\, P^{iJ}\, \mathring{\epsilon}_{JKL}\, \mathring{\mathbf{e}}^K \wedge \mathring{\mathbf{e}}^L. \tag{8.182}$$

To signal that the form (8.182) is defined in the referential manifold, we use a "$^\circ$." When "fed" a referential surface (8.181) spanned by two vectors $\mathring{\mathbf{u}}$ and $\mathring{\mathbf{w}}$, the two-form $\mathring{\mathbf{P}}^i$ returns the ith component of the traction on the corresponding spatial surface:

$$\mathring{\mathbf{P}}^i(\mathring{\mathbf{u}},\mathring{\mathbf{w}}) = P^{iJ}\, \mathring{\epsilon}_{JKL}\, u^K\, w^L. \tag{8.183}$$

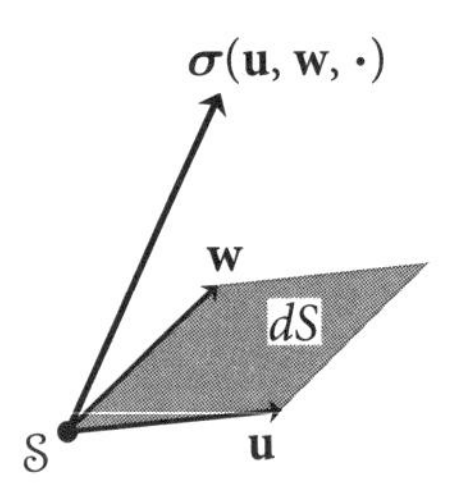

Figure 8.5: In the tangent space at point $\mathcal{S}$ in the spatial manifold, two nonparallel vectors $\mathbf{u}$ and $\mathbf{w}$ define a surface. The action of the stress tensor on this surface returns the traction vector: $\boldsymbol{\sigma}(\mathbf{u}, \mathbf{w}, \cdot) \equiv \boldsymbol{\sigma}^i(\mathbf{u}, \mathbf{w}) \otimes \mathbf{e}_i = \sigma^{ij}\, \epsilon_{jk\ell}\, u^k\, w^\ell\, \mathbf{e}_i = \sigma^{ij}\, \hat{n}_j\, dS\, \mathbf{e}_i$, where $\hat{n}_i\, dS = \epsilon_{ijk}\, u^j\, w^k$. The three traction two-forms consist of one normal traction two-form, $\boldsymbol{\sigma}^i\, \hat{n}_i$, which captures the normal stress on the oriented surface with normal $\hat{n}_i$ and area dS, and two shear traction two-forms, $\boldsymbol{\sigma}^j\, (\delta^i{}_j - \hat{n}^i\, \hat{n}_j)$, which capture the shear stresses on the oriented surface. Traction must be represented as a *two-form* because, according to the generalization of Stokes's theorem, only two-forms can be integrated over a surface, for example, in the balance of linear momentum.

Therefore, we refer to $\mathring{\mathbf{P}}^i$ as the vector-valued *first Piola–Kirchhoff stress two-forms*, which takes two vectors in the *referential* manifold and returns the ith component of the traction on the corresponding *spatial* surface element. Thus, the first Piola–Kirchhoff stress two-form exhibits its "two-point" nature. In terms of the pull-back, using relationship (6.31) between the spatial and referential volume forms, we have

$$\mathring{\mathbf{P}}^i = \varphi^* \boldsymbol{\sigma}^i, \tag{8.184}$$

in which the pullback is on the covariant components of the Cauchy stress two-form and does not involve the ith contravariant Eulerian component.

A much more appealing view of the first Piola–Kirchhoff stress may be obtained by expressing (8.182) in the form

$$\mathring{\mathbf{P}}^i = \tfrac{1}{2}\, P^i_{JK}\, \mathring{\mathbf{e}}^J \wedge \mathring{\mathbf{e}}^K, \tag{8.185}$$

where we have defined the components of Eulerian vector-valued first Piola–Kirchhoff two-forms as

$$\begin{aligned} P^i_{JK} &\equiv P^{iL}\, \mathring{\epsilon}_{LJK} \\ &= J^{-1}\, p^{iL}\, \epsilon_{LJK}, \end{aligned} \tag{8.186}$$

In the last equality we used relationship (6.31) between the referential and spatial volume forms. Upon comparing (8.186) to expression (8.176) for the Eulerian vector-valued components of the Cauchy stress two-forms, we realize that they are the same:

$$P^i_{JK} = \sigma^i_{JK}. \tag{8.187}$$

Thus, the Cauchy stress is an Eulerian valued two-form which may be expressed in either an Eulerian or Lagrangian form basis as

$$\begin{aligned}\boldsymbol{\sigma}^i &= \tfrac{1}{2}\,\sigma^i_{jk}\,\mathbf{e}^j \wedge \mathbf{e}^k \\ &= \tfrac{1}{2}\,P^i_{JK}\,\mathbf{e}^J \wedge \mathbf{e}^K.\end{aligned} \tag{8.188}$$

Notice how this view of the Cauchy stress as an Eulerian vector-valued two-form expressible in either Eulerian or Lagrangian components does not require the introduction of a referential manifold. The components of the first Piola–Kirchhoff stress naturally appear when the two-form is expressed in Lagrangian components: $\sigma^i_{JK} = P^i_{JK}$. Thus, expression (8.188) combines the traditional Cauchy and first Piola–Kirchhoff stresses in a single entity, $\boldsymbol{\sigma}^i$.

Finally, we view the second Piola–Kirchhoff stress as a vector-valued (pseudo) two-form in the referential manifold:

$$\mathring{\mathbf{S}}^I \equiv \tfrac{1}{2}\,\mathring{S}^{IJ}\,\mathring{\epsilon}_{JKL}\,\mathring{\mathbf{e}}^K \wedge \mathring{\mathbf{e}}^L. \tag{8.189}$$

When "fed" a referential surface spanned by two vectors $\mathring{\mathbf{u}}$ and $\mathring{\mathbf{w}}$, the two-form $\mathbf{S}^I$ returns the Ith component of traction on the referential surface:

$$\mathring{\mathbf{S}}^I(\mathring{\mathbf{u}}, \mathring{\mathbf{w}}) = \mathring{S}^{IJ}\,\mathring{\epsilon}_{JKL}\,u^K\,w^L. \tag{8.190}$$

Therefore, we refer to $\mathring{\mathbf{S}}^I$ as the vector-valued *second Piola–Kirchhoff stress two-form*, which takes two vectors in the referential manifold and returns the Ith component of the traction on this referential surface element. Thus, the second Piola–Kirchhoff stress two-form lives entirely in the referential manifold.

The coordinate-free second Piola–Kirchhoff stress tensor may be defined as

$$\mathring{\mathbf{S}} \equiv \mathring{\mathbf{S}}^I \otimes \mathring{\mathbf{e}}_I. \tag{8.191}$$

Using the pullback, we find the relationship

$$\mathring{\mathbf{S}} = \varphi^* \boldsymbol{\sigma}. \tag{8.192}$$

Compared to relationship (6.44) between the second Piola–Kirchhoff stress and the Cauchy stress, we do not need a factor J, because the pullback of the spatial volume form involves precisely this factor, as expressed in equation (6.31).

Again, we may obtain a much more appealing view of the second Piola–Kirchhoff stress by expressing it in the form

$$\mathring{\mathbf{S}}^I = \tfrac{1}{2}\,\mathring{S}^I_{JK}\,\mathring{\mathbf{e}}^J \wedge \mathring{\mathbf{e}}^K. \tag{8.193}$$

Here we have defined the components of the Lagrangian vector-valued second Piola–Kirchhoff stress two-forms as

$$\begin{aligned} \mathring{S}^I_{JK} &\equiv \mathring{S}^{IL}\,\mathring{\epsilon}_{LJK} \\ &= (F^{-1})^I{}_i\,P^{iL}\,\mathring{\epsilon}_{LJK} \\ &= (F^{-1})^I{}_i\,P^i_{JK}, \end{aligned} \tag{8.194}$$

where in the second equality we used relationship (6.50) between the first and second Piola–Kirchhoff stresses, and in the last equality, we used (8.186). We conclude that the second Piola–Kirchhoff stress is just the Cauchy stress expressed as Lagrangian vector-valued two-form:

$$\begin{aligned} \boldsymbol{\sigma}^I &= (F^{-1})^I{}_i\,\boldsymbol{\sigma}^i \\ &= \tfrac{1}{2}\,(F^{-1})^I{}_i\,P^i_{JK}\,\mathbf{e}^J\wedge\mathbf{e}^K \\ &= \tfrac{1}{2}\,\mathring{S}^I_{JK}\,\mathbf{e}^J\wedge\mathbf{e}^K. \end{aligned} \tag{8.195}$$

Notice how this view of the Cauchy stress as a Lagrangian vector-valued two-form does not require the introduction of a referential manifold. The components of the second Piola–Kirchhoff stress naturally appear when the Cauchy stress two-form is expressed as a Lagrangian vector-valued two-form: $\sigma^I_{JK}=\mathring{S}^I_{JK}$. The symmetry of the second Piola–Kirchhoff stress is captured by the expression

$$\boldsymbol{\sigma}^I\wedge\mathbf{e}^J=\boldsymbol{\sigma}^J\wedge\mathbf{e}^I, \tag{8.196}$$

which is just a Lagrangian version of (8.173).

In summary, the Eulerian and Lagrangian versions of the Cauchy stress two-forms naturally combine the traditional Cauchy and first and second Piola–Kirchhoff stresses in a single entity without the need for a referential manifold. Specifically, the elements σ^i_{jk} represent the Cauchy stress, the elements $\sigma^i_{JK}=P^i_{JK}$ represent the first Piola–Kirchhoff stress, and the elements $\sigma^I_{JK}=\mathring{S}^I_{JK}$ represent the second Piola–Kirchhoff stress.

Example 8.7 Strain as a One-Form-Valued Two-Vector

In example 6.4, we discussed various descriptions of strain based on maps between the spatial and referential manifolds of continuum mechanics. In this example, we formulate a description of strain in terms of two-vectors.

Thermodynamically, stress and strain form a *conjugate pair*. To capture conservation of linear momentum geometrically, as discussed in example 8.20, we must express the Cauchy stress as a *two-form*, representing traction on a surface, as discussed in example 8.6. This is because only *forms* can be integrated, as explained in section 8.11. Thus, strain must be expressed as a *two-vector*, making its Hodge dual, a one-form, conjugate to stress, as demonstrated in this example. Based on these considerations, we seek to express a generic strain tensor in terms of one-form-valued two-vectors, which are discussed in section 8.3. Specifically, we introduce strain as a one-form-valued two-vector:

$$\begin{aligned}\mathbf{E}_i &\equiv \tfrac{1}{2}\, E_{ij}\, \epsilon^{jk\ell}\, \mathbf{e}_k \wedge \mathbf{e}_\ell \\ &= \tfrac{1}{2}\, E_i^{jk}\, \mathbf{e}_j \wedge \mathbf{e}_k,\end{aligned} \tag{8.197}$$

where we have defined the components of the Eulerian one-form-valued *strain two-vectors* as

$$E_i^{jk} \equiv E_{i\ell}\, \epsilon^{\ell jk}. \tag{8.198}$$

Note that the elements E_i^{jk} of these one-form-valued two-vectors have three one-form components i, and, because of the antisymmetry in their upper two indices, there are three independent combinations of j and k, for a total of nine independent components, which reflect the nine components of the strain E_{ij}. Using property (5.155), the inverse of relationship (8.198) is

$$E_{i\ell} = \tfrac{1}{2}\, \epsilon_{\ell jk}\, E_i^{jk}. \tag{8.199}$$

The Eulerian "one-form-value" of the two-vector (8.197) is signaled by the covariant index i.

If we take the wedge product of the strain two-vector (8.197) with a frame $\mathbf{e}_j$, we obtain

$$\begin{aligned}\mathbf{E}_i \wedge \mathbf{e}_j &= \tfrac{1}{2}\, E_{im}\, \epsilon^{mk\ell}\, \mathbf{e}_k \wedge \mathbf{e}_\ell \wedge \mathbf{e}_j \\ &= E_{ik}\, \delta^k{}_j\, \boldsymbol{\epsilon} \\ &= E_{ij}\, \boldsymbol{\epsilon},\end{aligned} \tag{8.200}$$

where, in the second equality, we used one of the properties of the three-dimensional three-vector. Symmetry of the strain, $E_{ij} = E_{ji}$, is thus captured by the requirement

$$\mathbf{E}_i \wedge \mathbf{e}_j = \mathbf{E}_j \wedge \mathbf{e}_i. \tag{8.201}$$

If we change the underlying Eulerian vector basis to a Lagrangian basis, we find

$$\mathbf{E}_i = \tfrac{1}{2}\, E_i^{JK}\, \mathbf{e}_J \wedge \mathbf{e}_K, \tag{8.202}$$

where we have defined the components of the Eulerian one-form-valued Lagrangian two-vector as

$$E_i^{JK} \equiv (F^{-1})^J{}_j\, (F^{-1})^K{}_k\, E_i^{jk}. \tag{8.203}$$

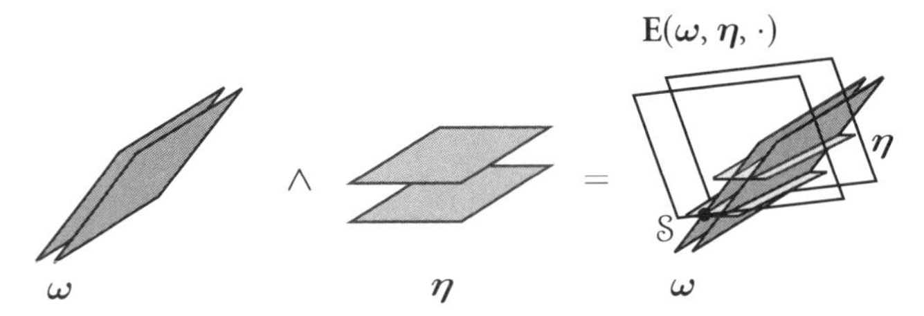

Figure 8.6: In the cotangent space at point $\mathbb{S}$ in the spatial manifold, two nonparallel one-forms $\boldsymbol{\omega}$ and $\boldsymbol{\eta}$ define a honeycomb structure $\boldsymbol{\omega}\wedge\boldsymbol{\eta}$ consisting of "circulation tubes." The action of the strain tensor on a tube returns the strain one-form: $\mathbf{E}(\boldsymbol{\omega},\boldsymbol{\eta},\,\cdot\,)\equiv\mathbf{E}_i(\boldsymbol{\omega},\boldsymbol{\eta})\otimes\mathbf{e}^i=E_{ij}\,\epsilon^{jk\ell}\,\omega_k\,\eta_\ell\,\mathbf{e}^i=E_{ij}\,\hat{n}^j\,dS\,\mathbf{e}^i$, where $\hat{n}^i\,dS=\epsilon^{ijk}\,\omega_j\,\eta_k$. The three strain two-vectors $\mathbf{E}_i$ consist of one *normal strain two-vector*, $\mathbf{E}_i\,\hat{n}^i$, which captures the normal strain on the tube with normal $\hat{n}^i$ and area dS, and two shear strain two-vectors, $\mathbf{E}_j\,(\delta_i{}^j-\hat{n}_i\,\hat{n}^j)$, which capture the shear strains on the tube.

As illustrated in figure 8.6, a circulation tube in the spatial manifold defined by two one-forms $\boldsymbol{\omega}$ and $\boldsymbol{\eta}$ is given in terms of a vector with elements

$$\hat{n}^i\,dS=\epsilon^{ijk}\,\omega_j\,\eta_k. \tag{8.204}$$

When "fed" a circulation tube defined by two one-forms $\boldsymbol{\omega}$ and $\boldsymbol{\eta}$ in the spatial manifold, the two-vector $\mathbf{E}_i$ returns the ith component of the *strain* on this tube:

$$\mathbf{E}_i(\boldsymbol{\omega},\boldsymbol{\eta})=E_i^{\,jk}\,\omega_j\,\eta_k=E_{ij}\,\epsilon^{jk\ell}\,\omega_k\,\eta_\ell=E_{ij}\,\hat{n}^j\,dS. \tag{8.205}$$

Therefore, we also refer to $\mathbf{E}_i$ as the *strain two-vectors*. The three strain two-vectors consist of one *normal strain two-vector*, $\mathbf{E}_i\,\hat{n}^i$, which captures the normal strain on the tube, and two shear strain two-vectors, $\mathbf{E}_j\,(\delta_i{}^j-\hat{n}_i\,\hat{n}^j)$, which capture the shear strains on the tube.

If we define the coordinate-free $(2,1)$ *strain tensor* as

$$\mathbf{E}\equiv\mathbf{E}_i\otimes\mathbf{e}^i, \tag{8.206}$$

then, when fed two one-forms $\boldsymbol{\omega}$ and $\boldsymbol{\eta}$, this tensor returns the strain one-form

$$\mathbf{E}(\boldsymbol{\omega},\boldsymbol{\eta},\,\cdot\,)\equiv\mathbf{E}_i(\boldsymbol{\omega},\boldsymbol{\eta})\otimes\mathbf{e}^i=E_{ij}\,\hat{n}^j\,dS\,\mathbf{e}^i, \tag{8.207}$$

as illustrated in figure 8.6.

Using definition (8.42), the Hodge duals of the strain two-vectors (8.197) are the *strain one-forms*

$$*\mathbf{E}_i=\tfrac{1}{2}\,\epsilon_{\ell jk}\,E_i^{\,jk}\,\mathbf{e}^\ell=E_{ij}\,\mathbf{e}^j, \tag{8.208}$$

where in the second equality we used relationship (8.199). The one-forms (8.208) are an alternative representation of strain.

8.9.2 Operations on Tensor-Valued Forms

One can perform several operations on tensor-valued forms that extend operations on ordinary forms. Here, we discuss the pullback and wedge product of these objects.[7]

Pullback

The pullback of a tensor-valued differential form $\boldsymbol{\alpha}$ is defined in the usual manner, expressed by equation (6.15), namely

$$(\varphi^* \boldsymbol{\alpha})(\mathbf{u}_1, \ldots, \mathbf{u}_k) = \boldsymbol{\alpha}(\varphi_* \mathbf{u}_1, \ldots, \varphi_* \mathbf{u}_k). \tag{8.209}$$

Exterior Product

The exterior or wedge product between two tensor-valued forms is the same as the wedge product between ordinary forms, defined by equation (8.24), except that the multiplication of the components is replaced by the tensor product. Specifically, let $\boldsymbol{\omega}$ be a tensor-valued k-form and $\boldsymbol{\eta}$ a tensor-valued ℓ-form. Their exterior product is a tensor-valued $(k+\ell)$-form given by

$$\boldsymbol{\omega} \wedge \boldsymbol{\eta} = \frac{1}{k!\,\ell!}\, \omega_{i_1 \cdots i_k}\, \eta_{j_1 \cdots j_\ell}\, \mathbf{e}^{i_1} \wedge \cdots \wedge \mathbf{e}^{i_k} \wedge \mathbf{e}^{j_1} \wedge \cdots \wedge \mathbf{e}^{j_\ell}. \tag{8.210}$$

As usual, the operation of multiplication is distributive, associative, and (anti)commutative. Concerning the latter, if $\boldsymbol{\omega}$ is a k-form and $\boldsymbol{\eta}$ is an ℓ-form, then

$$\boldsymbol{\omega} \wedge \boldsymbol{\eta} = (-1)^{k\ell}\, \boldsymbol{\eta} \wedge \boldsymbol{\omega}. \tag{8.211}$$

Contraction

We may contract tensor-valued forms with other tensors. For example, for a $(2,0)$ tensor, $\mathbf{T}$, and a vector-valued two-form, $\boldsymbol{\omega}^i$, we define their double-dot product,

$$\mathbf{T} : \boldsymbol{\omega}^i = T^{jk}\, \omega^i_{jk}, \tag{8.212}$$

which are the components of a vector.

Example 8.8 Form Version of Hooke's law

In example 8.6, we introduced the traction two-form $\boldsymbol{\sigma} = \boldsymbol{\sigma}^i \otimes \mathbf{e}_i$, and in example 8.7 we introduced the strain two-vector $\mathbf{E} = \mathbf{E}_i \otimes \mathbf{e}^i$. The generic component form of Hooke's law relates the Cauchy stress to a measure of strain:

$$\sigma^{ij} = c^{ijk\ell}\, E_{k\ell}. \tag{8.213}$$

[7] In this section, the specific tensor values of a form are omitted but may be general.

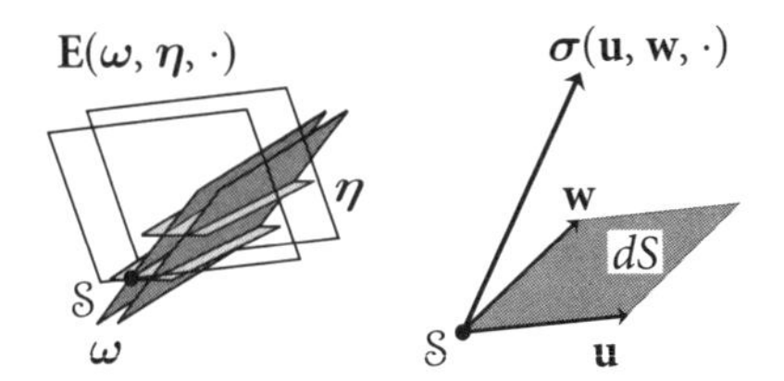

Figure 8.7: *Left*: In the cotangent space at point $\mathcal{S}$ in the spatial manifold, two nonparallel forms $\boldsymbol{\omega}$ and $\boldsymbol{\eta}$ define a circulation tube. The action of the strain tensor on this circulation tube returns the strain one-form: $\mathbf{E}(\boldsymbol{\omega},\boldsymbol{\eta},\,\cdot\,) \equiv \mathbf{E}_i(\boldsymbol{\omega},\boldsymbol{\eta}) \otimes \mathbf{e}^i = E_{ij}\,\hat{n}^j\,dS\,\mathbf{e}^i$. *Right*: In the tangent space at point $\mathcal{S}$ in the spatial manifold, two nonparallel vectors $\mathbf{u}$ and $\mathbf{w}$ define a surface. The action of the stress tensor on this surface returns the traction vector: $\boldsymbol{\sigma}(\mathbf{u},\mathbf{w},\,\cdot\,) \equiv \boldsymbol{\sigma}^i(\mathbf{u},\mathbf{w}) \otimes \mathbf{e}_i = \sigma^{ij}\,\hat{n}_j\,dS\,\mathbf{e}_i$. Geometrically, when we feed the elastic tensor, $\mathbf{c}$, a strained circulation tube, $\mathbf{E}$, in the spatial manifold at location $\mathcal{S}$, it returns the traction vector, $\boldsymbol{\sigma}$, at that location, as expressed in Hooke's law (8.218).

To express Hooke's law as a relationship between two-forms and two-vectors, we introduce the fourth-order tensor

$$\begin{aligned} \mathbf{c}^{ij} &\equiv \tfrac{1}{8}\, c^{ikjn}\, \epsilon_{k\ell m}\, \epsilon_{npq}\, \mathbf{e}^\ell \wedge \mathbf{e}^m \otimes \mathbf{e}^p \wedge \mathbf{e}^q \\ &= \tfrac{1}{8}\, c^{ij}{}_{k\ell mn}\, \mathbf{e}^k \wedge \mathbf{e}^\ell \otimes \mathbf{e}^m \wedge \mathbf{e}^n, \end{aligned} \tag{8.214}$$

where we have defined the components

$$c^{ij}{}_{k\ell mn} \equiv c^{ipjq}\, \epsilon_{pk\ell}\, \epsilon_{qmn}. \tag{8.215}$$

Now we define the coordinate-free $(2,4)$ elastic tensor as

$$\mathbf{c} \equiv \mathbf{e}_i \otimes \mathbf{c}^{ij} \otimes \mathbf{e}_j. \tag{8.216}$$

Then, the form version of Hooke's law is

$$\boldsymbol{\sigma}^i = \mathbf{c}^{ij} : \mathbf{E}_j, \tag{8.217}$$

or, contracting the three slots of the $(2,1)$ strain tensor $\mathbf{E}$ with the last three slots of the $(2,4)$ elastic tensor $\mathbf{c}$,

$$\boldsymbol{\sigma} = \mathbf{c} \,\vdots\, \mathbf{E}. \tag{8.218}$$

Geometrically, when we feed the elastic tensor, $\mathbf{c}$, a strained circulation tube, $\mathbf{E}$, in the spatial manifold at location $\mathcal{S}$, it returns the traction vector, $\boldsymbol{\sigma}$, at that location, as illustrated in Figure 8.7.

If the stress tensor is symmetric, as expressed by (8.174), then the elastic tensor has the symmetry

$$\mathbf{c}^{ik} \wedge \mathbf{e}^j = \mathbf{c}^{jk} \wedge \mathbf{e}^i. \tag{8.219}$$

Similarly, if the strain tensor is symmetric, as expressed by (8.201), then the elastic tensor has the symmetry

$$\mathbf{c}^{ij} \wedge \mathbf{e}^k = \mathbf{c}^{ik} \wedge \mathbf{e}^j. \tag{8.220}$$

For elastic materials, the internal energy per unit volume, e, takes the quadratic form

$$\begin{aligned} \mathrm{e} &= \tfrac{1}{2}\,\mathbf{E}_i : \mathbf{c}^{ij} : \mathbf{E}_j \\ &= \tfrac{1}{2}\,\mathbf{E} \vdots \mathbf{c} \vdots \mathbf{E} \\ &= \tfrac{1}{2}\,E_{ij}\, c^{ijk\ell}\, E_{k\ell}, \end{aligned} \tag{8.221}$$

which implies the final symmetry

$$\mathbf{c}^{ij} = \mathbf{c}^{ji}. \tag{8.222}$$

These collective symmetries reduce the number of independent parameters of the elastic tensor from 81 to 21. The number of independent parameters may be reduced further based on the consideration of additional symmetries under rotations and reflections, as discussed in example 5.17.

8.9.3 Connection One-Forms

To extend the exterior derivative, discussed in section 8.7, from forms to tensor-valued forms, we need to introduce the concept of a *connection one-form*. As discussed in section 7.1.4, the change of vector field $\mathbf{w}$ in the direction defined by vector field $\mathbf{u}$ is captured by the directional derivative $\nabla_{\mathbf{u}}\mathbf{w}$, which defines a third vector in the tangent space, Thus, the directional derivative of a non-coordinate basis vector $\mathbf{e}_i$ can be expressed as a linear combination of basis vectors:

$$\nabla_{\mathbf{u}}\mathbf{e}_i = \omega_i^{\ j}(\mathbf{u})\,\mathbf{e}_j. \tag{8.223}$$

Because the coefficients $\omega_i^{\ j}$ are linear functionals of $\mathbf{u} = u^k\,\mathbf{e}_k$, such that

$$\omega_i^{\ j}(\mathbf{u}) = u^k\,\omega_i^{\ j}(\mathbf{e}_k), \tag{8.224}$$

they are known as *connection one-forms*. Connection one-forms depend on a choice of basis and are, therefore, despite their name, not tensorial objects. We may write the connection one-form in terms of the one-form basis as

$$\omega_i^{\ j} = \omega_i^{\ j}(\mathbf{e}_k)\,\mathbf{e}^k. \tag{8.225}$$

Thus far, the basis $\mathbf{e}_i$ is general, that is, not tied to a particular coordinate system, as in the tetrad formalism discussed in section 5.3.1. If we identify the one-form basis $\mathbf{e}^i$ with the coordinate basis $\mathrm{d}x^i$, then we recover equation (7.19) for the directional derivative of a one-form coordinate basis, namely

$$\omega_i^j(\mathbf{e}_k) = \Gamma_{ki}^j. \tag{8.226}$$

Here we see that the connection one-forms, $\omega_i^j(\mathbf{e}_k)$, fulfill the role of the connection coefficients, Γ_{ki}^j, in a non-coordinate basis.

In a coordinate basis, the directional derivative of a vector $\mathbf{w}$ is expressed as

$$\nabla_{\mathbf{u}}\mathbf{w} = u^k\,(\partial_k w^j + \Gamma_{ki}^j\,w^i)\,\partial_j. \tag{8.227}$$

We may rewrite the terms $u^k\,\partial_k w^j$ in terms of the exterior derivative as $\mathrm{d}w^j(\mathbf{u})$, and thus a coordinate-free expression for the directional derivative is

$$\begin{aligned}\nabla_{\mathbf{u}}\mathbf{w} &= [\mathrm{d}w^j(\mathbf{u}) + w^i\,\omega_i^j(\mathbf{u})]\,\mathbf{e}_j\\ &= [\mathrm{d}(\,\mathbf{w}(\mathbf{e}^j)\,)(\mathbf{u}) + \mathbf{w}(\mathbf{e}^i)\,u^k\,\omega_i^j(\mathbf{e}_k)]\,\mathbf{e}_j,\end{aligned} \tag{8.228}$$

where we used the fact that $w^j = \mathbf{w}(\mathbf{e}^j)$. For the directional derivative of a one-form $\boldsymbol{\alpha}$ we have

$$\nabla_{\mathbf{u}}\boldsymbol{\alpha} = u^k\,(\partial_k\alpha_j - \Gamma_{kj}^i\,\alpha_i)\,\mathrm{d}x^j. \tag{8.229}$$

The coordinate-free expression for this directional derivative is

$$\begin{aligned}\nabla_{\mathbf{u}}\boldsymbol{\alpha} &= [\mathrm{d}\alpha_j(\mathbf{u}) - \alpha_i\,\omega_j^i(\mathbf{u})]\,\mathbf{e}^j\\ &= [\mathrm{d}\alpha_j(\mathbf{u}) - \alpha_i\,u^k\,\omega_j^i(\mathbf{e}_k)]\,\mathbf{e}^j.\end{aligned} \tag{8.230}$$

If this form is the gradient of a function, $\boldsymbol{\alpha} = \mathrm{d}f$, then $\alpha_j = \partial_j f = \mathrm{d}f(\mathbf{e}_j)$, and so

$$\nabla_{\mathbf{u}}\mathrm{d}f = [\mathrm{d}(\mathrm{d}f(\mathbf{e}_j))(\mathbf{u}) - \mathrm{d}f(\mathbf{e}_i)\,u^k\,\omega_j^i(\mathbf{e}_k)]\,\mathbf{e}^j. \tag{8.231}$$

8.9.4 Torsion Two-Forms

The concept of tensor-valued forms enables us to revisit notions of torsion and curvature without the introduction of local coordinates. For example, torsion is defined via the map (7.57), namely

$$\mathbf{t}(\mathbf{u},\mathbf{v}) \equiv \nabla_{\mathbf{u}}\mathbf{v} - \nabla_{\mathbf{v}}\mathbf{u} - [\mathbf{u},\mathbf{v}], \tag{8.232}$$

as illustrated in figure 8.8. Relative to a basis $\mathbf{e}_i$ its components are

$$\mathbf{t}(\mathbf{e}_i,\mathbf{e}_k) = \left[\omega_k^j(\mathbf{e}_i) - \omega_i^j(\mathbf{e}_k) - \tau_{ik}^j\right]\mathbf{e}_j, \tag{8.233}$$

where the structure coefficients τ_{ik}^j, given by equation (4.73), define the Lie bracket $[\mathbf{e}_i,\mathbf{e}_k]$. Recognizing the skewsymmetry of the torsion tensor in its two covariant indices, we express

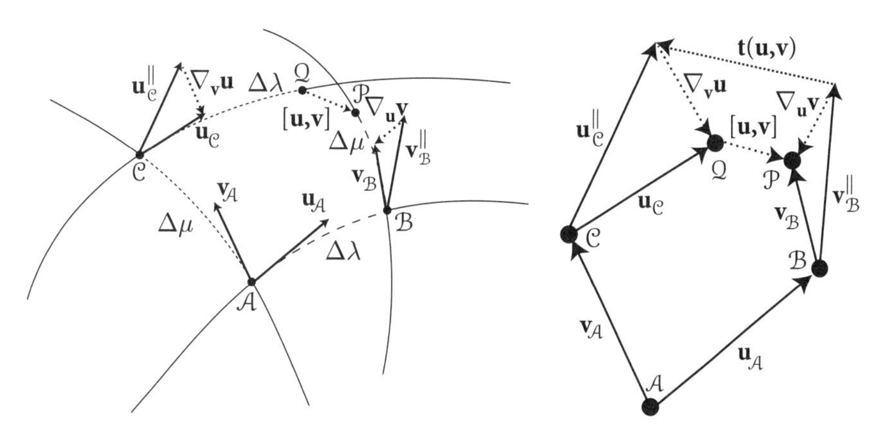

Figure 8.8: *Left*: The congruence of vector field $\mathbf{u}$ is determined by the parameter λ, and the congruence of vector field $\mathbf{v}$ is determined by the parameter μ. At point $\mathcal{A}$, the vectors are labeled $\mathbf{u}_{\mathcal{A}}$ and $\mathbf{v}_{\mathcal{A}}$. If we travel along the long-dashed integral curve defined by $\mathbf{u}_{\mathcal{A}}$ a step $\Delta\lambda$, we arrive at location $\mathcal{B}$, and if we continue from there along the long-dashed integral curve defined by $\mathbf{v}_{\mathcal{B}}$ a step $\Delta\mu$, we arrive at location $\mathcal{P}$. In contrast, if we travel along the short-dashed integral curve defined by $\mathbf{v}_{\mathcal{A}}$ a step $\Delta\mu$, we arrive at location $\mathcal{C}$, and if we continue from there along the short-dashed integral curve defined by $\mathbf{u}_{\mathcal{C}}$ a step $\Delta\lambda$, we arrive at location $\mathcal{Q}$. In the limit $\Delta\lambda \to 0$ and $\Delta\mu \to 0$, the "gap" between points $\mathcal{P}$ and $\mathcal{Q}$ corresponds to the Lie bracket, $[\mathbf{u},\mathbf{v}]$, as previously illustrated in figure 4.9. Vector $\mathbf{v}_{\mathcal{B}}^{\|}$ is parallel transported from $\mathcal{A}$ to $\mathcal{B}$ where it can be compared to $\mathbf{v}_{\mathcal{B}}$, thereby defining the directional derivative $\nabla_{\mathbf{u}}\mathbf{v}$. Similarly, vector $\mathbf{u}_{\mathcal{C}}^{\|}$ is parallel transported from $\mathcal{A}$ to $\mathcal{C}$ where it can be compared to $\mathbf{u}_{\mathcal{C}}$, thereby defining the directional derivative $\nabla_{\mathbf{v}}\mathbf{u}$. *Right*: "Shrunken" version of the image on the left, setting $\Delta\lambda = \Delta\mu = 1$, illustrating definition (8.232) for the torsion $\mathbf{t}(\mathbf{u},\mathbf{v})$ in terms of the directional derivatives $\nabla_{\mathbf{u}}\mathbf{v}$ and $\nabla_{\mathbf{v}}\mathbf{u}$ and the Lie bracket $[\mathbf{u},\mathbf{v}]$, namely, $\mathbf{t}(\mathbf{u},\mathbf{v}) = \nabla_{\mathbf{u}}\mathbf{v} - \nabla_{\mathbf{v}}\mathbf{u} - [\mathbf{u},\mathbf{v}]$.

it as

$$\begin{aligned} \mathbf{t} &= \tfrac{1}{2}\left[\omega^j_{\ k}(\mathbf{e}_i) - \omega^j_{\ i}(\mathbf{e}_k) - \tau^j_{\ ik}\right] \mathbf{e}_j \otimes \mathbf{e}^i \wedge \mathbf{e}^k \\ &= \mathbf{e}_j \otimes \left[\omega^j_{\ k} \wedge \mathbf{e}^k - \tfrac{1}{2}\,\tau^j_{\ ik}\,\mathbf{e}^i \wedge \mathbf{e}^k\right], \end{aligned} \tag{8.234}$$

where we have used equation (8.225). We have from equation (8.94)

$$\mathrm{d}\mathbf{e}^j + \tfrac{1}{2}\,\tau^j_{\ ik}\,\mathbf{e}^i \wedge \mathbf{e}^k = \mathbf{0}. \tag{8.235}$$

Thus we find that the torsion is given by

$$\mathbf{t} = \mathbf{e}_j \otimes (\mathrm{d}\mathbf{e}^j + \omega^j_{\ k} \wedge \mathbf{e}^k). \tag{8.236}$$

The object

$$\mathbf{t}^i = \mathrm{d}\mathbf{e}^i + \omega^i_{\ j} \wedge \mathbf{e}^j, \tag{8.237}$$

defines the vector-valued *torsion two-form* $\mathbf{t}^i$ relative to a basis.

In a nonholonomic coordinate basis, using relations (8.235), (8.225), and (8.226), we have

$$\begin{aligned} \mathbf{t}^i &= \mathrm{d}\mathbf{e}^i + \boldsymbol{\omega}^i_j \wedge \mathbf{e}^j \\ &= (\Gamma^i_{kj}\, \mathbf{e}^k) \wedge \mathbf{e}^j - \tfrac{1}{2}\, \tau^i_{jk}\, \mathbf{e}^j \wedge \mathbf{e}^k \\ &= \tfrac{1}{2}\, \mathbf{e}^k \wedge \mathbf{e}^j\, (\Gamma^i_{kj} - \Gamma^i_{jk} - \tau^i_{kj}), \end{aligned} \tag{8.238}$$

where we have used the skew symmetry of τ^i_{kj} in its covariant indices. In this case the torsion tensor has components

$$\mathbf{t}^i(\mathbf{e}_k, \mathbf{e}_j) = t_{kj}{}^i = \Gamma^i_{kj} - \Gamma^i_{jk} + \tau^i_{kj}, \tag{8.239}$$

in agreement with equation (7.58). In a holonomic coordinate basis, this reduces to

$$\mathbf{t}^i(\mathbf{e}_k, \mathbf{e}_j) = t_{kj}{}^i = \Gamma^i_{kj} - \Gamma^i_{jk}, \tag{8.240}$$

in agreement with equation (7.55).

8.9.5 Exterior Covariant Derivative

The introduction of connection one-forms in section 8.9.3 enables us to introduce a generalization of the exterior derivative, d, namely the *exterior covariant derivative*, denoted by D. In general, the exterior covariant derivative of a vector-valued k-form $\boldsymbol{\alpha}^i$ is defined by

$$\mathrm{D}\boldsymbol{\alpha}^i \equiv \mathrm{d}\boldsymbol{\alpha}^i + \boldsymbol{\omega}^i_j \wedge \boldsymbol{\alpha}^j. \tag{8.241}$$

Other attributes of the exterior covariant derivative are linearity, commutativity with the metric (since $\mathrm{D}\, g_{ij} = 0$ in the absence of nonmetricity, as discussed in section 8.9.10),

$$g_{ij}\, \mathrm{D}\boldsymbol{\alpha}^j = \mathrm{D}\boldsymbol{\alpha}_i, \tag{8.242}$$

and the relation

$$\mathrm{D}(\boldsymbol{\omega}^j_k \wedge \boldsymbol{\alpha}) \equiv (\mathrm{D}\boldsymbol{\omega}^j_k) \wedge \boldsymbol{\alpha} - \boldsymbol{\omega}^j_k \wedge \mathrm{d}\boldsymbol{\alpha}. \tag{8.243}$$

Its effect on a basis one-form $\mathbf{e}^j$ is

$$\mathrm{D}\mathbf{e}^j = \mathrm{d}\mathbf{e}^j + \boldsymbol{\omega}^j_k \wedge \mathbf{e}^k. \tag{8.244}$$

Upon comparison of the last result with equation (8.237), we conclude that the exterior covariant derivative of a basis one-form is the torsion two-form:

$$\mathbf{t}^j = \mathrm{D}\mathbf{e}^j. \tag{8.245}$$

For example, for a vector-valued zero-form u^i—that is, the component of a vector—we find from (8.241) that application of the exterior covariant derivative produces the vector-valued one-form

$$\begin{aligned}
\mathrm{D}u^i &= \mathrm{d}u^i + \omega^i_{\ k}\, u^k \\
&= \partial_j u^i\, \mathbf{e}^j + \omega^i_{\ k}(\mathbf{e}_j)\, u^k\, \mathbf{e}^j \\
&= (\partial_j u^i + \Gamma^i_{jk}\, u^k)\, \mathbf{e}^j \\
&= \nabla_j u^i\, \mathbf{e}^j,
\end{aligned} \tag{8.246}$$

where we used equation (8.225) in the second equality, and where we used equation (8.226)—valid in a local coordinate basis—in the third equality. This is precisely the covariant derivative of a vector (7.12)!

For a general tensor-valued k-form $\boldsymbol{\alpha}$ one adds terms $\boldsymbol{\omega}\wedge\boldsymbol{\alpha}$ and $-\boldsymbol{\omega}\wedge\boldsymbol{\alpha}$ depending on whether the tensor index is contravariant or covariant. For example, for a $(1,1)$-tensor-valued k-form, we have

$$\mathrm{D}\alpha^i_{\ j} \equiv \mathrm{d}\alpha^i_{\ j} + \omega^i_{\ k}\wedge\alpha^k_{\ j} - \omega^k_{\ j}\wedge\alpha^i_{\ k}. \tag{8.247}$$

To illustrate this generalization, in three dimensions, let us consider the one-form-valued two-form

$$\mathbf{T}_i \otimes \mathbf{e}^i = \mathbf{T}\cdot\boldsymbol{\epsilon} = T_i{}^j\, \epsilon_{jk\ell}\, \mathbf{e}^i\otimes\mathbf{e}^k\otimes\mathbf{e}^\ell = \tfrac{1}{2}\, T_i{}^j\, \epsilon_{jk\ell}\, \mathbf{e}^i\otimes\mathbf{e}^k\wedge\mathbf{e}^\ell. \tag{8.248}$$

We find that

$$\begin{aligned}
\mathrm{D}\mathbf{T}_i &= \mathrm{d}(\tfrac{1}{2}\, T_i{}^j\, \epsilon_{jk\ell}\, \mathbf{e}^k\wedge\mathbf{e}^\ell) - \omega^m_{\ i}\wedge(\tfrac{1}{2}\, T_m{}^j\, \epsilon_{jk\ell}\, \mathbf{e}^k\wedge\mathbf{e}^\ell) \\
&= \partial_n(\tfrac{1}{2}\, T_i{}^j\, \epsilon_{jk\ell})\, \mathbf{e}^n\wedge\mathbf{e}^k\wedge\mathbf{e}^\ell - \tfrac{1}{2}\, T_m{}^j\, \epsilon_{jk\ell}\, \Gamma^m_{ni}\, \mathbf{e}^n\wedge\mathbf{e}^k\wedge\mathbf{e}^\ell \\
&= \tfrac{1}{2}\, \partial_n(T_i{}^j\, \epsilon_{jk\ell} - T_m{}^j\, \epsilon_{jk\ell}\, \Gamma^m_{ni})\, \mathbf{e}^n\wedge\mathbf{e}^k\wedge\mathbf{e}^\ell \\
&= \tfrac{1}{2}\, (\partial_n T_i{}^j + T_i{}^j\, \Gamma^m_{nm} - \Gamma^m_{ni}\, T_m{}^j)\, \epsilon_{jk\ell}\, \mathbf{e}^n\wedge\mathbf{e}^k\wedge\mathbf{e}^\ell,
\end{aligned} \tag{8.249}$$

where in the last equality we have used the fact that $\partial_n\epsilon_{jk\ell} = \partial_n(\bar{g}\,\underline{\epsilon}_{jk\ell}) = \Gamma^m_{nm}\,\epsilon_{jk\ell}$. But, using the result (8.51),

$$\epsilon_{jk\ell}\, \mathbf{e}^n\wedge\mathbf{e}^k\wedge\mathbf{e}^\ell = 2\,\delta_j{}^n\, \epsilon_{123}\, \mathbf{e}^1\wedge\mathbf{e}^2\wedge\mathbf{e}^3 = 2\,\delta_j{}^n\, \boldsymbol{\epsilon}, \tag{8.250}$$

and so we find that (8.249) may be rewritten in the form

$$\begin{aligned}
\mathrm{D}\mathbf{T}_i &= (\partial_j T_i{}^j + T_i{}^j\, \Gamma^m_{jm} - \Gamma^m_{ji}\, T_m{}^j)\, \boldsymbol{\epsilon} \\
&= \nabla_j T_i{}^j\, \boldsymbol{\epsilon} \\
&\equiv \boldsymbol{\epsilon}\, \boldsymbol{\nabla}\cdot\mathbf{T}_i,
\end{aligned} \tag{8.251}$$

where we have used the definition of the covariant derivative of a tensor (7.21).

Example 8.9 Eulerian and Lagrangian Exterior Covariant Derivatives

For the exterior derivative of a vector-valued form α^i, we have the following tranformation between Eulerian and Lagrangian vector values:

$$\begin{aligned} \mathrm{D}\alpha^i &= \mathrm{d}\alpha^i + \boldsymbol{\Gamma}^i_j \wedge \alpha^j \\ &= F^i{}_I \, (\mathrm{d}\alpha^I + \boldsymbol{\Gamma}^I_J \wedge \alpha^J) \\ &= F^i{}_I \, \mathrm{D}\alpha^I, \end{aligned} \tag{8.252}$$

where we have defined the Eulerian and Lagrangian connection one-forms

$$\boldsymbol{\Gamma}^i_j \equiv \Gamma^i_{kj} \, \mathbf{e}^k, \tag{8.253}$$

$$\boldsymbol{\Gamma}^I_J \equiv \Gamma^I_{KJ} \, \mathbf{e}^K, \tag{8.254}$$

which are related via a form version of expression (7.33):

$$F^i{}_I \boldsymbol{\Gamma}^I_J = F^j{}_J \boldsymbol{\Gamma}^i_j + \mathrm{d}F^i{}_J. \tag{8.255}$$

Now we can raise and lower indices, for example, for a Lagrangian vector-valued form $\mathbf{q}^I$, we have

$$g_{IJ} \, \mathrm{D}\mathbf{q}^J = \mathrm{D}\mathbf{q}_I. \tag{8.256}$$

Example 8.10 Mixed Exterior Covariant Derivative

Using the tetrad formulation and notation of general relativity introduced in example 5.8, the exterior covariant derivative of a mixed tensor-valued form $\alpha^{\mu\cdots}{}_{\nu\dots}{}^{a\cdots}{}_{b\dots}$, involving both coordinate indices $\mu, \nu, \dots$ and frame indices $a, b, \dots$, is defined as

$$\begin{aligned} \tilde{\mathrm{D}}\alpha^{\mu\cdots}{}_{\nu\dots}{}^{a\cdots}{}_{b\dots} \equiv\ & \mathrm{d}\alpha^{\mu\cdots}{}_{\nu\dots}{}^{a\cdots}{}_{b\dots} \\ & + \Gamma^\mu_\tau \wedge \alpha^{\tau\cdots}{}_{\nu\dots}{}^{a\cdots}{}_{b\dots} + \cdots \\ & - \Gamma^\tau_\nu \wedge \alpha^{\mu\cdots}{}_{\tau\dots}{}^{a\cdots}{}_{b\dots} - \cdots \\ & + \omega^a_c \wedge \alpha^{\mu\cdots}{}_{\nu\dots}{}^{c\cdots}{}_{b\dots} + \cdots \\ & - \omega^c_b \wedge \alpha^{\mu\cdots}{}_{\nu\dots}{}^{a\cdots}{}_{c\dots} - \cdots, \end{aligned} \tag{8.257}$$

where we have defined the coordinate connection one-forms

$$\Gamma^\mu_\nu \equiv \Gamma^\nu_{\tau\mu} \, e^\tau, \tag{8.258}$$

and, using the tetrad connection coefficients $\omega^c_{\mu a}$, introduced in section 7.1.9, the tetrad connection one-forms

$$\omega^b_a \equiv \omega^c_{\mu a} \, e^\mu. \tag{8.259}$$

The mixed exterior covariant derivative of the tetrad transformation matrices vanishes:

$$\tilde{\mathrm{D}} e_a{}^{\mu} = 0, \qquad \tilde{\mathrm{D}} e_{\mu}{}^{a} = 0. \tag{8.260}$$

These expressions are equivalent to equations (7.108) and (7.109) in terms of the mixed covariant derivative.

Example 8.11 Spin Connection One-Forms

In a tetrad basis, the mixed exterior covariant derivative (8.257) of the components of the metric tensor is

$$\tilde{\mathrm{D}} g_{ab} = \mathrm{d} g_{ab} - \omega_a^c \, g_{cb} - \omega_b^c \, g_{ac} = 0, \tag{8.261}$$

where, in the last equality, we have assumed metricity. For orthogonal frames, as discussed in section 5.10.4,

$$g_{ab} = \eta_{ab}, \tag{8.262}$$

where η_{ab} denotes the constant metric of Minkowski space. In such a case, $\mathrm{d} g_{ab} = 0$, and thus we find from (8.261) that

$$\omega_{ab} = -\omega_{ab}, \tag{8.263}$$

where we used the notation $\omega_a^c \, g_{cb} = \omega_{ab}$. We conclude that the connection one-forms of an orthogonal frame are antisymmetric. Such connection one-forms are called *spin connection one-forms* (e.g., Carroll, 2004).

8.9.6 Covariant Lie Derivative

Tensor-valued forms require a generalization of the exterior derivative to accommodate the connection, leading to the introduction of the exterior covariant derivative discussed in Chapter 8.9.5. This derivative enables raising and lowering indices on expressions involving exterior covariant differentiation, as expressed in (8.242). In this section, we seek a similar generalization of the Lie derivative of tensor-valued forms. Such a generalization may be obtained by replacing the exterior derivatives in the Cartan magic formula (8.145) with exterior covariant derivatives. To illustrate the procedure, consider a tensor-valued form α^i_j, and define its covariant Lie derivative as

$$\begin{aligned}
\mathfrak{L}_{\mathbf{u}} \alpha^i_j &\equiv \mathbf{i}_{\mathbf{u}} \mathrm{D} \alpha^i_j + \mathrm{D}(\mathbf{i}_{\mathbf{u}} \alpha^i_j) \\
&= \mathbf{i}_{\mathbf{u}} \mathrm{d} \alpha^i_j + \omega^i_k(\mathbf{u}) \, \alpha^k_j - \omega^k_j(\mathbf{u}) \, \alpha^i_k + \mathrm{d}(\mathbf{i}_{\mathbf{u}} \alpha^i_j) + \omega^i_k \wedge \mathbf{i}_{\mathbf{u}} \alpha^k_j - \omega^k_j \wedge \mathbf{i}_{\mathbf{u}} \alpha^i_k \\
&= \mathcal{L}_{\mathbf{u}} \alpha^i_j + \omega^i_k(\mathbf{u}) \, \alpha^k_j - \omega^k_j(\mathbf{u}) \, \alpha^i_k + \omega^i_k \wedge \mathbf{i}_{\mathbf{u}} \alpha^k_j - \omega^k_j \wedge \mathbf{i}_{\mathbf{u}} \alpha^i_k,
\end{aligned} \tag{8.264}$$

where we used the exterior covariant derivative (8.247). Expression (8.264) demonstrates how the Lie derivative of a tensor-valued form needs to be modified for each of its tensor components to obtain its covariant version.

As special cases, for a tensor-valued n-form $\boldsymbol{\alpha}^i_j$ in an n-dimensional manifold we have $\mathrm{D}\boldsymbol{\alpha}^i_j = 0$ and thus

$$\mathfrak{L}_{\mathbf{u}}\boldsymbol{\alpha}^i_j = \mathrm{D}(\mathbf{i}_{\mathbf{u}}\boldsymbol{\alpha}^i_j)\,, \tag{8.265}$$

whereas for a tensor-valued scalar function α^i_j we have $\mathbf{i}_{\mathbf{u}}\alpha^i_j = 0$ and thus

$$\mathfrak{L}_{\mathbf{u}}\alpha^i_j = \mathbf{i}_{\mathbf{u}}\mathrm{D}\alpha^i_j\,. \tag{8.266}$$

Consequently, upon regarding the components of the metric tensor, g_{ij}, as a tensor-valued zero-form, we have $\mathfrak{L}_{\mathbf{u}}\, g_{ij} = \mathbf{i}_{\mathbf{u}}\,\mathrm{D}g_{ij} = 0$. Thus, we can raise and lower indices through the covariant Lie derivative, e.g.,

$$g_{ik}\, g^{j\ell}\, \mathfrak{L}_{\mathbf{u}}\boldsymbol{\alpha}^i_j = \mathfrak{L}_{\mathbf{u}}\boldsymbol{\alpha}_k^{\ell}\,, \tag{8.267}$$

which complements expression (8.242) for the exterior covariant derivative.

Example 8.12 Eulerian and Lagrangian Covariant Lie derivatives

To relate covariant expressions for the Lie derivative of Eulerian and Lagrangian tensor-valued forms, we consider the following transformation:

$$\begin{aligned}
\mathfrak{L}_{\mathbf{v}}\mathbf{q}^i &= \mathfrak{L}_{\mathbf{v}}(F^i{}_I\,\mathbf{q}^I) \\
&= F^i{}_I\,\mathfrak{L}_{\mathbf{v}}\mathbf{q}^I + \mathbf{q}^I\,\mathfrak{L}_{\mathbf{v}}F^i{}_I \\
&= F^i{}_I\,\mathfrak{L}_{\mathbf{v}}\mathbf{q}^I + \mathbf{q}^I\,\mathbf{i}_{\mathbf{v}}\mathrm{D}F^i{}_I \\
&= F^i{}_I\,\mathfrak{L}_{\mathbf{v}}\mathbf{q}^I\,,
\end{aligned} \tag{8.268}$$

where in the third equality we have used the fact that for a scalar $\mathbf{i}_{\mathbf{v}}F^i{}_I = 0$, and in the fourth equality the fact that, according to (8.255),

$$\mathrm{D}F^i{}_J = \mathrm{d}F^i{}_J + F^j{}_J\,\boldsymbol{\Gamma}^i_j - F^i{}_I\,\boldsymbol{\Gamma}^I_J = 0\,. \tag{8.269}$$

Expression (8.268) for the transformation of the covariant Lie derivative of an Eulerian vector-valued form to a Lagrangian vector-valued form complements equation (8.252) for the transformation of the exterior covariant derivative.

Thus, we can raise and lower Lagrangian indices through the covariant Lie derivative, e.g.,

$$g_{IJ}\, \mathfrak{L}_{\mathbf{v}}\mathbf{q}^J = \mathfrak{L}_{\mathbf{v}}\mathbf{q}_I, \tag{8.270}$$

which complements expression (8.256) for the exterior covariant derivative.

8.9.7 Curvature Two-Forms

In this section, we explore the connection between exterior covariant differentiation and curvature. To establish this connection, we apply the exterior covariant derivative to the torsion $\mathbf{t}^j = \mathrm{D}\mathbf{e}^j$, given in equation (8.244). We find that

$$\begin{aligned}
\mathrm{D}\mathbf{t}^j &= \mathrm{D}^2\mathbf{e}^j \\
&= \mathrm{D}\mathrm{d}\mathbf{e}^j + \mathrm{D}(\omega^j_k \wedge \mathbf{e}^k) \\
&= \mathrm{d}^2\mathbf{e}^j + \omega^j_k \wedge \mathrm{d}\mathbf{e}^k + (\mathrm{D}\omega^j_k) \wedge \mathbf{e}^k - \omega^j_k \wedge \mathrm{d}\mathbf{e}^k \\
&= \mathrm{D}\omega^j_k \wedge \mathbf{e}^k \\
&= (\mathrm{d}\omega^j_k + \omega^j_\ell \wedge \omega^\ell_k) \wedge \mathbf{e}^k,
\end{aligned} \tag{8.271}$$

where we have used the fact that $\mathrm{d}^2\mathbf{e}^i = \mathbf{0}$. What is the entity $\mathrm{d}\omega^j_k + \omega^j_\ell \wedge \omega^\ell_k$?

Consider the definition of curvature in terms of the map (7.64), namely

$$\mathbf{r}(\mathbf{u},\mathbf{v})\,\mathbf{w} = \boldsymbol{\nabla}_{\mathbf{u}}\,\boldsymbol{\nabla}_{\mathbf{v}}\,\mathbf{w} - \boldsymbol{\nabla}_{\mathbf{v}}\boldsymbol{\nabla}_{\mathbf{u}}\,\mathbf{w} - \boldsymbol{\nabla}_{[\mathbf{u},\mathbf{v}]}\,\mathbf{w}, \tag{8.272}$$

as illustrated in figure 8.9. Using the repeated application of the coordinate-free definition of the directional derivative given by equation (8.228), we have

$$\begin{aligned}
\boldsymbol{\nabla}_{\mathbf{u}}\boldsymbol{\nabla}_{\mathbf{v}}\mathbf{w} &= \boldsymbol{\nabla}_{\mathbf{u}}[\mathrm{d}w^j(\mathbf{v}) + w^i\,\omega^j_i(\mathbf{v})]\,\mathbf{e}_j] \\
&= \{\mathrm{d}[\mathrm{d}w^j(\mathbf{v}) + w^i\,\omega^j_i(\mathbf{v})](\mathbf{u}) + [\mathrm{d}w^i(\mathbf{v}) + w^k\,\omega^i_k(\mathbf{v})]\,\omega^j_i(\mathbf{u})\}\,\mathbf{e}_j \\
&= \{\mathrm{d}(\mathrm{d}w^j(\mathbf{v}))(\mathbf{u}) + \mathrm{d}w^i(\mathbf{u})\,\omega^j_i(\mathbf{v}) + w^i(\mathrm{d}\omega^j_i(\mathbf{v}))(\mathbf{u}) \\
&\quad + [\mathrm{d}w^i(\mathbf{v}) + w^k\,\omega^i_k(\mathbf{v})]\,\omega^j_i(\mathbf{u})\}\,\mathbf{e}_j.
\end{aligned} \tag{8.273}$$

Using the definition of the Lie bracket (4.74) and the coordinate-free directional derivative (8.228), we have

$$\boldsymbol{\nabla}_{[\mathbf{u},\mathbf{v}]}\,\mathbf{w} = \{\mathrm{d}w^j([\mathbf{u},\mathbf{v}]) + w^i\omega^j_i([\mathbf{u},\mathbf{v}])\}\,\mathbf{e}_j. \tag{8.274}$$

Therefore, we find that

$$\begin{aligned}
\mathbf{r}(\mathbf{u},\mathbf{v})\,\mathbf{w} = \{&\mathrm{d}(\mathrm{d}w^j(\mathbf{v}))(\mathbf{u}) + w^i(\mathrm{d}\omega^j_i(\mathbf{v}))(\mathbf{u}) + w^i\,\omega^k_i(\mathbf{v})\,\omega^j_k(\mathbf{u}) - \mathrm{d}(\mathrm{d}w^j(\mathbf{u}))(\mathbf{v}) \\
&- w^i\,(\mathrm{d}\omega^j_i(\mathbf{u}))(\mathbf{v}) - w^i\,\omega^k_i(\mathbf{u})\,\omega^j_k(\mathbf{v}) - \mathrm{d}w^j([\mathbf{u},\mathbf{v}]) - w^i\,\omega^j_i([\mathbf{u},\mathbf{v}])\}\,\mathbf{e}_j.
\end{aligned} \tag{8.275}$$

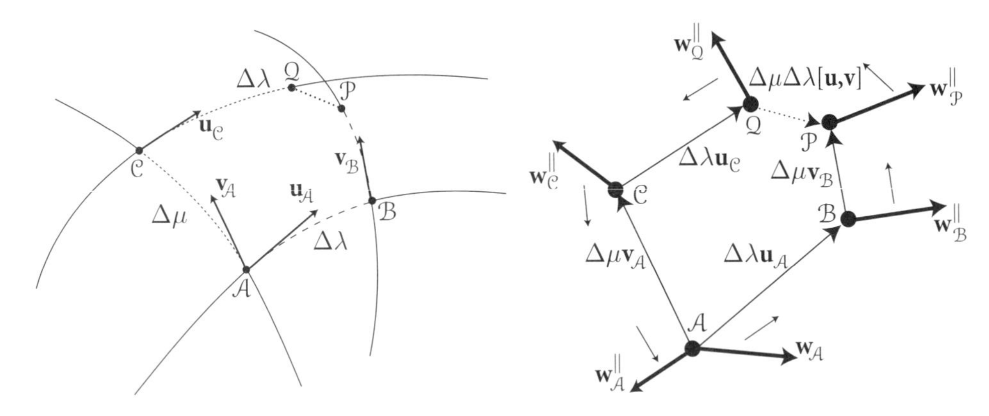

Figure 8.9: *Left*: The congruence of vector field $\mathbf{u}$ is determined by the parameter λ, and the congruence of vector field $\mathbf{v}$ is determined by the parameter μ. At point $\mathcal{A}$, the vectors are labeled $\mathbf{u}_\mathcal{A}$ and $\mathbf{v}_\mathcal{A}$. If we travel along the long-dashed integral curve defined by $\mathbf{u}_\mathcal{A}$ a step $\Delta\lambda$, we arrive at location $\mathcal{B}$, and if we continue from there along the long-dashed integral curve defined by $\mathbf{v}_\mathcal{B}$ a step $\Delta\mu$, we arrive at location $\mathcal{P}$. In contrast, if we travel along the short-dashed integral curve defined by $\mathbf{v}_\mathcal{A}$ a step $\Delta\mu$, we arrive at location $\mathcal{C}$, and if we continue from there along the short-dashed integral curve defined by $\mathbf{u}_\mathcal{C}$ a step $\Delta\lambda$, we arrive at location $\mathcal{Q}$. In the limit $\Delta\lambda \to 0$ and $\Delta\mu \to 0$, the "gap" between points $\mathcal{P}$ and $\mathcal{Q}$ corresponds to the Lie bracket, $[\mathbf{u}, \mathbf{v}]$, as previously illustrated in figure 4.9. *Right*: "Shrunken" version of the image on the left, illustrating definition (8.272) for the curvature $\mathbf{r}(\mathbf{u}, \mathbf{v})\,\mathbf{w}$. Vector $\mathbf{w}_\mathcal{A}$ at point $\mathcal{A}$ is parallel transported a distance $\Delta\lambda\,\mathbf{u}_\mathcal{A}$ to point $\mathcal{B}$ where it is labeled $\mathbf{w}^{||}_\mathcal{B}$, followed by parallel transport over a distance $\Delta\mu\,\mathbf{v}_\mathcal{B}$ to point $\mathcal{P}$ where it is labeled $\mathbf{w}^{||}_\mathcal{P}$, past the Lie bracket a distance $\Delta\mu\,\Delta\lambda\,[\mathbf{u}, \mathbf{v}]$ to point $\mathcal{Q}$ where it is labeled $\mathbf{w}^{||}_\mathcal{Q}$, followed by parallel transport over a distance $-\Delta\lambda\,\mathbf{u}_\mathcal{C}$ to point $\mathcal{C}$ where it is labeled $\mathbf{w}^{||}_\mathcal{C}$, concluded by parallel transport over a distance $-\Delta\mu\,\mathbf{v}_\mathcal{A}$ to point $\mathcal{A}$ where it is labeled $\mathbf{w}^{||}_\mathcal{A}$. The curvature is captured by the difference between $\mathbf{w}_\mathcal{A}$ and its parallel transported version around the loop $\mathbf{w}^{||}_\mathcal{A}$.

But according to the coordinate-free definition of the exterior derivative of a one-form (8.92), we have

$$\mathrm{dd}w^j(\mathbf{u}, \mathbf{v}) = \mathrm{d}(\mathrm{d}w^j(\mathbf{v}))(\mathbf{u}) - \mathrm{d}(\mathrm{d}w^j(\mathbf{u}))(\mathbf{v}) - \mathrm{d}w^j([\mathbf{u}, \mathbf{v}]) = 0, \tag{8.276}$$

and

$$\mathrm{d}\omega_i^{\ j}(\mathbf{u}, \mathbf{v}) = \mathrm{d}(\omega_i^{\ j}(\mathbf{v}))(\mathbf{u}) - \mathrm{d}(\omega_i^{\ j}(\mathbf{u}))(\mathbf{v}) - \omega_i^{\ j}([\mathbf{u}, \mathbf{v}]), \tag{8.277}$$

and thus we obtain a beautiful expression for the curvature:

$$\begin{aligned}\mathbf{r}(\mathbf{u}, \mathbf{v})\,\mathbf{w} &= w^i\,[\mathrm{d}\omega_i^{\ j}(\mathbf{u}, \mathbf{v}) + \omega_k^{\ j}(\mathbf{u})\,\omega_i^{\ k}(\mathbf{v}) - \omega_i^{\ k}(\mathbf{u})\,\omega_k^{\ j}(\mathbf{v})]\,\mathbf{e}_j \\ &= w^i\,(\mathrm{d}\omega_i^{\ j} + \omega_k^{\ j} \wedge \omega_i^{\ k})(\mathbf{u}, \mathbf{v})\,\mathbf{e}_j.\end{aligned} \tag{8.278}$$

This leads us to define the $(1, 1)$-tensor-valued *curvature two-form*

$$\mathbf{r}_i^{\ j} \equiv \mathrm{d}\omega_i^{\ j} + \omega_k^{\ j} \wedge \omega_i^{\ k}. \tag{8.279}$$

In components, we have

$$\begin{aligned}
\mathbf{r}^k_\ell(\mathbf{e}_i,\mathbf{e}_j) &= \mathrm{d}\boldsymbol{\omega}^k_\ell(\mathbf{e}_i,\mathbf{e}_j) + \boldsymbol{\omega}^m_\ell(\mathbf{e}_j)\,\boldsymbol{\omega}^k_m(\mathbf{e}_i) - \boldsymbol{\omega}^m_\ell(\mathbf{e}_i)\,\boldsymbol{\omega}^k_m(\mathbf{e}_j) \\
&= \mathrm{d}(\boldsymbol{\omega}^k_\ell(\mathbf{e}_j))(\mathbf{e}_i) - \mathrm{d}(\boldsymbol{\omega}^k_\ell(\mathbf{e}_i))(\mathbf{e}_j) - \boldsymbol{\omega}^k_\ell([\mathbf{e}_i,\mathbf{e}_j]) \\
&\quad + \boldsymbol{\omega}^m_\ell(\mathbf{e}_j)\,\boldsymbol{\omega}^k_m(\mathbf{e}_i) - \boldsymbol{\omega}^m_\ell(\mathbf{e}_i)\,\boldsymbol{\omega}^k_m(\mathbf{e}_j) \\
&= \partial_i\Gamma^k_{j\ell} - \partial_j\Gamma^k_{i\ell} - \tau^m_{ij}\,\Gamma^k_{m\ell} + \Gamma^m_{j\ell}\,\Gamma^k_{im} - \Gamma^m_{i\ell}\,\Gamma^k_{jm}.
\end{aligned} \tag{8.280}$$

Using expression (7.65) for the curvature tensor obtained in the context of covariant differentiation, we see that the curvature two-form and the Riemann tensor are related via

$$\mathbf{r}^k_\ell(\mathbf{e}_i,\mathbf{e}_j) = r_{ij\ell}{}^k. \tag{8.281}$$

Finally, circling back to the first result in the section, we have established the fact that the exterior covariant derivative of the vector-valued torsion two-form is the $(1,1)$-tensor-valued curvature two-form:

$$\mathrm{D}\mathbf{t}^j = \mathbf{r}^j_i \wedge \mathbf{e}^i. \tag{8.282}$$

For a general vector-valued k-form we find

$$\begin{aligned}
\mathrm{D}^2\boldsymbol{\alpha}^i &= \mathrm{D}(\mathrm{D}\boldsymbol{\alpha}^i) \\
&= \mathrm{d}(\mathrm{D}\boldsymbol{\alpha}^i) + \boldsymbol{\omega}^i_j \wedge (\mathrm{D}\boldsymbol{\alpha}^j) \\
&= \mathrm{d}(\mathrm{d}\boldsymbol{\alpha}^i + \boldsymbol{\omega}^i_j \wedge \boldsymbol{\alpha}^j) + \boldsymbol{\omega}^i_j \wedge (\mathrm{d}\boldsymbol{\alpha}^j + \boldsymbol{\omega}^j_k \wedge \boldsymbol{\alpha}^k) \\
&= \mathrm{d}\boldsymbol{\omega}^i_j \wedge \boldsymbol{\alpha}^j - \boldsymbol{\omega}^i_j \wedge \mathrm{d}\boldsymbol{\alpha}^j + \boldsymbol{\omega}^i_j \wedge \mathrm{d}\boldsymbol{\alpha}^j + \boldsymbol{\omega}^i_j \wedge \boldsymbol{\omega}^j_k \wedge \boldsymbol{\alpha}^k \\
&= (\mathrm{d}\boldsymbol{\omega}^i_k + \boldsymbol{\omega}^i_j \wedge \boldsymbol{\omega}^j_k) \wedge \boldsymbol{\alpha}^k \\
&= \mathbf{r}^i_k \wedge \boldsymbol{\alpha}^k,
\end{aligned} \tag{8.283}$$

where we have used (8.279) in the last equality. So unlike d^2, D^2 does not vanish in general.

8.9.8 Commutator of Covariant Lie and Exterior Covariant Derivatives

In this section, we investigate the commutativity of the covariant Lie derivative, $\mathfrak{L}_\mathbf{u}$, and the exterior covariant derivative, D. To illustrate the problem, consider a vector-valued form $\boldsymbol{\alpha}^i$. According to the covariant version of the Cartan magic formula (8.264), its covariant Lie derivative is

$$\mathfrak{L}_\mathbf{u}\boldsymbol{\alpha}^i = \mathbf{i}_\mathbf{u}\mathrm{D}\boldsymbol{\alpha}^i + \mathrm{D}(\mathbf{i}_\mathbf{u}\boldsymbol{\alpha}^i), \tag{8.284}$$

and the exterior covariant derivative of this expression is

$$\mathrm{D}\mathfrak{L}_{\mathbf{u}}\alpha^i = \mathrm{D}(\mathbf{i}_{\mathbf{u}}\mathrm{D}\alpha^i) + \mathrm{D}^2(\mathbf{i}_{\mathbf{u}}\alpha^i). \tag{8.285}$$

Similarly, the covariant Lie derivative of its exterior covariant derivative is

$$\mathfrak{L}_{\mathbf{u}}\mathrm{D}\alpha^i = \mathbf{i}_{\mathbf{u}}\mathrm{D}^2\alpha^i + \mathrm{D}(\mathbf{i}_{\mathbf{u}}\mathrm{D}\alpha^i). \tag{8.286}$$

Thus, using expression (8.283), we find that

$$\begin{aligned} \mathrm{D}\mathfrak{L}_{\mathbf{u}}\alpha^i - \mathfrak{L}_{\mathbf{u}}\mathrm{D}\alpha^i &= \mathrm{D}^2(\mathbf{i}_{\mathbf{u}}\alpha^i) - \mathbf{i}_{\mathbf{u}}\mathrm{D}^2\alpha^i \\ &= \mathbf{r}^i_j \wedge \mathbf{i}_{\mathbf{u}}\alpha^j - \mathbf{i}_{\mathbf{u}}\mathbf{r}^i_j \wedge \alpha^j, \end{aligned} \tag{8.287}$$

We conclude that when the curvature vanishes, the covariant Lie derivative and exterior covariant derivative commute.

8.9.9 Bianchi Identities Revisited

The vector-valued torsion two-form $\mathbf{t}^i$ defined by equation (8.237) and the $(1,1)$-tensor-valued curvature two-form $\mathbf{r}^i_j$ given in equation (8.279)—expressing the torsion and curvature two-forms in terms of connection one-forms—illustrate that the torsion and curvature two-forms are tensors, even though the connection one-forms are not, just like in a coordinate basis, the connection coefficients are not tensorial, whereas the torsion and curvature tensors are. The Bianchi identities (7.66) and (7.67) may be rewritten in terms of the exterior covariant derivative in the concise form

$$\mathrm{D}\mathbf{r}^i_j = \mathbf{0}, \tag{8.288}$$

$$\mathrm{D}\mathbf{t}^i = \mathbf{r}^i_j \wedge \mathbf{e}^j. \tag{8.289}$$

We have already demonstrated the second Bianchi identity in equation (8.271). To obtain the first identity, start with equation (8.247) for the exterior covariant derivative of a $(1,1)$-tensor-valued form and substitute expression (8.279):

$$\begin{aligned} \mathrm{D}\mathbf{r}^i_j &= \mathrm{d}\mathbf{r}^i_j + \omega^i_k \wedge \mathbf{r}^k_j - \omega^k_j \wedge \mathbf{r}^i_k \\ &= \mathrm{d}(\mathrm{d}\omega^i_j + \omega^i_k \wedge \omega^k_j) + \omega^i_k \wedge (\mathrm{d}\omega^k_j + \omega^k_\ell \wedge \omega^\ell_j) \\ &\quad - \omega^k_j \wedge (\mathrm{d}\omega^i_k + \omega^i_\ell \wedge \omega^\ell_k) \\ &= \mathrm{d}\omega^i_k \wedge \omega^k_j - \omega^i_k \wedge \mathrm{d}\omega^k_j + \omega^i_k \wedge (\mathrm{d}\omega^k_j + \omega^k_\ell \wedge \omega^\ell_j) \\ &\quad - \omega^k_j \wedge (\mathrm{d}\omega^i_k + \omega^i_\ell \wedge \omega^\ell_k) \\ &= \mathrm{d}\omega^i_k \wedge \omega^k_j + \omega^i_k \wedge \omega^k_\ell \wedge \omega^\ell_j - \omega^k_j \wedge \mathrm{d}\omega^i_k - \omega^k_j \wedge \omega^i_\ell \wedge \omega^\ell_k \\ &= \mathrm{d}\omega^i_k \wedge \omega^k_j - \omega^k_j \wedge \mathrm{d}\omega^i_k \\ &= \mathbf{0}. \end{aligned} \tag{8.290}$$

The last equality follows from property (8.211) for the wedge product of a two-form, $d\omega^i_k$, with a one-form, ω^k_j.

Example 8.13 Bianchi Identities in General Relativity

In Einstein–Cartan general relativity, torsion

$$\mathbf{t}^\mu = d\mathbf{e}^\mu + \omega^\mu_\nu \wedge \mathbf{e}^\nu \tag{8.291}$$

and curvature

$$\mathbf{r}^\mu_\nu = d\omega^\mu_\nu + \omega^\mu_\tau \wedge \omega^\tau_\nu \tag{8.292}$$

are linked via the Bianchi identities

$$D\mathbf{t}^\mu = \mathbf{r}^\mu_\nu \wedge \mathbf{e}^\nu, \tag{8.293}$$

$$D\mathbf{r}^\mu_\nu = \mathbf{0}. \tag{8.294}$$

In the absence of torsion, in Einstein general relativity, we have

$$D\mathbf{r}^\mu_\nu = \mathbf{0}, \tag{8.295}$$

$$\mathbf{r}^\mu_\nu \wedge \mathbf{e}^\nu = \mathbf{0}. \tag{8.296}$$

8.9.10 Nonmetricity Revisited

In section 7.1.13 we introduced the notion of nonmetricity, which was captured by the nonmetricity tensor (7.144). In the context of differential forms, we capture nonmetricity by introducing the *nonmetricity one-forms*

$$\mathbf{Q}_{ij} \equiv Dg_{ij} = dg_{ij} - \omega^k_i\, g_{kj} - \omega^k_j\, g_{ik}. \tag{8.297}$$

Now it may be shown that

$$D\mathbf{Q}_{ij} = -\mathbf{r}^k_i\, g_{kj} - \mathbf{r}^k_j\, g_{ik}, \tag{8.298}$$

which may be regarded as a third Bianchi identity in the presence of nonmetricity (Hehl et al., 1995).

Example 8.14 Nonmetricity in General Relativity

In metric-affine gravitation theory, nonmetricity

$$\mathbf{Q}_{\mu\nu} = dg_{\mu\nu} - \omega^\tau_\mu\, g_{\tau\nu} - \omega^\tau_\nu\, g_{\mu\tau} \tag{8.299}$$

$$\mathbf{t}^\mu = \mathrm{d}\mathbf{e}^\mu + \boldsymbol{\omega}^\mu_\nu \wedge \mathbf{e}^\nu$$
$$\mathbf{r}^\mu_\nu = \mathrm{d}\boldsymbol{\omega}^\mu_\nu + \boldsymbol{\omega}^\mu_\tau \wedge \boldsymbol{\omega}^\tau_\nu$$
$$\mathbf{Q}_{\mu\nu} = \mathrm{d}g_{\mu\nu} - \boldsymbol{\omega}^\tau_\mu g_{\tau\nu} - \boldsymbol{\omega}^\tau_\nu g_{\mu\tau}$$

$$\mathrm{D}\mathbf{t}^\mu = \mathbf{r}^\mu_\nu \wedge \mathbf{e}^\nu$$
$$\mathrm{D}\mathbf{r}^\mu_\nu = \mathbf{0}$$
$$\mathrm{D}\mathbf{Q}_{\mu\nu} = -\mathbf{r}_{\mu\nu} - \mathbf{r}_{\nu\mu}$$

Table 8.1: Geometry of metric-affine gravitation theory (e.g., Hehl et al., 1995; Vitagliano et al., 2011). When the nonmetricity tensor vanishes, $\mathbf{Q}_{\mu\nu} = \mathbf{0}$, these equations reduce to Einstein–Cartan geometry (e.g., Hehl et al., 1976; Trautman, 2006). If, furthermore, the torsion vanishes, $\mathbf{t}^\mu = \mathbf{0}$, we obtain the geometry of Einstein general relativity. *Top*: The geometry of metric-affine gravitation theory is captured by the torsion two-forms $\mathbf{t}^\mu$, the curvature two-forms $\mathbf{r}^\mu_\nu$, and the nonmetricity one-forms $\mathbf{Q}_{\mu\nu}$. These quantities determine the field strengths in this framework. *Bottom*: The dynamics of metric-affine gravitation theory are captured by the *post-Riemannian Bianchi identities* linking torsion, curvature, and nonmetricity. It is interesting to note that defects in the field of continuum mechanics can be explained by *intrinsic* torsion, curvature, and nonmetricity, which correspond to dislocations, disclinations, and point defects. The material's intrinsic geometry is also regulated by a group of post-Riemannian Bianchi identities (see, e.g., Roychowdhury & Gupta, 2013; Tromp, 2025).

is governed by a third Bianchi identity,

$$\mathrm{D}\mathbf{Q}_{\mu\nu} = -\mathbf{r}_{\mu\nu} - \mathbf{r}_{\nu\mu}, \tag{8.300}$$

where $\mathbf{r}_{\mu\nu} \equiv \mathbf{r}^\tau_\mu g_{\tau\nu}$. In the Einstein and Einstein–Cartan versions of general relativity, $\mathbf{Q}_{\mu\nu} = \mathbf{0}$ and the curvature two-forms exhibit the antisymmetry

$$\mathbf{r}_{\mu\nu} = -\mathbf{r}_{\nu\mu}. \tag{8.301}$$

The geometry of the various flavors of general relativity is summarized in table 8.1.

8.10 Integration of Forms

8.10.1 Line Integrals

We discuss in chapter 6 that k-forms may be integrated because they may be pulled back by a map φ. This is not true for vectors or general tensors, whose pullback requires both φ and its inverse φ^{-1}, which may not exist. Consider a one-form $\boldsymbol{\omega} = \omega_i \,\mathrm{d}x^i$. We evaluate this form along a parameterized curve C defined in terms of a generating parameter λ, $\lambda_0 \leq \lambda \leq \lambda_1$, by $\{x^j(\lambda)\}$. We write the integral of the one-form along this curve as

$$\int_C \boldsymbol{\omega} = \int_C \omega_i \,\mathrm{d}x^i. \tag{8.302}$$

Since the action of a one-form on a vector produces a real number, let us feed $\mathrm{d}\omega = \omega_i\, \mathrm{d}x^i$ with a small vector, $(d\lambda)\, \mathrm{d}/\mathrm{d}\lambda$, which is tangent to curve C:

$$\mathrm{d}\omega(d\lambda\, \mathrm{d}/\mathrm{d}\lambda) = \omega_i\, \mathrm{d}x^i(d\lambda\, \mathrm{d}/\mathrm{d}\lambda) = \omega_i \frac{\mathrm{d}x^i}{\mathrm{d}\lambda}\, d\lambda. \tag{8.303}$$

Thus, we make the association[8]

$$\int_C \omega \equiv \int_{\lambda_0}^{\lambda_1} \omega_i(x^i(\lambda)) \frac{\mathrm{d}x^i}{\mathrm{d}\lambda}\, d\lambda = \int_{\lambda_0}^{\lambda_1} \omega_i(x^j(\lambda))\, v^i(x^j(\lambda))\, d\lambda. \tag{8.304}$$

The last equality results in the expression for integrating along a curve based on the speed, $v^i = \mathrm{d}x^i/\mathrm{d}\lambda$, along this curve.

Example 8.15 Line Integral

Let us evaluate the integral of a one-form in Cartesian coordinates in $\mathbb{R}^3$. Consider $\omega = x^2\, \mathrm{d}x + x\, y\, \mathrm{d}y + \mathrm{d}z$ along path $c\,(\lambda)$, where $x\,(\lambda) = \lambda$, $y\,(\lambda) = \lambda^2$, and $z\,(\lambda) = 1$, between $\lambda_1 = 0$ and $\lambda_2 = 1$. We write

$$\begin{aligned} \int \omega &= \int x^2\, \mathrm{d}x + x\, y\, \mathrm{d}y + \mathrm{d}z \\ &= \int_0^1 \left(\lambda^2 \frac{\mathrm{d}x}{\mathrm{d}\lambda} + \lambda^3 \frac{\mathrm{d}y}{\mathrm{d}\lambda} + 1 \frac{\mathrm{d}z}{\mathrm{d}\lambda} \right) d\lambda. \end{aligned} \tag{8.305}$$

Since

$$\frac{\mathrm{d}x}{\mathrm{d}\lambda} = 1, \qquad \frac{\mathrm{d}y}{\mathrm{d}\lambda} = 2\,\lambda, \qquad \frac{\mathrm{d}z}{\mathrm{d}\lambda} = 0, \tag{8.306}$$

we obtain

$$\int \omega = \int_0^1 \left(\lambda^2 + 2\,\lambda^4 \right) d\lambda = \frac{11}{15}. \tag{8.307}$$

8.10.2 Surface Integrals

Consider a two-form, $\omega = \sum_{i<j} \omega_{ij}\, \mathrm{d}x^i \wedge \mathrm{d}x^j$, on a two-dimensional surface S. We evaluate this form by parameterizing points on the surface in terms of coordinates $\lambda_0 \le \lambda \le \lambda_1$ and $\mu_0 \le \mu \le \mu_1$, such that $\{x^k(\lambda, \mu)\}$ maps S. We may write the integral of the two-form over the surface as

$$\int_S \omega = \int_S \sum_{i<j} \omega_{ij}\, \mathrm{d}x^i \wedge \mathrm{d}x^j. \tag{8.308}$$

[8] Note the distinction between a small curve element, which is denoted by the Italic "d," and the exterior derivative, which is denoted by the Roman "d."

As we did in the previous section, we feed the two-form $\boldsymbol{\omega}$ two small vectors $(d\lambda)\,\partial/\partial\lambda$ and $(d\mu)\,\partial/\partial\mu$ tangent to curves defined by λ and μ, respectively. We find

$$\int_S \omega = \int_{\lambda_0}^{\lambda_1} \int_{\mu_0}^{\mu_1} \sum_{i<j} \omega_{ij}(x^k(\lambda,\mu)) \, \frac{\partial(x^i,x^j)}{\partial(\lambda,\mu)} \, d\lambda \, d\mu, \tag{8.309}$$

where we introduced the determinants of 2×2 matrices:

$$\frac{\partial(x^i,x^j)}{\partial(\lambda,\mu)} = \det \begin{pmatrix} \frac{\partial x^i}{\partial\lambda} & \frac{\partial x^j}{\partial\lambda} \\ \frac{\partial x^i}{\partial\mu} & \frac{\partial x^j}{\partial\mu} \end{pmatrix}. \tag{8.310}$$

In a three-dimensional manifold, there would only be three such determinants, associated with the three two-forms $\mathrm{d}x^1 \wedge \mathrm{d}x^2$, $\mathrm{d}x^1 \wedge \mathrm{d}x^3$, and $\mathrm{d}x^2 \wedge \mathrm{d}x^3$.

Example 8.16 Surface Integral

In Cartesian components in $\mathbb{R}^3$, consider the two-form

$$\omega = x\,\mathrm{d}x \wedge \mathrm{d}y + y\,\mathrm{d}y \wedge \mathrm{d}z. \tag{8.311}$$

Let surface S be defined by

$$x(u,v) = u+v, \qquad y(u,v) = u^2 - v^2, \qquad z(u,v) = uv, \tag{8.312}$$

where $u, v \in [0,1]$. We have

$$\frac{\partial(x,y)}{\partial(u,v)} = -2(u+v), \qquad \frac{\partial(z,x)}{\partial(u,v)} = -u+v,$$

$$\frac{\partial(y,z)}{\partial(u,v)} = 2(u^2+v^2). \tag{8.313}$$

Thus we find

$$\begin{aligned} \int_S \omega &= \int_S x\,\mathrm{d}x \wedge \mathrm{d}y + y\,\mathrm{d}y \wedge \mathrm{d}z \\ &= \int_0^1 \int_0^1 \left[x(u,v) \frac{\partial(x,y)}{\partial(u,v)} + y(u,v) \frac{\partial(y,z)}{\partial(u,v)} \right] du\,dv \end{aligned}$$

$$= -2\int_0^1\int_0^1 [(u+v)(u+v)-(u^2-v^2)(u^2+v^2)]du\,dv$$
$$= -\frac{7}{3},$$

a real number assigned to a surface.

8.10.3 Volume Integrals

Finally, consider a three-form $\omega = \sum_{i<j<k} \omega_{ijk}\, \mathrm{d}x^i \wedge \mathrm{d}x^j \wedge \mathrm{d}x^k$ over a three-dimensional volume V. We parameterize the volume in terms of coordinates $\lambda_0 \le \lambda \le \lambda_1$, $\mu_0 \le \mu \le \mu_1$, and $\tau_0 \le \tau \le \tau_1$, such that $\{x^\ell(\lambda,\mu,\tau)\}$ describes points in the volume. We may write the integral of the three-form over the volume as

$$\int_V \omega = \int_V \sum_{i<j<k} \omega_{ijk}\, \mathrm{d}x^i \wedge \mathrm{d}x^j \wedge \mathrm{d}x^k. \tag{8.314}$$

Upon feeding the three-form ω three small vectors $(d\lambda)\,\partial/\partial\lambda$, $(d\mu)\,\partial/\partial\mu$, and $(d\tau)\,\partial/\partial\tau$ tangent to curves defined by λ, μ, and τ, respectively, we find

$$\int_V \omega = \int_{\lambda_0}^{\lambda_1}\int_{\mu_0}^{\mu_1}\int_{\tau_0}^{\tau_1} \sum_{i<j<k} \omega_{ijk}(x^\ell(\lambda,\mu,\tau))\, \frac{\partial(x^i,x^j,x^k)}{\partial(\lambda,\mu,\tau)}\, d\lambda\, d\mu\, d\tau,$$

where we introduced the determinants of 3×3 matrices

$$\frac{\partial(x^i,x^j,x^k)}{\partial(\lambda,\mu,\tau)} = \det\begin{pmatrix} \frac{\partial x^i}{\partial\lambda} & \frac{\partial x^j}{\partial\lambda} & \frac{\partial x^k}{\partial\lambda} \\ \frac{\partial x^i}{\partial\mu} & \frac{\partial x^j}{\partial\mu} & \frac{\partial x^k}{\partial\mu} \\ \frac{\partial x^i}{\partial\tau} & \frac{\partial x^j}{\partial\tau} & \frac{\partial x^k}{\partial\tau} \end{pmatrix}. \tag{8.315}$$

In a three-dimensional manifold, there exists only one volume form, namely, $\mathrm{d}x^1 \wedge \mathrm{d}x^2 \wedge \mathrm{d}x^3$, and only one determinant, namely, $\partial(x^1,x^2,x^3)/\partial(\lambda,\mu,\tau)$.

Example 8.17 Volume Integral

Consider the three-form

$$\omega = (x+z)\,\mathrm{d}x \wedge \mathrm{d}y \wedge \mathrm{d}z, \tag{8.316}$$

and the volume $V = [0,1]\times[0,1]\times[0,1]$. We have

$$\int_V \omega = \int_V (x+z)\,\mathrm{d}x \wedge \mathrm{d}y \wedge \mathrm{d}z = \int_0^1\int_0^1\int_0^1 (x+z)\,dx\,dy\,dz = 1.$$

8.11 Stokes's Theorem

As stated in the preface of the book by Spivak (1965),

> Stokes' Theorem may be considered as a case study in the value of generalization.

Spivak notes that, despite the "horde of difficult definitions," the statement and proof of the theorem are relatively simple. This book takes a similar approach by emphasizing definitions and notations, which may not be immediately accessible to readers but provide a solid foundation for deeper understanding.

Let ω denote an $(n-1)$-form on an n-dimensional manifold, V, which we can view as a volume. *Stokes's theorem* states that the integral of the exterior derivative of this form, $\mathrm{d}\omega$, over V is equal to the integral of the form itself, ω, over ∂V, which is the boundary of V, and can be viewed as an $(n-1)$-dimensional hypersurface (e.g., Spivak, 1965; Frankel, 2004)

$$\int_V \mathrm{d}\omega = \int_{\partial V} \omega. \tag{8.317}$$

Given the properties of the exterior derivative, on the left-hand side, we have an n-dimensional integral of an n-form, and on the right-hand side, an $(n-1)$-dimensional integral of an $(n-1)$-form. A standard requirement of integrability is satisfied: integration of a k-form is defined only on a k-dimensional manifold. The coordinate independence of identity (8.317) is ensured if both are definite integrals: the integration results in a number, not in another form; this is consistent with standard definitions of integrations of such forms: only definite integrals are allowed, (see, e.g., Schutz, 1980).

The volume V and its surface ∂V must agree in the sense that if $\{x^i\}$ are n-coordinates on a portion of V, with x^1 chosen such that $x^1 = 0$ on the boundary ∂V, then for a point $\mathcal{S}$ on the boundary the orientation $\partial_1 \mathcal{S} \wedge \partial_2 \mathcal{S} \wedge \cdots \wedge \partial_n \mathcal{S}$ on V implies the orientation $\partial_2 \mathcal{S} \wedge \cdots \wedge \partial_n \mathcal{S}$ on ∂V. The reason one can integrate differential forms but not general tensors is that, as discussed in sectionr 6.2.1, one can always *pullback* a form from one manifold to another manifold, but not vectors or general tensors.

Let us explore Stokes's theorem in $\mathbb{R}^3$, where it encompasses the traditional Green's, Gauss's, and Stokes's theorems.

8.11.1 Fundamental Theorem of Calculus

Consider a one-dimensional domain given by an interval $V = [x_1, x_2]$ on the x-axis. Let f be a differentiable zero-form, and let $\mathcal{S}_1 = \mathcal{S}(x_1)$ and $\mathcal{S}_2 = \mathcal{S}(x_2)$. In view of equation (8.317), we may write

$$\int_V \mathrm{d}\,\mathrm{f} = \mathrm{f}(\mathcal{S}_2) - \mathrm{f}(\mathcal{S}_1). \tag{8.318}$$

Equivalently, in coordinates we may write

$$\int_{x_1}^{x_2} f'(x)\, dx = f(x_2) - f(x_1), \tag{8.319}$$

with $f'(x) \equiv df/dx$, which is the standard form of the fundamental theorem of calculus.

In equation (8.319), the integrand represents a one-form expressed in coordinates x. If we change the coordinates, the one-form remains the same in the sense that its integral value stays constant, although its functional dependence may change. To emphasize this coordinate independence, we denote the integrand by $\mathrm{d}\,\mathrm{f}$, as in expression (8.318). This expression is a statement of Stokes's theorem (8.317), where V is the interval $[x_1, x_2]$ and ∂V is its oriented boundary, consisting of the ordered points x_1 and x_2. The notation used in (8.317), introduced by Élie Cartan, is known as the *fundamental theorem of exterior calculus* and allows us to generalize integral theorems of vector calculus. We discuss these generalizations in sections 8.11.2 and 8.11.3.

Looking at the right-hand side of (8.317) in light of (8.319), we see a zero-form integrated over a zero-dimensional boundary: values at discrete points. On the other hand, the left-hand side shows a one-form integrated over a one-dimensional region. This is a general pattern: a p-form, $\boldsymbol{\omega}$, integrated along a p-dimensional boundary, is equal to a $(p+1)$-form integrated over a $(p+1)$-dimensional region.

We require the boundaries and regions to be *compact* and *oriented*. Compactness means there are no infinite extensions or "holes" in the region. For example, $\mathbb{R}$ and $[-1,0) \cup (0,1]$ are not compact due to extension and removal of zero, respectively, but $[-1,1]$ is compact. Compactness also implies that the region is simply connected, meaning any closed curve can be shrunk to a point by an infinitesimal deformation within the domain.

As for orientation, a zero-dimensional manifold, which is a set of discrete points, is oriented if we can assign a positive or negative sign to each point. A one-dimensional manifold, a curve, is oriented if it has a positive and negative direction of displacement. A two-dimensional manifold is oriented if it has a positive and negative direction of rotation of a tangent vector at every point. A formal definition of orientability is introduced in section 3.5.

A boundary, ∂N, of a compact and oriented $(n+1)$-manifold, N, consists of points of N that do not lie in its interior. The boundary of a boundary is empty: $\partial\,\partial N = \partial^2 N = \emptyset$; in other words, $\partial^2 = 0$, which corresponds to $\mathrm{d}^2 = \mathbf{0}$, stated in expression (8.86).

For a disc, which is two-dimensional, the boundary is a one-dimensional compact and oriented region; for a volume, which is three-dimensional, the boundary is a two-dimensional compact and oriented region. Such cases are discussed in sections 8.11.2 and 8.11.3, respectively.

8.11.2 Green's Theorem

In $\mathbb{R}^2$, consider an area D with edge ∂D and the one-form

$$\boldsymbol{\omega} = \omega_x\,\mathrm{d}x + \omega_y\,\mathrm{d}y. \tag{8.320}$$

Then $\mathrm{d}\boldsymbol{\omega}$ is the two-form

$$\mathrm{d}\boldsymbol{\omega} = (\partial_x\omega_y - \partial_y\omega_x)\,\mathrm{d}x \wedge \mathrm{d}y. \tag{8.321}$$

Thus Stokes's theorem (8.317) implies

$$\int_D (\partial_x\omega_y - \partial_y\omega_x)\,\mathrm{d}x \wedge \mathrm{d}y = \int_{\partial D} \omega_x\,\mathrm{d}x + \omega_y\,\mathrm{d}y. \tag{8.322}$$

Using the results in sections 8.10.2 and 8.10.1, we deduce that the last equation is Green's theorem.

8.11.3 Gauss's Theorem

Next, in $\mathbb{R}^3$, consider a volume V with surface ∂V and the interior product (8.31) of a vector $\mathbf{u}$ with the volume form, that is, the flux $\omega = \mathbf{i_u}\epsilon = \mathbf{u} \cdot \epsilon$. According to the formulation on page 150, the exterior derivative, $\mathrm{d}\omega$, is the divergence of the vector field times the volume form: $\epsilon \operatorname{div} \mathbf{u}$. Thus, we find immediately Gauss's theorem:

$$\int_{\partial V} \mathbf{u} \cdot \epsilon = \int_V \epsilon \operatorname{div} \mathbf{u}, \tag{8.323}$$

or, alternatively, using the interior product and exterior derivative,

$$\int_{\partial V} \mathbf{i_u}\epsilon = \int_V \mathrm{d}(\mathbf{u} \cdot \epsilon). \tag{8.324}$$

8.11.4 Stokes's Theorem

In $\mathbb{R}^3$, the results in section 8.2 immediately imply the classical Stokes's theorem:

$$\int_S \mathrm{d}\omega = \int_S \operatorname{curl} \omega \cdot \epsilon = \int_{\partial S} \omega, \tag{8.325}$$

where S is a surface and ∂S its edge.

8.11.5 Variational Principles

Much of contemporary physics is based on the concept of *variational principles*. These principles employ the calculus of variations to solve mathematical and physical problems. Later in this book, we explore the use of variational principles in general relativity and continuum mechanics. In this section, we consider variational principles in their most generic form. Let V denote n-dimensional spacetime with $(n-1)$-dimensional boundary ∂V, and let φ^μ, $\mu = 1, \ldots, n$, denote a generic set of n k-form fields. For example, these fields might represent the scalar particle motion in continuum mechanics. The *Lagrangian density* $\boldsymbol{L}$ is a differential n-form which depends on the fields φ^μ and their (four-dimensional) exterior covariant derivatives $\mathrm{D}\varphi^\mu$:

$$\boldsymbol{L} = \boldsymbol{L}(\varphi^\mu, \mathrm{D}\varphi^\mu). \tag{8.326}$$

The *action* is defined by the spacetime integral

$$I = \int_V \boldsymbol{L}. \tag{8.327}$$

Thus, given fields φ^μ and their exterior covariant derivatives $\mathrm{D}\varphi^\mu$, the calculation of the action returns a number. The objective is to find fields φ^μ such that the action I is stationary with respect to perturbations in the fields, $\delta\varphi^\mu$. Using the calculus of variations, we have

$$
\begin{aligned}
\delta I &= \int_V \delta \boldsymbol{L} \\
&= \int_V \delta\varphi^\mu \wedge \frac{\partial \boldsymbol{L}}{\partial \varphi^\mu} + \delta \mathrm{D}\varphi^\mu \wedge \frac{\partial \boldsymbol{L}}{\partial \mathrm{D}\varphi^\mu} \\
&= \int_V \delta\varphi^\mu \wedge \frac{\partial \boldsymbol{L}}{\partial \varphi^\mu} + \mathrm{D}\delta\varphi^\mu \wedge \frac{\partial \boldsymbol{L}}{\partial \mathrm{D}\varphi^\mu} \\
&= \int_V \delta\varphi^\mu \wedge \left[\frac{\partial \boldsymbol{L}}{\partial \varphi^\mu} - (-1)^k \mathrm{D}\frac{\partial \boldsymbol{L}}{\partial \mathrm{D}\varphi^\mu}\right] + \mathrm{d}\left[\delta\varphi^\mu \wedge \frac{\partial \boldsymbol{L}}{\partial \mathrm{D}\varphi^\mu}\right] \\
&= \int_V \delta\varphi^\mu \wedge \left[\frac{\partial \boldsymbol{L}}{\partial \varphi^\mu} - (-1)^k \mathrm{D}\frac{\partial \boldsymbol{L}}{\partial \mathrm{D}\varphi^\mu}\right] + \int_{\partial V} \delta\varphi^\mu \wedge \frac{\partial \boldsymbol{L}}{\partial \mathrm{D}\varphi^\mu},
\end{aligned}
\tag{8.328}
$$

where in the third equality we used the fact that the variation of $\mathrm{D}\varphi^\mu$ is determined by the variation of $\delta\varphi^\mu$, that is, $\delta\mathrm{D}\varphi^\mu = \mathrm{D}\delta\varphi^\mu$, and in the last equality, we used Stokes's theorem (8.317). Stationarity of the action,

$$\delta I = 0, \tag{8.329}$$

implies the *Euler–Lagrange equations*

$$\frac{\partial \boldsymbol{L}}{\partial \varphi^\mu} - (-1)^k \mathrm{D}\frac{\partial \boldsymbol{L}}{\partial \mathrm{D}\varphi^\mu} = \mathbf{0}, \qquad \text{in } V, \tag{8.330}$$

and the boundary conditions

$$\frac{\partial \boldsymbol{L}}{\partial \mathrm{D}\varphi^\mu} = \mathbf{0}, \qquad \text{in } \partial V. \tag{8.331}$$

Sometimes the surface integral in the last equality in (8.328) is assumed to vanish by restricting the variations $\delta\varphi^\mu$ to be zero in ∂V.

The results in this section illustrate that once a suitable Lagrangian density (8.326) has been identified, the governing equations follow naturally in the form of the Euler–Lagrange equations (8.330). Snell's law, Fermat's principle, Hamilton's principle, the Rayleigh–Ritz method, and the Einstein–Hilbert and Palatini actions in general relativity are examples of applications of variational principles.

8.11.6 Noether's Theorem

Noether's theorem (Noether, 1918) states that every differentiable symmetry of the action of a physical system with conservative forces has a corresponding conservation law. Our goal in this section is to investigate possible symmetries of the action (8.327), and connect those symmetries to a conservation law. A *continuous symmetry* is a *transformation* of the k-form fields, $\delta\varphi^\mu$, which makes the Lagrangian density n-form $\boldsymbol{L}$ transform as

$$\delta \boldsymbol{L} = -\,\mathrm{d}\mathbf{K} \tag{8.332}$$

where $\mathbf{K}$ denotes a generic $(n-1)$-form which either vanishes on the boundary or the transformation of the fields, $\delta\varphi^\mu$, is such that it vanishes on the boundary. Then, in the transformation of the action (8.327), the exterior derivative $\mathbf{dK}$ integrates out, and the action remains stationary under this specific transformation.

According to Noether's theorem, for every continuous symmetry, there exists a *Noether current*: an $(n-1)$-form $\mathbf{J}$ that is *conserved* in the following sense. We compute the transformation of the Lagrangian density (8.326) induced by transformations of the fields φ^μ:

$$\begin{aligned}\delta\boldsymbol{L} &= \delta\varphi^\mu \wedge \frac{\partial\boldsymbol{L}}{\partial\varphi^\mu} + \delta\mathrm{D}\varphi^\mu \wedge \frac{\partial\boldsymbol{L}}{\partial\mathrm{D}\varphi^\mu} \\ &= \delta\varphi^\mu \wedge \left[\frac{\partial\boldsymbol{L}}{\partial\varphi^\mu} - (-1)^k\,\mathrm{D}\frac{\partial\boldsymbol{L}}{\partial\mathrm{D}\varphi^\mu}\right] + (-1)^k\,\delta\varphi^\mu \wedge \mathrm{D}\frac{\partial\boldsymbol{L}}{\partial\mathrm{D}\varphi^\mu} \\ &\quad + \mathrm{D}\delta\varphi^\mu \wedge \frac{\partial\boldsymbol{L}}{\partial\mathrm{D}\varphi^\mu} \\ &= \delta\varphi^\mu \wedge \left[\frac{\partial\boldsymbol{L}}{\partial\varphi^\mu} - (-1)^k\,\mathrm{D}\frac{\partial\boldsymbol{L}}{\partial\mathrm{D}\varphi^\mu}\right] + \mathrm{d}\left[\delta\varphi^\mu \wedge \frac{\partial\boldsymbol{L}}{\partial\mathrm{D}\varphi^\mu}\right].\end{aligned} \tag{8.333}$$

Upon equating the final equality with (8.332), we find that

$$\mathrm{d}\mathbf{J} = -\,\delta\varphi^\mu \wedge \left[\frac{\partial\boldsymbol{L}}{\partial\varphi^\mu} - (-1)^k\,\mathrm{D}\frac{\partial\boldsymbol{L}}{\partial\mathrm{D}\varphi^\mu}\right], \tag{8.334}$$

where we have defined the *Noether current*

$$\mathbf{J} \equiv \delta\varphi^\mu \wedge \frac{\partial\boldsymbol{L}}{\partial\mathrm{D}\varphi^\mu} + \mathbf{K}. \tag{8.335}$$

We conclude that, for fields φ^μ that satisfy the Euler–Lagrange equations (8.330), thus making the right-hand side of expression (8.334) vanish, if a transformation $\delta\varphi^\mu$ leaves the action (8.327) unchanged, then there exists a corresponding Noether current defined by (8.335) which is a closed form:

$$\mathrm{d}\mathbf{J} = \mathbf{0}. \tag{8.336}$$

This closed form is the *conservation law* that corresponds to the symmetric transformation $\delta\varphi^\mu$. According to Poincaré's lemma, discussed in section 8.7, if an $(n-1)$-form $\mathbf{J}$ is closed, then there exists an $(n-2)$-form $\boldsymbol{\omega}$ such that $\mathbf{J} = \mathrm{d}\boldsymbol{\omega}$, and $\boldsymbol{\omega}$ may be regarded as the potential giving rise to the Noether current.

Diffeomorphisms

Let us consider transformations of the k-form fields, $\delta\varphi^\mu$, that are *diffeomorphisms*, that is, smooth, invertible functions that map one differentiable manifold to another. In this case, we consider *infinitesimal* transformations of the form

$$\delta\varphi^\mu = -\,\mathfrak{L}_{\boldsymbol{\xi}}\varphi^\mu = -\,\mathbf{i}_{\boldsymbol{\xi}}\mathrm{D}\varphi^\mu - \mathrm{D}\mathbf{i}_{\boldsymbol{\xi}}\varphi^\mu, \tag{8.337}$$

where in the second equality we used the covariant "Cartan magic formula" (8.264). The transformations (8.337) are generated by infinitesimal vector fields $\boldsymbol{\xi}$ which we express in the form

$$\boldsymbol{\xi} \equiv \xi^{\nu} \lambda^{\mu}{}_{\nu} \, \mathbf{e}_{\mu}, \tag{8.338}$$

such that for infinitesimal translations in time and space $\lambda^{\mu}{}_{\nu} = \delta^{\mu}{}_{\nu}$, whereas for infinitesimal spatial rotations $\lambda^{i}{}_{k} = \epsilon^{i}{}_{jk} \, x^{j}$ and $\lambda^{0}{}_{\nu} = 0$.

The Lagrangian density $\boldsymbol{L}$ is an n-form in the cotangent space of n-dimensional spacetime V, and thus $\mathrm{d}\boldsymbol{L} = \mathbf{0}$. Consequently, the infinitesimal perturbation of the Lagrangian density under the vector field $\boldsymbol{\xi}$ is

$$\delta \boldsymbol{L} = -\mathcal{L}_{\boldsymbol{\xi}} \boldsymbol{L} = -\mathrm{d}\mathbf{i}_{\boldsymbol{\xi}} \boldsymbol{L}, \tag{8.339}$$

which is an exterior derivative of the form (8.332), with $\mathbf{K} = \mathbf{i}_{\boldsymbol{\xi}} \boldsymbol{L}$. We conclude that for a diffeomorphism the Noether current (8.335) takes the form

$$\mathbf{J} = -\mathfrak{L}_{\boldsymbol{\xi}} \varphi^{\mu} \wedge \frac{\partial \boldsymbol{L}}{\partial \mathrm{D}\varphi^{\mu}} + \mathbf{i}_{\boldsymbol{\xi}} \boldsymbol{L}. \tag{8.340}$$

By selecting infinitesimal translations in time and space or infinitesimal rotations based on (8.338), we may, respectively, obtain expressions of conservation of energy, linear momentum, and angular momentum by substituting the corresponding Noether current (8.340) in expression (8.336).

Finally, if the fields φ^{μ} are *scalar* fields, then expression (8.337) reduces to $\delta\varphi^{\mu} = -\mathfrak{L}_{\boldsymbol{\xi}} \varphi^{\mu} = -\mathbf{i}_{\boldsymbol{\xi}} \mathrm{D}\varphi^{\mu}$, and we may express the Noether current (8.340) in the form

$$\mathbf{J} = -\mathbf{i}_{\boldsymbol{\xi}} \left(\mathrm{D}\varphi^{\mu} \wedge \frac{\partial \boldsymbol{L}}{\partial \mathrm{D}\varphi^{\mu}} - \boldsymbol{L} \right). \tag{8.341}$$

8.11.7 Applications

We conclude our discussion of Stokes's theorem by presenting several examples of its application in continuum mechanics and general relativity. The complete set of conservation laws in continuum mechanics is summarized in table 8.2.

Example 8.18 Reynolds Transport Theorem

In this example, we investigate the rate of change of some extensive mechanical or thermodynamical scalar quantity with volumetric density q contained in a *comoving volume* V with boundary ∂V, which both always consist of the same matter. Thus, matter in a comoving volume moves along with the velocity, $\mathbf{v}$. This situation is analogous to *Leibniz's rule* in one dimension. We introduce the three-form

$$\mathbf{q} \equiv \mathrm{q} \, \boldsymbol{\epsilon} = q \, \tfrac{1}{3!} \, \epsilon_{ijk} \, \mathbf{e}^{i} \wedge \mathbf{e}^{j} \wedge \mathbf{e}^{k}. \tag{8.342}$$

The corresponding volume transport of "q-stuff" may be defined in terms of the velocity as

$$\mathbf{v}\cdot\mathbf{q} \equiv q\, v^i\, \tfrac{1}{2}\, \epsilon_{ijk}\, \mathbf{e}^j \wedge \mathbf{e}^k. \tag{8.343}$$

Reynolds transport theorem states that

$$\frac{d}{dt}\int_V \mathbf{q} = \int_V \mathrm{d}_t\mathbf{q} + \int_{\partial V} \mathbf{v}\cdot\mathbf{q}. \tag{8.344}$$

The first term on the right-hand side of equation (8.344) captures local changes in q-stuff within the comoving volume V based on the Euler derivative discussed in section 7.2, whereas the second term captures the flux of q-stuff advected along with the comoving surface ∂V. Using the generalization of Stokes's theorem (8.317), Reynolds theorem becomes

$$\frac{d}{dt}\int_V \mathbf{q} = \int_V [\mathrm{d}_t\mathbf{q} + \mathrm{d}(\mathbf{v}\cdot\mathbf{q})]. \tag{8.345}$$

We may obtain an alternative form of Reynolds theorem based on relationship (6.31) between the referential and spatial volume forms and definition (7.216) of the Lie derivative with respect to the flow of matter. Let V^0 denote the comoving volume V at the referential time $t=0$. We have

$$\begin{aligned}\frac{d}{dt}\int_V \mathbf{q} &= \frac{d}{dT}\int_{V^0} \varphi^*\mathbf{q}\\ &= \int_{V^0} \partial_T\varphi^*\mathbf{q}\\ &= \int_V \varphi_*\partial_T\varphi^*\mathbf{q}\\ &= \int_V \mathcal{L}_{\mathbf{v}}\mathbf{q}.\end{aligned} \tag{8.346}$$

What if the field q takes on tensor values, for example, in the form of the vector-valued three-form $\mathbf{q}^i \equiv q^i\,\epsilon$? As long as these components refer to a constant *Cartesian* basis $\{\mathbf{e}_i\}$, Reynolds theorem remains valid, because the constant basis can be taken outside the integrals, leading to the component-by-component requirement (for further discussions, see, e.g., Misner et al., 2017, Box 5.4 II)

$$\begin{aligned}\frac{d}{dt}\int_V \mathbf{q}^i &= \int_V [\mathrm{d}_t\mathbf{q}^i + \mathrm{d}(\mathbf{v}\cdot\mathbf{q}^i)]\\ &= \int_V \mathcal{L}_{\mathbf{v}}\mathbf{q}^i.\end{aligned} \tag{8.347}$$

Mass	$\mathcal{L}_{\mathbf{v}}\boldsymbol{\rho} = \mathbf{0}$
Linear momentum	$\mathfrak{L}_{\mathbf{v}}\mathbf{p}^i = \mathrm{D}\boldsymbol{\sigma}^i + \mathbf{f}^i$
Angular momentum	$\boldsymbol{\sigma}^i \wedge \mathbf{e}^j - \boldsymbol{\sigma}^j \wedge \mathbf{e}^i = \mathbf{0}$
First law of thermodynamics	$\mathcal{L}_{\mathbf{v}}\mathbf{e} = \boldsymbol{\sigma}_i \wedge \mathrm{D}v^i + \mathbf{h} - \mathrm{d}\mathbf{H}$
Second law of thermodynamics	$\boldsymbol{\rho}\,\theta\,\mathcal{L}_{\mathbf{v}}\mathrm{S} \geq \mathcal{L}_{\mathbf{v}}\mathbf{e} - \boldsymbol{\sigma}_i \wedge \mathrm{D}v^i + \mathrm{d}\log\theta \wedge \mathbf{H}$
Caloric equation of state	$\mathbf{e} = \mathbf{e}(\,\mathrm{D}\varphi^i, \mathrm{S}, \xi\,)$
Thermodynamic stress	$\boldsymbol{\sigma}_i = \frac{\partial \mathbf{e}}{\partial \mathrm{D}\varphi^i}$
Material velocity	$v^i = \mathfrak{L}_{\mathbf{v}}\varphi^i$

Mass	$\mathcal{L}_{\mathbf{v}}\boldsymbol{\rho} = \mathbf{0}$
Linear momentum	$\mathfrak{L}_{\mathbf{v}}\mathbf{p}^i = \mathrm{D}\boldsymbol{\sigma}^i + \mathbf{f}^i$
Angular momentum	$\mathfrak{L}_{\mathbf{v}}\boldsymbol{\theta}^{ij} = \mathrm{D}\boldsymbol{\mu}^{ij} + \mathbf{c}^{ij} + \boldsymbol{\sigma}^i \wedge \mathbf{e}^j - \boldsymbol{\sigma}^j \wedge \mathbf{e}^i$
First law of thermodynamics	$\mathcal{L}_{\mathbf{v}}\mathbf{e} = \boldsymbol{\sigma}_i \wedge \mathrm{D}v^i - \varpi_{ij}\,\boldsymbol{\sigma}^i \wedge \mathbf{e}^j$
	$+\,\frac{1}{2}\,\boldsymbol{\mu}_i{}^j \wedge \mathrm{D}\varpi^i{}_j + \mathbf{h} - \mathrm{d}\mathbf{H}$
Second law of thermodynamics	$\boldsymbol{\rho}\,\theta\,\mathcal{L}_{\mathbf{v}}\mathrm{S} \geq \mathcal{L}_{\mathbf{v}}\mathbf{e} - \boldsymbol{\sigma}_i \wedge \mathrm{D}v^i + \varpi_{ij}\,\boldsymbol{\sigma}^i \wedge \mathbf{e}^j$
	$-\,\frac{1}{2}\,\boldsymbol{\mu}_i{}^j \wedge \mathrm{D}\varpi^i{}_j + \mathrm{d}\log\theta \wedge \mathbf{H}$
Caloric equation of state	$\mathbf{e} = \mathbf{e}(\,\Psi^i{}_j\,\mathrm{D}\varphi^j, (\Psi^t)^i{}_k\,\mathrm{D}\Psi^k{}_j, \mathrm{S}, \xi\,)$
Thermodynamic stress	$\boldsymbol{\sigma}_i \equiv \frac{\partial \mathbf{e}}{\partial(\Psi^j{}_k\,\mathrm{D}\varphi^k)}\,\Psi^j{}_i$
Thermodynamic couple stress	$\boldsymbol{\mu}_i{}^j \equiv -\,\frac{\partial \mathbf{e}}{\partial[(\Psi^t)^i{}_k\,\mathrm{D}\Psi^k{}_j]}$
Material velocity	$v^i = \mathfrak{L}_{\mathbf{v}}\varphi^i$
Material total angular velocity	$\varpi^i{}_j = -\,(\Psi^t)^i{}_k\,\mathfrak{L}_{\mathbf{v}}\Psi^k{}_j$

Table 8.2: Complete set of the laws of continuum mechanics in tensor-valued forms. The top panel excludes complications due to intrinsic particle rotation, whereas the bottom panel includes those complications. The indices i and j refer to general coordinates for the vector/tensor values of the forms, and the Lie and exterior derivatives of such tensor-valued forms are covariant, that is, $\mathfrak{L}_{\mathbf{v}}$ instead of $\mathcal{L}_{\mathbf{v}}$ and D instead of d, as discussed in examples 8.9 and 8.12. The underlying form basis is completely general, including both Eulerian and convected Lagrangian forms. The former capture Eulerian descriptions of the conservation laws of continuum mechanics, whereas the latter capture Lagrangian descriptions. The laws involve the mass three-form $\boldsymbol{\rho}$, the vector-valued linear momentum three-forms $\mathbf{p}^i$, the vector-valued Cauchy-stress two-forms $\boldsymbol{\sigma}^i$, the vector-valued volume force three-forms $\mathbf{f}^i$, the tensor-valued material spin three-forms $\boldsymbol{\theta}^{ij}$, the tensor-valued couple two-forms $\boldsymbol{\mu}^{ij}$, the tensor-valued volume torque three-forms $\mathbf{c}^{ij}$, the internal energy three-form $\mathbf{e}$, the heat production three-form $\mathbf{h}$, the heat flux two-form $\mathbf{H}$, the specific entropy density three-form S, thermodynamic internal state variables ξ, and the absolute temperature θ. The quantity $\varpi_{ij} = W_{ij} + w_{ij}$ captures the total angular velocity of a particle, which includes its vorticity W_{ij} as well as its intrinsic material angular velocity w_{ij}. Particle motion is denoted by φ^i and total particle rotation by $\Psi^i{}_j$.

Example 8.19 Conservation of Mass

Suppose the thermodynamical scalar quantity q introduced in example 8.18 is the mass density ρ. In that case, we consider the *mass three-form*

$$\boldsymbol{\rho} \equiv \rho\, \boldsymbol{\epsilon} = \rho\, \tfrac{1}{3!}\, \epsilon_{ijk}\, \mathbf{e}^i \wedge \mathbf{e}^j \wedge \mathbf{e}^k. \tag{8.348}$$

Upon defining the *mass flux two-form* or *linear momentum two-form*

$$\mathbf{v} \cdot \boldsymbol{\rho} = \tfrac{1}{2}\, \rho\, v^i\, \epsilon_{ijk}\, \mathbf{e}^j \wedge \mathbf{e}^k \tag{8.349}$$

and using Reynolds theorem (8.345), *conservation of mass* may be expressed as

$$\frac{d}{dt} \int_V \boldsymbol{\rho} = \int_V [\mathrm{d}_t \boldsymbol{\rho} + \mathrm{d}(\mathbf{v} \cdot \boldsymbol{\rho})] = 0. \tag{8.350}$$

Since the final result must hold for any comoving volume, we have the *continuity equation*

$$\mathrm{d}_t \boldsymbol{\rho} + \mathrm{d}(\mathbf{v} \cdot \boldsymbol{\rho}) = \mathbf{0}. \tag{8.351}$$

Using the alternative form of Reynolds theorem (8.346), conservation of mass may be expressed in terms of the Lie derivative with respect to the flow of matter as

$$\mathcal{L}_{\mathbf{v}} \boldsymbol{\rho} = \mathbf{0}. \tag{8.352}$$

The equivalence of expressions (8.353) and (8.54) is a result of the Cartan magic formula (8.138) for the mass three-form $\boldsymbol{\rho}$.

Example 8.20 Conservation of Linear Momentum

To develop a form version of *conservation of linear momentum* in continuum mechanics, in Cartesian Eulerian spatial coordinates, we define the vector-valued *linear momentum three-forms* in terms of the linear momentum two-form (8.349) as

$$\mathbf{p}^i \equiv \mathbf{v} \cdot \boldsymbol{\rho} \wedge \mathbf{e}^i = v^i\, \boldsymbol{\rho}, \tag{8.353}$$

and we introduce the vector-valued *volume force three-forms*

$$\mathbf{f}^i \equiv g^i\, \boldsymbol{\rho}, \tag{8.354}$$

where g^i measures the components of an applied force per unit mass. As discussed in example 8.6, tractions may be captured by regarding the Cauchy stress as a Cartesian

Eulerian vector-valued two-form, which may be expressed in either an Eulerian or Lagrangian coframe as

$$\begin{aligned}\sigma^i &= \tfrac{1}{2}\,\sigma^i_{jk}\,\mathbf{e}^j \wedge \mathbf{e}^k \\ &= \tfrac{1}{2}\,P^i_{JK}\,\mathbf{e}^J \wedge \mathbf{e}^K.\end{aligned} \tag{8.355}$$

Conservation of linear momentum for a comoving volume V with surface S then implies a balance between the rate of change of total linear momentum and the total applied surface and body forces:

$$\frac{d}{dt}\int_V \mathbf{p}^i = \int_S \sigma^i + \int_V \mathbf{f}^i. \tag{8.356}$$

The form version of Reynolds transport theorem, discussed in example 8.18, states that the rate of change of total linear momentum of a comoving volume V with surface S is determined by

$$\frac{d}{dt}\int_V \mathbf{p}^i = \int_V \mathrm{d}_t\mathbf{p}^i + \int_S \mathbf{v}\cdot\mathbf{p}^i, \tag{8.357}$$

which implies that the balance of total linear momentum (8.356) may be rewritten in the form

$$\int_V \mathrm{d}_t\mathbf{p}^i = \int_S (\sigma^i - \mathbf{v}\cdot\mathbf{p}^i) + \int_V \mathbf{f}^i. \tag{8.358}$$

The generalization of Stokes's theorem (8.317) then implies that

$$\int_V \mathrm{d}_t\mathbf{p}^i = \int_V [\mathrm{d}(\sigma^i - \mathbf{v}\cdot\mathbf{p}^i) + \mathbf{f}^i]. \tag{8.359}$$

Since this result needs to hold for any comoving volume V, we find the conservation law of linear momentum

$$\mathrm{d}_t\mathbf{p}^i + \mathrm{d}(\mathbf{v}\cdot\mathbf{p}^i) = \mathrm{d}\sigma^i + \mathbf{f}^i. \tag{8.360}$$

In terms of the Lie derivative, invoking (8.347), we have

$$\mathcal{L}_{\mathbf{v}}\mathbf{p}^i = \mathrm{d}\sigma^i + \mathbf{f}^i. \tag{8.361}$$

Results (8.360) and (8.361) are valid provided the index i refers to a Cartesian, constant spatial reference frame. In general spatial coordinates, these equations need to be modified by replacing the Lie and exterior derivatives of forms with their covariant counterparts, as discussed in sections 8.9.5 and 8.9.6. Thus, the invariant forms of conservation of linear momentum are

$$\mathrm{d}_t \mathbf{p}^i + \mathrm{D}(\mathbf{v} \cdot \mathbf{p}^i) = \mathrm{D}\boldsymbol{\sigma}^i + \mathbf{f}^i, \tag{8.362}$$

and

$$\mathfrak{L}_{\mathbf{v}} \mathbf{p}^i = \mathrm{D}\boldsymbol{\sigma}^i + \mathbf{f}^i. \tag{8.363}$$

By expressing equations (8.362) and (8.363) in either Eulerian or Lagrangian coframes, using the transformations discussed in examples 8.9 and 8.12, Tromp (2025) obtains the classical Eulerian and Lagrangian conservation laws in terms of the Cauchy stress and the first Piola–Kirchhoff stress, respectively. In example 8.26, we obtain these equations of motion using a variational approach.

Example 8.21 Conservation of Angular Momentum

In this example, we explore a version of continuum mechanics in which particles may have an *intrinsic spin*. A continuum that supports spin is referred to as a *Cosserat micropolar medium*, in reference to the Cosserat brothers (Cosserat & Cosserat, 1896, 1907, 1909), who inspired Cartan's work on torsion and spin. The Cosserat equations are discussed by Truesdell & Toupin (1960), and more modern descriptions of the theory are presented by Malvern (1969) and Nowacki (1986); the latter also discusses the physical properties of micropolar media. Maugin & Metrikine (2010) provide a full overview of both Cosserat theory and its generalizations, with Maugin (2010) providing a useful conceptual introduction.

Working in Cartesian Eulerian spatial coordinates $\{r^i\}$, we define the tensor-valued *angular momentum three-forms* as

$$\mathbf{l}^{ij} \equiv (v^i r^j - v^j r^i)\,\boldsymbol{\rho} + \boldsymbol{\theta}^{ij}, \tag{8.364}$$

where $\boldsymbol{\theta}^{ij} = -\boldsymbol{\theta}^{ji}$ captures a form of intrinsic *spin angular momentum*:

$$\boldsymbol{\theta}^{ij} \equiv \theta^{ij}\,\boldsymbol{\epsilon}. \tag{8.365}$$

Next, we introduce the tensor-valued *volume torque three-forms* as

$$\mathbf{n}^{ij} \equiv (g^i r^j - r^i g^j)\,\boldsymbol{\rho} + \mathbf{c}^{ij}, \tag{8.366}$$

where the antisymmetric three-forms $\mathbf{c}^{ij} = -\mathbf{c}^{ji}$ capture applied body torques:

$$\mathbf{c}^{ij} \equiv c^{ij}\,\boldsymbol{\epsilon}. \tag{8.367}$$

Surface torques are captured in terms of the vector-valued two-forms

$$\nu^{ij} \equiv \sigma^i r^j - \sigma^j r^i + \mu^{ij}, \tag{8.368}$$

where the Cauchy stress two-forms are given by (8.170), and where $\mu^{ij} = -\mu^{ij}$ capture intrinsic *couple stresses*:

$$\begin{aligned}\mu^{ij} &\equiv \tfrac{1}{2}\,\mu^{ijk}\,\epsilon_{k\ell m}\,\mathbf{e}^\ell \wedge \mathbf{e}^m \\ &= \tfrac{1}{2}\,\mu^{ijk}\,(F^{-1})^K{}_k\,\epsilon_{KLM}\,\mathbf{e}^L \wedge \mathbf{e}^M.\end{aligned} \tag{8.369}$$

When fed two vectors $\mathbf{u}$ and $\mathbf{w}$ defining a surface, the couple-stress two-forms return the torques on the corresponding surface. Conservation of angular momentum for a comoving volume V with surface S then implies a balance between the rate of change of total angular momentum and the total applied surface and body torques:

$$\frac{d}{dt}\int_V \mathbf{l}^{ij} = \int_S \nu^{ij} + \int_V \mathbf{n}^{ij}. \tag{8.370}$$

The form version of Reynolds transport theorem, discussed in example 8.18, states that the rate of change of total angular momentum of a comoving volume V with surface S is determined by

$$\frac{d}{dt}\int_V \mathbf{l}^{ij} = \int_V \mathrm{d}_t\mathbf{l}^{ij} + \int_S \mathbf{v}\cdot\mathbf{l}^{ij}, \tag{8.371}$$

which implies that the balance of total angular momentum (8.370) may be rewritten in the form

$$\int_V \mathrm{d}_t\mathbf{l}^{ij} = \int_S (\nu^{ij} - \mathbf{v}\cdot\mathbf{l}^{ij}) + \int_V \mathbf{n}^{ij}. \tag{8.372}$$

The generalization of Stokes's theorem then implies that

$$\int_V \mathrm{d}_t\mathbf{l}^{ij} = \int_V [\mathrm{d}(\nu^{ij} - \mathbf{v}\cdot\mathbf{l}^{ij}) + \mathbf{n}^{ij}]. \tag{8.373}$$

Since this result needs to hold for any comoving volume V, we find the conservation law of angular momentum

$$\mathrm{d}_t\mathbf{l}^{ij} + \mathrm{d}(\mathbf{v}\cdot\mathbf{l}^{ij}) = \mathcal{L}_\mathbf{v}\mathbf{l}^{ij} = \mathrm{d}\nu^{ij} + \mathbf{n}^{ij}, \tag{8.374}$$

where we have used (8.347). Upon substituting equations (8.364), (8.366), and (8.368) and using the form version of conservation of linear momentum (8.361), we find the form version of conservation of angular momentum:

$$\mathcal{L}_\mathbf{v}\theta^{ij} = \mathrm{d}\mu^{ij} + \mathbf{c}^{ij} + \sigma^i \wedge \mathbf{e}^j - \sigma^j \wedge \mathbf{e}^i. \tag{8.375}$$

In general spatial coordinates, these equations need to be modified by replacing the Lie and exterior derivatives of forms with their covariant counterparts, as discussed in sections 8.9.5 and 8.9.6. By making the substitutions, we obtain an expression valid in general spatial coordinates:

$$\mathfrak{L}_{\mathbf{v}}\boldsymbol{\theta}^{ij} = \mathrm{D}\boldsymbol{\mu}^{ij} + \mathbf{c}^{ij} + \boldsymbol{\sigma}^i \wedge \mathbf{e}^j - \boldsymbol{\sigma}^j \wedge \mathbf{e}^i. \tag{8.376}$$

In example 8.26, we obtain these dynamical equations using a variational approach.

In the absence of intrinsic spin, applied volume torques, and couple stress, conservation of angular momentum reduces to

$$\boldsymbol{\sigma}^i \wedge \mathbf{e}^j = \boldsymbol{\sigma}^j \wedge \mathbf{e}^i, \tag{8.377}$$

implying the symmetry of the Cauchy stress, as discussed in example 8.6.

Example 8.22 First Law of Thermodynamics

The formulation of the *first law of thermodynamics*, which is a statement of *conservation of energy*, must recognize heat as a form of energy. This necessitates the introduction of several thermodynamic concepts, as follows. Working in Cartesian Eulerian spatial coordinates, we introduce the *kinetic energy density* as the three-form

$$\tfrac{1}{2} v^i \mathbf{p}_i, \tag{8.378}$$

where $\mathbf{p}_i$ denotes the linear momentum three-form (8.353) and v^i a component of the material velocity. For now, we ignore kinetic energy in the form of intrinsic spin, but we consider this complication in example 8.23. Next, we introduce the *internal energy three-form* $\mathbf{e}$, which captures thermomechanical energy. Thus, the total energy of a comoving volume is

$$\int_V \left(\tfrac{1}{2} v^i \mathbf{p}_i + \mathbf{e}\right). \tag{8.379}$$

The rate of change of the total energy is balanced by work done by the body forces

$$\int_V v^i \mathbf{f}_i, \tag{8.380}$$

surface tractions

$$\int_S v^i \boldsymbol{\sigma}_i, \tag{8.381}$$

heating due to the *rate of internal heating*,

$$\mathbf{h} \equiv \mathrm{h}\,\epsilon, \tag{8.382}$$

such that the total heating equals

$$\int_V \mathbf{h}, \tag{8.383}$$

and thermal conduction captured by the *heat flux two-form*

$$\mathbf{H} = \tfrac{1}{2}\, H^i\, \epsilon_{ijk}\, \mathbf{e}^j \wedge \mathbf{e}^k, \tag{8.384}$$

such that the total heat flux equals

$$\int_S \mathbf{H}. \tag{8.385}$$

Thus, we have the balance

$$\frac{d}{dt}\int_V (\tfrac{1}{2}\, v^i\, \mathbf{p}_i + \mathbf{e}) = \int_V v^i\, \mathbf{f}_i + \int_S v^i\, \boldsymbol{\sigma}_i + \int_V \mathbf{h} - \int_S \mathbf{H}. \tag{8.386}$$

Using the form version of Reynolds transport theorem, discussed in example 8.18, we find

$$\int_V \mathrm{d}_t(\tfrac{1}{2}\, v^i\, \mathbf{p}_i + \mathbf{e}) = \int_V v^i\, \mathbf{f}_i + \int_S [v^i\, \boldsymbol{\sigma}_i - \mathbf{H} - \mathbf{v}\cdot(\tfrac{1}{2}\, v^i\, \mathbf{p}_i + \mathbf{e})]. \tag{8.387}$$

The generalization of Stokes's theorem then implies that

$$\int_V \mathrm{d}_t(\tfrac{1}{2}\, v^i\, \mathbf{p}_i + \mathbf{e}) = \int_V (v^i\, \mathbf{f}_i + \mathbf{h}) + \mathrm{d}[v^i\, \boldsymbol{\sigma}_i - \mathbf{H} - \mathbf{v}\cdot(\tfrac{1}{2}\, v^i\, \mathbf{p}_i + \mathbf{e})].$$

Since this result needs to hold for any comoving volume V, we find the energy conservation law

$$\mathcal{L}_{\mathbf{v}}(\tfrac{1}{2}\, v^i\, \mathbf{p}_i + \mathbf{e}) = v^i\, \mathbf{f}_i + \mathbf{h} + \mathrm{d}(v^i\, \boldsymbol{\sigma}_i - \mathbf{H}). \tag{8.388}$$

Writing $\mathbf{p}_i = v_i\, \boldsymbol{\rho}$ and invoking the form version of conservation of mass (8.351), we obtain the final form version of conservation of energy:

$$\mathcal{L}_{\mathbf{v}}\mathbf{e} = \boldsymbol{\sigma}_i \wedge \mathrm{D}v^i + \mathbf{h} - \mathrm{d}\mathbf{H}. \tag{8.389}$$

In this case, the change from an exterior derivative d to an exterior covariant derivative D occurs naturally.

Example 8.23 First Law of Thermodynamics with Spin

In this example, we consider a form version of conservation of energy in the presence of *intrinsic spin*, a complication introduced in example 8.21. Working in Cartesian Eulerian spatial coordinates, we express the kinetic energy density as the three-form (8.378). In addition to vorticity, $W_{ij} = \frac{1}{2}(\nabla_j v_i - \nabla_i v_j)$, introduced in example 7.2, a particle is allowed to have a *material angular velocity* captured by the antisymmetric elements $w_{ij} = -w_{ji}$, such that the *total angular velocity* of a particle is the sum of its vorticity and material angular velocity:

$$\varpi_{ij} \equiv W_{ij} + w_{ij}. \tag{8.390}$$

Thus, the total angular velocity vector of a particle is $\varpi_i = \frac{1}{2}\epsilon_{ijk}\varpi^{jk}$, and if the *intrinsic spin angular momentum vector* is defined in terms of the components θ^{ij} of the intrinsic spin angular momentum form (8.365) as $\theta^i = \frac{1}{2}\epsilon^{ijk}\theta_{jk}$, then the rotational energy is

$$\tfrac{1}{2}\varpi_i\theta^i = \tfrac{1}{4}\varpi^{jk}\theta_{jk}. \tag{8.391}$$

The *rotational kinetic energy density* of a particle is therefore expressed as the three-form

$$\tfrac{1}{4}\varpi_{ij}\boldsymbol{\theta}^{ij}, \tag{8.392}$$

where $\boldsymbol{\theta}^{ij}$ are the *spin angular momentum three-forms* defined in (8.365). We express the internal energy density as the three-form $\mathbf{e}$, such that the total energy of a comoving volume is

$$\int_V \left[\tfrac{1}{2}(v^i\mathbf{p}_i + \tfrac{1}{2}\varpi_{ij}\boldsymbol{\theta}^{ij}) + \mathbf{e}\right]. \tag{8.393}$$

Work done by body forces (8.380) and surface forces (8.381) plus body and surface torques may be expressed as

$$\int_V (v^i\mathbf{f}_i + \tfrac{1}{2}\varpi_{ij}\mathbf{c}^{ij}), \tag{8.394}$$

and

$$\int_S (v^i\boldsymbol{\sigma}_i + \tfrac{1}{2}\varpi_{ij}\boldsymbol{\mu}^{ij}), \tag{8.395}$$

where $\mathbf{c}^{ij}$ denotes the applied volume torque three-forms (8.367) and $\boldsymbol{\mu}^{ij}$ the surface torque two-forms (8.368). Heating due to the rate of internal heating three-form $\mathbf{h}$ is represented by expression (8.383), and thermal conduction is captured by the integrated heat flux (8.385). Conservation of energy is expressed by the balance

$$\frac{d}{dt}\int_V \left[\tfrac{1}{2}\,(v^i\,\mathbf{p}_i + \tfrac{1}{2}\,\varpi_{ij}\,\boldsymbol{\theta}^{ij}) + \mathbf{e}\right] = \int_V (v^i\,\mathbf{f}_i + \tfrac{1}{2}\,\varpi_{ij}\,\mathbf{c}^{ij} + \mathbf{h}) + \int_S (v^i\,\boldsymbol{\sigma}_i + \tfrac{1}{2}\,\varpi_{ij}\,\boldsymbol{\mu}^{ij} - \mathbf{H}).$$

Using the form version of Reynolds transport theorem and the generalization of Stokes's theorem, we find the local law

$$\mathcal{L}_{\mathbf{v}}\left[\tfrac{1}{2}\,(v^i\,\mathbf{p}_i + \tfrac{1}{2}\,\varpi_{ij}\,\boldsymbol{\theta}^{ij}) + \mathbf{e}\right] = v^i\,\mathbf{f}_i + \tfrac{1}{2}\,\varpi_{ij}\,\mathbf{c}^{ij} + \mathbf{h} + \mathrm{d}(v^i\,\boldsymbol{\sigma}_i + \tfrac{1}{2}\,\varpi_{ij}\,\boldsymbol{\mu}^{ij}) - \mathrm{d}\mathbf{H}.$$

Substitute $\mathbf{p}^i = v^i\,\rho$ and use conservation of mass, $\mathcal{L}_{\mathbf{v}}\rho = \mathbf{0}$, and conservation of linear momentum (8.361) to find

$$\mathcal{L}_{\mathbf{v}}\left(\tfrac{1}{4}\,\varpi_{ij}\,\boldsymbol{\theta}^{ij} + \mathbf{e}\right) = \boldsymbol{\sigma}_i \wedge \mathrm{D}v^i + \tfrac{1}{2}\,\varpi_{ij} \wedge (\mathrm{D}\boldsymbol{\mu}^{ij} + \mathbf{c}^{ij}) + \tfrac{1}{2}\,\boldsymbol{\mu}^{ij} \wedge \mathrm{D}\varpi_{ij} + \mathbf{h} - \mathrm{d}\mathbf{H}.$$

Again, the change from an exterior derivative d to an exterior covariant derivative D occurs naturally.

Assume that, like linear momentum density $\mathbf{p}^i$ is related to the material velocity v^i, spin angular momentum density $\boldsymbol{\theta}^{ij}$ is related to the total angular velocity ϖ_{ij} via the inertia density tensor

$$\boldsymbol{\theta}^{ij} = \tfrac{1}{2}\,\mathbf{j}^{ijk\ell}\,\varpi_{k\ell}, \tag{8.396}$$

where $\mathbf{j}^{ijk\ell} = \mathbf{j}^{k\ell ij} = -\,\mathbf{j}^{jik\ell} = -\,\mathbf{j}^{ij\ell k}$ denote the *inertia density three-forms,*

$$\mathbf{j}^{ijk\ell} \equiv j^{ijk\ell}\,\boldsymbol{\epsilon}. \tag{8.397}$$

Then we are left with

$$\begin{aligned}\tfrac{1}{8}\,\varpi_{ij}\,(\mathcal{L}_{\mathbf{v}}\mathbf{j}^{ijk\ell})\,\varpi_{k\ell} + \mathcal{L}_{\mathbf{v}}\mathbf{e} &= \boldsymbol{\sigma}_i \wedge \mathrm{D}v^i + \tfrac{1}{2}\,\boldsymbol{\mu}_i{}^j \wedge \mathrm{D}\varpi^i{}_j \\ &+ \mathbf{h} - \mathrm{d}\mathbf{H} + \tfrac{1}{2}\,\varpi_{ij}\,(\mathrm{D}\boldsymbol{\mu}^{ij} + \mathbf{c}^{ij} - \mathcal{L}_{\mathbf{v}}\boldsymbol{\theta}^{ij}).\end{aligned} \tag{8.398}$$

We expect that the inertia density three-forms are conserved in the sense

$$\varpi_{ij}\,(\mathcal{L}_{\mathbf{v}}\mathbf{j}^{ijk\ell})\,\varpi_{k\ell} = \mathbf{0}, \tag{8.399}$$

much like mass is conserved in the sense $\mathcal{L}_{\mathbf{v}}\rho = \mathbf{0}$. Thus, upon involving conservation of angular momentum (8.376), we are left with

$$\mathcal{L}_{\mathbf{v}}\mathbf{e} = \boldsymbol{\sigma}_i \wedge (\mathrm{D}v^i - \varpi^i{}_j\,\mathbf{e}^j) + \tfrac{1}{2}\,\boldsymbol{\mu}_i{}^j \wedge \mathrm{D}\varpi^i{}_j + \mathbf{h} - \mathrm{d}\mathbf{H}. \tag{8.400}$$

Using the surface traction two-forms (8.170), we have

$$\begin{aligned}\boldsymbol{\sigma}_i \wedge \mathrm{D}v^i &= \boldsymbol{\sigma}_i \wedge \mathbf{e}^j\,(\nabla_j v^i)\\ &= \sigma^{ij}\,(W_{ij}+D_{ij})\,\epsilon\\ &= \widetilde{\sigma}^{ij}\,W_{ij}\,\epsilon + \widehat{\sigma}^{ij}\,D_{ij}\epsilon\end{aligned} \tag{8.401}$$

and

$$\begin{aligned}\varpi_{ij}\,\boldsymbol{\sigma}^i \wedge \mathbf{e}^j &= \tfrac{1}{2}\,(W_{ij}+w_{ij})\,(\sigma^{ij}-\sigma^{ji})\,\epsilon\\ &= (W_{ij}+w_{ij})\,\widetilde{\sigma}^{ij}\,\epsilon.\end{aligned} \tag{8.402}$$

Thus, conservation of energy is expressed by (see, e.g., Malvern, 1969, Section 5.4, where intrinsic spin is ignored)

$$\mathcal{L}_{\mathbf{v}}\mathbf{e} = \widehat{\boldsymbol{\sigma}}_i \wedge \mathrm{D}v^i - w^i{}_j\,\widetilde{\boldsymbol{\sigma}}_i \wedge \mathbf{e}^j + \tfrac{1}{2}\,\boldsymbol{\mu}_i{}^j \wedge \mathrm{D}\varpi^i{}_j + \mathbf{h} - \mathrm{d}\mathbf{H}. \tag{8.403}$$

Example 8.24 Second Law of Thermodynamics

The first law of thermodynamics (8.389) or, with spin, (8.403) states that energy cannot be created or destroyed, only transferred or converted. However, it does not ensure that heat will flow naturally from hot to cold. The *second law of thermodynamics* is necessary to ensure that heat always flows in the right direction. To achieve this, the theory introduces the concept of *entropy*, which is a measure of the disorder or randomness in a system. Specifically, it defines the entropy per unit mass or the *specific entropy density* S. Under the assumption of incrementally isothermal conditions, the production of entropy per unit volume is determined by $\mathbf{h}/\theta$, where $\mathbf{h}$ is the rate of internal heating (8.382) and θ is the *absolute temperature*. The entropy flux is determined by $\mathbf{H}/\theta$, where $\mathbf{H}$ denotes the heat flux two-form (8.384).[a] The second law of thermodynamics stipulates that the rate of change of entropy in a comoving volume V is greater than or equal to the entropy generated in V plus the entropy flux into V[b] through its surface S:

$$\frac{d}{dt}\int_V \rho\,\mathrm{S} \geq \int_V \frac{\mathbf{h}}{\theta} - \int_S \frac{\mathbf{H}}{\theta}. \tag{8.404}$$

Using the form version of Reynolds transport theorem, discussed in example 8.18, and conservation of mass (8.352), we find

$$\int_V \rho\,\mathcal{L}_{\mathbf{v}}\mathrm{S} \geq \int_V \left[\frac{\mathbf{h}}{\theta} - \mathrm{d}\left(\frac{\mathbf{H}}{\theta}\right)\right], \tag{8.405}$$

and since this result has to hold for any comoving volume, we obtain the local law known as the *Clausius–Duhem inequality*

$$\rho\,\theta\,\mathcal{L}_{\mathbf{v}}\mathrm{S} \geq \mathbf{h} - \mathrm{d}\mathbf{H} + \mathrm{d}\log\theta \wedge \mathbf{H}. \tag{8.406}$$

For an *adiabatic process*, $\mathcal{L}_{\mathbf{v}}\mathrm{S} = 0$, that is, entropy is constant. We can use the first law of thermodynamics (8.389) to rewrite the second law in the alternative form

$$\rho\,\theta\,\mathcal{L}_{\mathbf{v}}\mathrm{S} \geq \mathcal{L}_{\mathbf{v}}\mathbf{e} - \boldsymbol{\sigma}_i \wedge \mathrm{D}v^i + \mathrm{d}\log\theta \wedge \mathbf{H}. \tag{8.407}$$

or, in the presence of spin, we use (8.400) to rewrite the second law as

$$\begin{aligned}\rho\,\theta\,\mathcal{L}_{\mathbf{v}}\mathrm{S} \geq{}& \mathcal{L}_{\mathbf{v}}\mathbf{e} - \boldsymbol{\sigma}_i \wedge (\mathrm{D}v^i + \varpi^i{}_j \wedge \mathbf{e}^j) \\ &- \tfrac{1}{2}\,\boldsymbol{\mu}_i{}^j \wedge \mathrm{D}\varpi^i{}_j + \mathrm{D}\log\theta \wedge \mathbf{H}.\end{aligned} \tag{8.408}$$

Only for a *reversible process* does the equal sign hold in (8.406), (8.407), and (8.408).

For a system in equilibrium, $\mathcal{L}_{\mathbf{v}}\mathrm{S} = 0$, $\mathcal{L}_{\mathbf{v}}\mathbf{e} = \mathbf{0}$, and $v^i = 0$, in which case the second law states that

$$\mathrm{d}\log\theta \wedge \mathbf{H} \leq \mathbf{0}, \tag{8.409}$$

implying that the heat flux, $\mathbf{H}$, must be in the direction of the negative temperature gradient, $-\,\mathrm{d}\theta$. In other words, heat must flow from hot to cold, as desired.

---

[a] In classical thermodynamics (e.g., Adkins, 1983; Callen, 1985; Borgnakke & Sonntag, 2012), using conventional notation, the first law is written as $dU = \delta Q + \delta W$, expressing the change in internal energy, dU, in terms of the difference between the infinitesimal amount of heat supplied, δQ, and the infinitesimal amount of work done on the system, δW. For a reversible heat transfer, the change in entropy, dS, due to an infinitesimal amount of heat at absolute temperature T is determined by $dS = \delta Q/T$, and the thermodynamic work done on the system may be expressed in terms of the pressure, P, and change in volume, dV, as $\delta W = -P\,dV$. Thus, conservation of energy may be expressed as $dU = T\,dS - P\,dV$, corresponding to an *equation of state* $U = U(S, V)$. For irreversible natural processes, the second law of thermodynamics requires that the change in entropy satisfies the inequality $dS \geq \delta Q/T$. Thus, the terms $\mathbf{h}/\theta$ and $\mathbf{H}/\theta$ are the continuum mechanical equivalent of the classical change in entropy $\delta Q/T$.

[b] Remember that $\mathbf{H}$ denotes the heat flux *out of* the comoving volume.

Example 8.25 Caloric Equations of State and Gibbs Relations

The *Helmholtz free energy* three-form $\boldsymbol{\Phi}$ combines the internal energy three-form $\mathbf{e}$ and the specific entropy density S:

$$\boldsymbol{\Phi} \equiv \mathbf{e} - \rho\,\mathrm{S}\,\theta. \tag{8.410}$$

Equation (8.410) is an example of a *Legendre transformation*, in which the internal energy three-form $\mathbf{e}$ is considered to be a function of the specific entropy density S, whereas the Helmholtz free energy three-form $\boldsymbol{\Phi}$ is considered to be a function of the absolute temperature θ. The fields S and θ are *conjugate variables*. Upon taking the Lie derivative of this expression, using conservation of mass (8.352), we find the relationship

$$\mathcal{L}_{\mathbf{v}}\boldsymbol{\Phi} = \mathcal{L}_{\mathbf{v}}\mathbf{e} - \rho\,\theta\,\mathcal{L}_{\mathbf{v}}\mathrm{S} - \rho\,\mathrm{S}\,\mathcal{L}_{\mathbf{v}}\theta. \tag{8.411}$$

In the absence of intrinsic spin, we may use (8.411) to rewrite the second law of thermodynamics (8.407) in the form

$$\boldsymbol{\sigma}_i \wedge \mathrm{D}v^i \geq \mathcal{L}_{\mathbf{v}}\boldsymbol{\Phi} + \rho\,\mathrm{S}\,\mathcal{L}_{\mathbf{v}}\theta + \mathrm{d}\log\theta \wedge \mathbf{H}. \tag{8.412}$$

The next step is to introduce a *caloric equation of state*. In expression (8.389) for conservation of energy, we see that stress is captured by a one-form-valued two-form, $\boldsymbol{\sigma}_i$. Internal energy, $\mathbf{e}$, is a three-form. We need to determine a thermodynamic mechanical parameter which is the *conjugate* of the stress. Geometrically, we are looking for vector-valued one-forms, let's call them $\boldsymbol{\vartheta}^i$, such that the derivative of the internal energy three-form with respect to these vector-valued one-forms, $\partial\mathbf{e}/\partial\boldsymbol{\vartheta}^i$, yields the stress one-form-valued two-forms, $\boldsymbol{\sigma}_i$.

At this point, we need to distinguish two types of deformation: *elastic deformation*, which is a *reversible process* wherein the continuum returns to its original equilibrium state once the stress is removed, and *plastic deformation*, which is irreversible. Plastic deformation is stress-free, so only the mechanical parameters $\boldsymbol{\vartheta}^i$ can produce stress.

Based on these considerations, the most general material is one whose Helmholtz free energy three-form $\boldsymbol{\Phi}$ is a function of the yet-to-be-determined one-forms $\boldsymbol{\vartheta}^i$, the local absolute temperature θ, and a collection of additional thermodynamic *internal state variables* ξ, which might include chemical composition or defect densities. Thus, based on the Helmholtz free energy, the *caloric equation of state* is

$$\boldsymbol{\Phi} = \boldsymbol{\Phi}(\boldsymbol{\vartheta}^i, \theta, \xi). \tag{8.413}$$

The rate of change of the Helmholtz free energy three-form is determined by the *Gibbs relation*

$$\begin{aligned}\mathcal{L}_{\mathbf{v}}\boldsymbol{\Phi} &= \left(\frac{\partial\boldsymbol{\Phi}}{\partial\boldsymbol{\vartheta}^i}\right)_{\theta,\xi} \wedge \mathfrak{L}_{\mathbf{v}}\boldsymbol{\vartheta}^i + \left(\frac{\partial\boldsymbol{\Phi}}{\partial\theta}\right)_{\boldsymbol{\vartheta}^i,\xi} \mathcal{L}_{\mathbf{v}}\theta \\ &\quad + \left(\frac{\partial\boldsymbol{\Phi}}{\partial\xi}\right)_{\boldsymbol{\vartheta}^i,\theta} \mathcal{L}_{\mathbf{v}}\xi,\end{aligned} \tag{8.414}$$

where we have used conservation of mass (8.352). We also used the covariant Lie derivative of the vector-valued one-forms, $\mathfrak{L}_{\mathbf{v}}\boldsymbol{\vartheta}^i$. Upon substituting the Gibbs relation (8.414) into (8.412), we find

$$\begin{aligned}\boldsymbol{\sigma}_i \wedge \mathrm{D}v^i - \left(\frac{\partial \boldsymbol{\Phi}}{\partial \boldsymbol{\vartheta}^i}\right)_{\theta,\xi} \wedge \mathfrak{L}_{\mathbf{v}}\boldsymbol{\vartheta}^i \geq \left[\boldsymbol{\rho}\,\mathrm{S} + \left(\frac{\partial \boldsymbol{\Phi}}{\partial \theta}\right)_{\boldsymbol{\vartheta}^i,\xi}\right] \mathcal{L}_{\mathbf{v}}\theta \\ + \left(\frac{\partial \boldsymbol{\Phi}}{\partial \xi}\right)_{\boldsymbol{\vartheta}^i,\theta} \mathcal{L}_{\mathbf{v}}\xi + \mathrm{d}\log\theta \wedge \mathbf{H}.\end{aligned} \tag{8.415}$$

Alternatively, the internal energy $\mathbf{e}$ is a function of the vector-valued one-forms $\boldsymbol{\vartheta}^i$, the local entropy S, and the internal state variables ξ. Thus, we have the caloric equation of state

$$\mathbf{e} = \mathbf{e}(\,\boldsymbol{\vartheta}^i, \mathrm{S}, \xi\,), \tag{8.416}$$

with the Gibbs relationship

$$\mathcal{L}_{\mathbf{v}}\mathbf{e} = \left(\frac{\partial \mathbf{e}}{\partial \boldsymbol{\vartheta}^i}\right)_{\mathrm{S},\xi} \wedge \mathfrak{L}_{\mathbf{v}}\boldsymbol{\vartheta}^i + \left(\frac{\partial \mathbf{e}}{\partial \mathrm{S}}\right)_{\boldsymbol{\vartheta}^i,\xi} \mathcal{L}_{\mathbf{v}}\mathrm{S} + \left(\frac{\partial \mathbf{e}}{\partial \xi}\right)_{\boldsymbol{\vartheta}^i,\mathrm{S}} \mathcal{L}_{\mathbf{v}}\xi. \tag{8.417}$$

We use relationship (8.411) to rewrite (8.417) in the form

$$\begin{aligned}\mathcal{L}_{\mathbf{v}}\boldsymbol{\Phi} &= \left(\frac{\partial \mathbf{e}}{\partial \boldsymbol{\vartheta}^i}\right)_{\mathrm{S},\xi} \wedge \mathfrak{L}_{\mathbf{v}}\boldsymbol{\vartheta}^i + \left[\left(\frac{\partial \mathbf{e}}{\partial \mathrm{S}}\right)_{\boldsymbol{\vartheta}^i,\xi} - \boldsymbol{\rho}\,\theta\right] \mathcal{L}_{\mathbf{v}}\mathrm{S} \\ &+ \left(\frac{\partial \mathbf{e}}{\partial \xi}\right)_{\boldsymbol{\vartheta}^i,\mathrm{S}} \mathcal{L}_{\mathbf{v}}\xi - \boldsymbol{\rho}\,\mathrm{S}\,\mathcal{L}_{\mathbf{v}}\theta.\end{aligned} \tag{8.418}$$

Upon comparing this expression to (8.414), we deduce the requirements

$$\left(\frac{\partial \boldsymbol{\Phi}}{\partial \boldsymbol{\vartheta}^i}\right)_{\theta,\xi} = \left(\frac{\partial \mathbf{e}}{\partial \boldsymbol{\vartheta}^i}\right)_{\mathrm{S},\xi}, \tag{8.419}$$

$$\boldsymbol{\rho}\,\theta = \left(\frac{\partial \mathbf{e}}{\partial \mathrm{S}}\right)_{\boldsymbol{\vartheta}^i,\xi}, \tag{8.420}$$

$$\boldsymbol{\rho}\,\mathrm{S} = -\left(\frac{\partial \boldsymbol{\Phi}}{\partial \theta}\right)_{\boldsymbol{\vartheta}^i,\xi}, \tag{8.421}$$

$$\left(\frac{\partial \mathbf{e}}{\partial \xi}\right)_{\boldsymbol{\vartheta}^i,\mathrm{S}} = \left(\frac{\partial \boldsymbol{\Phi}}{\partial \xi}\right)_{\boldsymbol{\vartheta}^i,\theta}. \tag{8.422}$$

Now, as we set out to do, we identify the stress one-form-valued two-forms $\boldsymbol{\sigma}_i$ with the partial derivatives of the energy three-forms with respect to the vector-valued one-forms $\boldsymbol{\vartheta}^i$, thereby rendering them conjugate thermodynamic parameters:

$$\boldsymbol{\sigma}_i \equiv \left(\frac{\partial \boldsymbol{\Phi}}{\partial \boldsymbol{\vartheta}^i}\right)_{\theta,\xi} = \left(\frac{\partial \mathbf{e}}{\partial \boldsymbol{\vartheta}^i}\right)_{\mathrm{S},\xi}. \tag{8.423}$$

Upon using equations (8.419), (8.421), (8.422), and (8.423) in the Clausius–Duhem inequality (8.415), we find that the second law of thermodynamics may be stated as

$$\boldsymbol{\sigma}_i \wedge \mathbf{D}_{\mathrm{p}}^i \geq \left(\frac{\partial \mathbf{e}}{\partial \xi}\right)_{\vartheta^i, S} \mathcal{L}_{\mathbf{v}} \xi + \mathrm{d} \log \theta \wedge \mathbf{H}. \tag{8.424}$$

The quantity

$$\mathbf{D}_{\mathrm{p}}^i \equiv \mathrm{D} v^i - \mathfrak{L}_{\mathbf{v}} \boldsymbol{\vartheta}^i \tag{8.425}$$

captures the *plastic deformation rate*, that is, irreversible deformation that does not produce stresses. Thus, the term on the left-hand side of the inequality captures the rate of work done against plastic deformation.

Elastic deformation is reversible, implying an equal sign in (8.424), and *adiabatic*, implying no heat production or heat flow. Thus, in the absence of complications due to chemistry or defects, we are led to the requirement that

$$\boldsymbol{\sigma}_i \wedge \mathbf{D}_{\mathrm{p}}^i = \mathbf{0}. \tag{8.426}$$

Since this needs to hold for any stress, we must impose the conditions

$$\mathbf{D}_{\mathrm{p}}^i = \mathbf{0}, \tag{8.427}$$

which, according to (8.425), imply that

$$\mathfrak{L}_{\mathbf{v}} \boldsymbol{\vartheta}^i = \mathrm{D} v^i. \tag{8.428}$$

Condition (8.428) finally enables us to determine the mechanical parameters $\boldsymbol{\vartheta}^i$. They are given in terms of the motion φ^i by

$$\boldsymbol{\vartheta}^i = \mathrm{D} \varphi^i. \tag{8.429}$$

To confirm this, we use the commutativity of the exterior Lie derivative, $\mathfrak{L}_{\mathbf{v}}$, and the exterior covariant derivative, D, expressed by (8.287), and the fact that curvature vanishes in continuum mechanics:

$$\begin{aligned} \mathfrak{L}_{\mathbf{v}} \boldsymbol{\vartheta}^i &= \mathfrak{L}_{\mathbf{v}} \mathrm{D} \varphi^i \\ &= \mathrm{D} \mathfrak{L}_{\mathbf{v}} \varphi^i \\ &= \mathrm{D} v^i. \end{aligned} \tag{8.430}$$

In the third equality, we have made the identification

$$v^i = \mathfrak{L}_{\mathbf{v}} \varphi^i = \mathcal{L}_{\mathbf{v}} \varphi^i + \boldsymbol{\Gamma}^i_j(\mathbf{v})\, \varphi^j, \tag{8.431}$$

which is a generalized, covariant, version of our previous definition of the material velocity (4.8). We can actually identify the mechanical parameters ϑ^i with the coframe $\mathbf{e}^i$ as follows:

$$\begin{aligned}\mathfrak{L}_{\mathbf{v}}\mathbf{e}^i &= \mathcal{L}_{\mathbf{v}}\mathbf{e}^i + \boldsymbol{\Gamma}^i_j(\mathbf{v})\,\mathbf{e}^j \\ &= (\partial_j v^i + \Gamma^i_{kj}\,v^k)\,\mathbf{e}^j \\ &= (\nabla_j v^i)\,\mathbf{e}^j \\ &= \mathrm{D}v^i \\ &= \mathfrak{L}_{\mathbf{v}}\vartheta^i.\end{aligned} \tag{8.432}$$

Thus we can use ϑ^i, $\mathbf{e}^i$, and $\mathrm{D}\varphi^i$ interchangeably:

$$\vartheta^i = \mathrm{D}\varphi^i = \mathbf{e}^i. \tag{8.433}$$

In conclusion, the desired caloric equation of state is

$$\mathbf{e} = \mathbf{e}(\,\mathrm{D}\varphi^i, \mathrm{S}, \xi\,), \tag{8.434}$$

such that the stress one-form-valued two-forms are determined by

$$\boldsymbol{\sigma}_i \equiv \left(\frac{\partial \boldsymbol{\Phi}}{\partial \mathrm{D}\varphi^i}\right)_{\theta,\xi} = \left(\frac{\partial \mathbf{e}}{\partial \mathrm{D}\varphi^i}\right)_{\mathrm{S},\xi}, \tag{8.435}$$

thereby forming a conjugate pair with the exterior covariant derivative of the motion $\mathrm{D}\varphi^i$.

To accommodate spin, we introduce a *particle rotation tensor* $\Psi^i{}_j$. Rotations are discussed in section 5.15, and have the important property that their inverse is equal to their transpose: $(\Psi^{-1})^i{}_j = (\Psi^t)^i{}_j$. We define the following vector-/tensor-valued one-forms in terms of the particle motion φ^i and rotation $\Psi^i{}_j$:

$$\vartheta^i \equiv \Psi^i{}_j\,\mathrm{D}\varphi^j = \Psi^i{}_j\,\mathbf{e}^j, \tag{8.436}$$

and

$$\boldsymbol{\kappa}^i{}_j \equiv \Psi^i{}_k\,\mathrm{D}(\Psi^t)^k{}_j. \tag{8.437}$$

Note the antisymmetry

$$\boldsymbol{\kappa}^{ij} = -\,\boldsymbol{\kappa}^{ji}. \tag{8.438}$$

Next, we modify the caloric equation of state (8.434) as

$$\mathbf{e} = \mathbf{e}(\,\vartheta^i, \boldsymbol{\kappa}^i{}_j, \mathrm{S}, \xi\,). \tag{8.439}$$

In this case, using (8.411) in the Clausius–Duhem inequality (8.408), invoking relationships (8.421), (8.422), and (8.423), we find

$$\begin{aligned}\boldsymbol{\sigma}_i \wedge (\mathrm{D}v^i - \varpi^i{}_j\, \mathbf{e}^j) + \tfrac{1}{2}\, \boldsymbol{\mu}_i{}^j \wedge \mathrm{D}\varpi^i{}_j \\ - \frac{\partial \mathbf{e}}{\partial \boldsymbol{\vartheta}^i} \wedge \mathfrak{L}_{\mathbf{v}}\boldsymbol{\vartheta}^i - \frac{\partial \mathbf{e}}{\partial \boldsymbol{\kappa}^i{}_j} \wedge \mathfrak{L}_{\mathbf{v}}\boldsymbol{\kappa}^i{}_j \\ \geq \frac{\partial \mathbf{e}}{\partial \xi}\, \mathfrak{L}_{\mathbf{v}}\xi + \mathrm{d}\log\theta \wedge \mathbf{H}.\end{aligned} \tag{8.440}$$

We define

$$\boldsymbol{\sigma}_i \equiv \frac{\partial \mathbf{e}}{\partial \boldsymbol{\vartheta}^j}\, \Psi^j{}_i, \tag{8.441}$$

and

$$\boldsymbol{\mu}_i{}^j \equiv 2\, \frac{\partial\, \mathbf{e}}{\partial \boldsymbol{\kappa}^k{}_\ell}\, \Psi^k{}_i\, (\Psi^t)^j{}_\ell. \tag{8.442}$$

Thanks to (8.438), we have the antisymmetry

$$\boldsymbol{\mu}^{ij} = -\, \boldsymbol{\mu}^{ij}. \tag{8.443}$$

The second law of thermodynamics with spin (8.440) may now be expressed in the form

$$\boldsymbol{\sigma}_i \wedge (\mathbf{D}^i_{\mathrm{p}} - \boldsymbol{\varpi}^i_{\mathrm{p}}) + \tfrac{1}{2}\, \boldsymbol{\mu}_i{}^j \wedge \boldsymbol{\Pi}^i_{\mathrm{p}j} \geq \frac{\partial \mathbf{e}}{\partial \xi}\, \mathfrak{L}_{\mathbf{v}}\xi + \mathrm{d}\log\theta \wedge \mathbf{H}, \tag{8.444}$$

where we have defined the plastic deformation rates

$$\mathbf{D}^i_{\mathrm{p}} \equiv \mathrm{D}v^i - \mathfrak{L}_{\mathbf{v}}\mathrm{D}\varphi^i, \tag{8.445}$$

$$\boldsymbol{\varpi}^i_{\mathrm{p}} \equiv \varpi^i{}_j\, \mathbf{e}^j + (\mathrm{D}\varphi^j)\, (\Psi^t)^i{}_k\, \mathfrak{L}_{\mathbf{v}}\Psi^k{}_j, \tag{8.446}$$

and

$$\boldsymbol{\Pi}^i_{\mathrm{p}j} \equiv \mathrm{D}\Psi^i{}_j - (\Psi^t)^i{}_\ell\, \Psi^m{}_j\, \mathfrak{L}_{\mathbf{v}}[\Psi^\ell{}_k\, \mathrm{D}(\Psi^t)^k{}_m]. \tag{8.447}$$

In the elastic case, in the absence of complications due to chemistry or defects, we are led to the requirement that $\mathbf{D}^i_{\mathrm{p}}$, $\boldsymbol{\varpi}^i_{\mathrm{p}}$, and $\boldsymbol{\Pi}^i_{\mathrm{p}j}$ vanish, leading to the connections between motion and rotation on one hand, and particle velocity and total angular velocity on the other:

$$\mathfrak{L}_{\mathbf{v}}\mathrm{D}\varphi^i = \mathrm{D}v^i, \tag{8.448}$$

$$(\Psi^t)^i{}_k\,\mathfrak{L}_{\mathbf{v}}\Psi^k{}_j = -\varpi^i{}_j, \tag{8.449}$$

$$(\Psi^t)^i{}_\ell\,\Psi^m{}_j\,\mathfrak{L}_{\mathbf{v}}[\Psi^\ell{}_k\,\mathrm{D}(\Psi^t)^k{}_m] = \mathrm{D}\varpi^i{}_j. \tag{8.450}$$

Now we make the identifications $\mathrm{D}\varphi^i = \mathbf{e}^i$ and $\mathfrak{L}_{\mathbf{v}}\mathbf{e}^i = \mathrm{D}v^i$, such that equality (8.448) is satisfied. Then, with some effort, expression (8.450) may be obtained by taking the exterior derivative of (8.449), implying the former is redundant. To obtain this consistency between (8.449) and (8.450), the caloric equation of state (8.439) must have the functional dependence $\Psi^i{}_k\,\mathrm{D}(\Psi^t)^k{}_j$.

Example 8.26 Action in Continuum Mechanics

Based on the results in example 8.25, we are now in a position to define a form version of *Hamilton's principle* in continuum mechanics, which is an application of the variational principles discussed in section 8.11.5. In the absence of spin, we express the *Lagrangian density* as the three-form

$$\boldsymbol{L}(\,\mathfrak{L}_{\mathbf{v}}\varphi^i, \mathrm{D}\varphi^i\,) = \tfrac{1}{2}\,\boldsymbol{\rho}\,g_{ij}\,(\mathfrak{L}_{\mathbf{v}}\varphi^i)(\mathfrak{L}_{\mathbf{v}}\varphi^j) - \mathbf{e}(\,\mathrm{D}\varphi^i\,), \tag{8.451}$$

where $\mathbf{e}$ denotes the isentropic version of the caloric equation of state (8.434). The *action* is given by

$$I = \iint_V \boldsymbol{L}(\,\mathfrak{L}_{\mathbf{v}}\varphi^i, \mathrm{D}\varphi^i\,), \tag{8.452}$$

where the four-dimensional integral is over time and a spatial volume V. The variation of this action is

$$\begin{aligned}
\delta I &= \iint_V \frac{\partial \boldsymbol{L}}{\partial \mathfrak{L}_{\mathbf{v}}\varphi^i}\,\delta\mathfrak{L}_{\mathbf{v}}\varphi^i + \frac{\partial \boldsymbol{L}}{\partial \mathrm{D}\varphi^i} \wedge \delta\mathrm{D}\varphi^i \\
&= \iint_V \mathfrak{L}_{\mathbf{v}}\left(\frac{\partial \boldsymbol{L}}{\partial \mathfrak{L}_{\mathbf{v}}\varphi^i}\,\delta\varphi^i\right) + \mathrm{d}\left(\frac{\partial \boldsymbol{L}}{\partial \mathrm{D}\varphi^i}\,\delta\varphi^i\right) \\
&\quad - \delta\varphi^i\left(\mathfrak{L}_{\mathbf{v}}\frac{\partial \boldsymbol{L}}{\partial \mathfrak{L}_{\mathbf{v}}\varphi^i} + \mathrm{D}\frac{\partial \boldsymbol{L}}{\partial \mathrm{D}\varphi^i}\right) \\
&= \iint_S \frac{\partial \boldsymbol{L}}{\partial \mathrm{D}\varphi^i}\,\delta\varphi^i - \iint_V \delta\varphi^i\left(\mathfrak{L}_{\mathbf{v}}\frac{\partial \boldsymbol{L}}{\partial \mathfrak{L}_{\mathbf{v}}\varphi^i} + \mathrm{D}\frac{\partial \boldsymbol{L}}{\partial \mathrm{D}\varphi^i}\right).
\end{aligned} \tag{8.453}$$

During the integration by parts from the first to the second equality, the desired exterior and exterior covariant derivatives naturally emerge. From the second to the third equality, the time integral of the Lie derivative integrates out under the assumption that the perturbations vanish at the starting and ending times, and the exterior derivative over the volume V integrates to an integral over its surface S based on Stokes's theorem (8.317).

Stationarity of the action, $\delta I = 0$, then leads to conservation of linear momentum of motion (8.363)

$$\mathfrak{L}_{\mathbf{v}} \frac{\partial \boldsymbol{L}}{\partial \mathfrak{L}_{\mathbf{v}} \varphi^i} + \mathrm{D} \frac{\partial \boldsymbol{L}}{\partial \mathrm{D}\varphi^i} = \mathbf{0}, \qquad \text{in } V, \tag{8.454}$$

subject to the boundary conditions

$$\frac{\partial \boldsymbol{L}}{\partial \mathrm{D}\varphi^i} = \mathbf{0}, \qquad \text{in } S. \tag{8.455}$$

We have, using (8.353),

$$\frac{\partial \boldsymbol{L}}{\partial \mathfrak{L}_{\mathbf{v}} \varphi^i} = \boldsymbol{\rho}\, g_{ij} \,(\mathfrak{L}_{\mathbf{v}} \varphi^j) = \boldsymbol{\rho}\, v_i = \mathbf{p}_i, \tag{8.456}$$

and, using (8.435),

$$\frac{\partial \boldsymbol{L}}{\partial \mathrm{D}\varphi^i} = -\frac{\partial \mathbf{e}}{\partial \mathrm{D}\varphi^i} = -\boldsymbol{\sigma}^i, \tag{8.457}$$

and thus we recover conservation of linear momentum (8.363):

$$\mathfrak{L}_{\mathbf{v}} \mathbf{p}_i = \mathrm{D}\boldsymbol{\sigma}_i. \tag{8.458}$$

Example 8.27 Action in Continuum Mechanics with Spin

In the presence of spin, to make things compact, let us consider a four-dimensional motion

$$x^\mu = \varphi^\mu (X^{\mu'}), \tag{8.459}$$

and let us introduce a four-dimensional *total rotation* $\Psi^\mu{}_\nu$. Based on the form of the internal energy density in the presence of spin expressed in (8.439), we introduce the Lagrangian density four-form

$$\boldsymbol{L} = \boldsymbol{L}(\Psi^\mu_\nu\, \mathrm{D}\varphi^\nu, \Psi^\mu_\tau\, \mathrm{D}(\Psi^t)^\tau{}_\nu), \tag{8.460}$$

with an associated four-dimensional action

$$I = \iint_V \boldsymbol{L}(\Psi^\mu{}_\nu\, \mathrm{D}\varphi^\nu, \Psi^\mu{}_\tau\, \mathrm{D}(\Psi^t)^\tau{}_\nu). \tag{8.461}$$

In this application of the variational principles discussed in section 8.11.5, there are two sets of fields capturing *motion* and *rotation*. Let us introduce the auxiliary one-forms

$$\vartheta^\mu \equiv \Psi^\mu{}_\nu \, \mathrm{D}\varphi^\nu, \tag{8.462}$$

and

$$\kappa^\mu{}_\nu \equiv \Psi^\mu{}_\tau \, \mathrm{D}(\Psi^t)^\tau{}_\nu, \tag{8.463}$$

where we note the antisymmetry

$$\kappa^{\mu\nu} = -\kappa^{\nu\mu}. \tag{8.464}$$

Then the variation of the action (8.461) may be expressed in the form

$$\begin{aligned}
\delta I &= \iint_V \delta(\mathrm{D}\varphi^\mu) \wedge \frac{\delta L}{\delta \mathrm{D}\varphi^\mu} + \delta\Psi^\mu{}_\nu \, \frac{\delta L}{\delta \Psi^\mu{}_\nu} \\
&\qquad + \delta[\mathrm{D}(\Psi^t)^\mu{}_\nu] \wedge \frac{\delta L}{\delta \mathrm{D}(\Psi^t)^\mu{}_\nu} \\
&= \iint_V -\mathrm{d}\left[\delta\varphi^\mu \, \boldsymbol{\sigma}_\mu + \tfrac{1}{2}\,\delta(\Psi^t)^\mu{}_\nu \, \Psi^\nu{}_\tau \, \boldsymbol{\mu}_\mu{}^\tau\right] + \delta\varphi^\mu \, \mathrm{D}\boldsymbol{\sigma}_\mu \\
&\qquad - (\Psi^t)^\mu{}_\tau \, \delta\Psi^\tau{}_\nu \left(\mathbf{e}^\nu \wedge \boldsymbol{\sigma}_\mu + \tfrac{1}{2}\,\mathrm{D}\boldsymbol{\mu}_\mu{}^\nu\right) \\
&= \iint_V \delta\varphi^\mu \, \mathrm{D}\boldsymbol{\sigma}_\mu - (\Psi^t)^\mu{}_\tau \, \delta\Psi^\tau{}_\nu \left(\mathbf{e}^\nu \wedge \boldsymbol{\sigma}_\mu + \tfrac{1}{2}\,\mathrm{D}\boldsymbol{\mu}_\mu{}^\nu\right) \\
&\qquad - \iint_S \mathrm{d}\left[\delta\varphi^\mu \, \boldsymbol{\sigma}_\mu + \tfrac{1}{2}\,\delta(\Psi^t)^\mu{}_\nu \, \Psi^\nu{}_\tau \, \boldsymbol{\mu}_\mu{}^\tau\right] + \delta\varphi^\mu \, \mathrm{D}\boldsymbol{\sigma}_\mu,
\end{aligned} \tag{8.465}$$

where we have defined the stress three-forms

$$\boldsymbol{\sigma}_\mu \equiv -\frac{\partial L}{\partial \vartheta^\nu} \, \Psi^\nu{}_\mu, \tag{8.466}$$

and the couple-stress three-forms

$$\boldsymbol{\mu}_\mu{}^\nu \equiv -2 \, \frac{\partial L}{\partial \kappa^\tau{}_\sigma} \, \Psi^\tau{}_\mu \, (\Psi^t)^\nu{}_\sigma, \tag{8.467}$$

such that, thanks to (8.464),

$$\boldsymbol{\mu}^{\mu\nu} = -\boldsymbol{\mu}^{\nu\mu}. \tag{8.468}$$

In the final equality, we used Stokes's theorem (8.317). Thus, stationarity of the action implies conservation of linear momentum

$$D\boldsymbol{\sigma}_{\mu} = \mathbf{0}, \qquad \text{in } V, \tag{8.469}$$

and conservation of angular momentum

$$D\boldsymbol{\mu}^{\mu\nu} = \boldsymbol{\sigma}^{\mu} \wedge \mathbf{e}^{\nu} - \boldsymbol{\sigma}^{\nu} \wedge \mathbf{e}^{\mu}, \qquad \text{in } V, \tag{8.470}$$

subject to the boundary conditions

$$\boldsymbol{\sigma}_{\mu} = \mathbf{0}, \qquad \text{in } S, \tag{8.471}$$

and

$$\boldsymbol{\mu}_{\mu}{}^{\nu} = \mathbf{0}, \qquad \text{in } S. \tag{8.472}$$

In this case, we recover a four-dimensional version of conservation of angular momentum (8.375).

Example 8.28 Burgers and Frank Vectors

In this example, we provide geometrical interpretations of the dislocation and disclination two-forms, which were introduced in example 8.5. Much of Kondo's work on dislocations and disclinations was based on such considerations from the outset (e.g., Kondo, 1952, 1955a,b,c; Takeo & Ito, 1997). As stated in example 8.18, we may integrate tensor-valued forms provided their tensor components refer to a constant Cartesian basis. For this reason, we transform the Lagrangian tensor-valued dislocation and disclination two-forms (8.135) and (8.136) to Cartesian Eulerian tensor-valued forms:

$$\boldsymbol{\alpha}^{i} = F^{i}{}_{I}\,\boldsymbol{\alpha}^{I} \qquad \text{and} \qquad \boldsymbol{\Xi}^{i}_{j} = F^{i}{}_{I}\,(F^{-1})^{J}{}_{j}\,\boldsymbol{\Xi}^{I}_{J}. \tag{8.473}$$

Upon integrating the dislocation two-forms $\boldsymbol{\alpha}^{i}$ over a small surface S, as illustrated in figure 8.10, we obtain the *Burgers vector* with Eulerian components b^{i}:

$$b^{i} \equiv \int_{S} \boldsymbol{\alpha}^{i}. \tag{8.474}$$

The Burgers vector indicates the direction in which a material element tends to move as a result of the dislocation. Dislocations may be regarded as capturing an *intrinsic torsion* of the material.

Similarly, the disclination two-forms $\boldsymbol{\Xi}^{i}_{j}$ capture a change in the orientation of a vector when parallel transported around an oriented loop defining a small surface S, as illustrated in figure 8.10. Let w^{j} denote the components at the start of the loop,

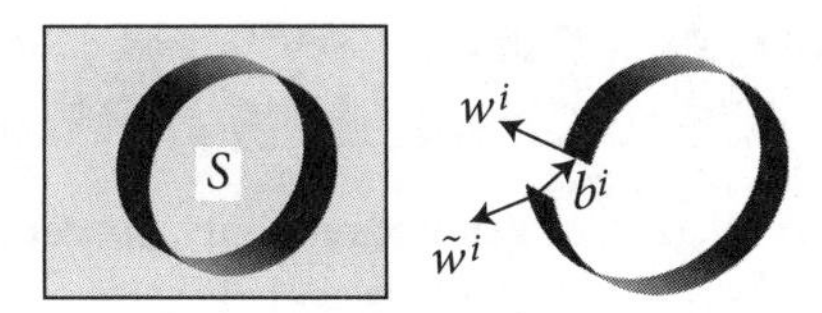

Figure 8.10: *Left*: A thin strip of material contained within a strained continuum with defects surrounds a surface S. *Right*: When the strip is removed from the strained continuum and cut, it takes on its natural, relaxed state. The distance between the two ends of the cut in the relaxed state corresponds to the *Burgers vector* with Cartesian components b^i. This vector is determined by integrating the dislocation two-form $\boldsymbol{\alpha}^i$ over the surface S, as expressed in (8.474), and captures an intrinsic torsion. The vector with Cartesian components w^j is parallel transported around the cut strip and becomes $\tilde{w}^j$ at its other end. The difference $\Delta w^i = \tilde{w}^i - w^i$ corresponds to the *Frank vector*, which is determined by integrating the quantity $\boldsymbol{\Xi}^i_j\, w^j$ over the surface S, as expressed in (8.475). The Frank vector captures an intrinsic curvature of the material. Dislocations and disclinations are perfectly captured by the cover artwork "Thetis Circle" by Beverly Pepper (1922–2020).

and $\tilde{w}^j$ the components at the end. Then

$$\Delta w^i \equiv \int_S \boldsymbol{\Xi}^i_j\, w^j \tag{8.475}$$

captures the disclination, $\Delta w^i = \tilde{w}^i - w^i$, between the orientation of the two vectors with components $\tilde{w}^i$ and w^i. When $w^j = 1$, equation (8.475) captures the *Frank vector*. Disclinations may be regarded as capturing an *intrinsic curvature* of the material.

The Cartesian Eulerian tensor-valued dislocation two-forms

$$\boldsymbol{\alpha}^i = \tfrac{1}{2}\, F^i{}_I\, \mathring{\alpha}_{KJ}{}^I\, \mathbf{e}^K \wedge \mathbf{e}^J, \tag{8.476}$$

and the Cartesian Eulerian tensor-valued disclination three-forms

$$\boldsymbol{\Xi}^i \equiv \boldsymbol{\Xi}^i_j \wedge \mathbf{e}^j = \tfrac{1}{2}\, F^i{}_I\, \mathring{\Xi}_{KLJ}{}^I\, \mathbf{e}^K \wedge \mathbf{e}^L \wedge \mathbf{e}^J, \tag{8.477}$$

satisfy the Bianchi identities (e.g., Edelen & Lagoudas, 1988; Tromp, 2025)

$$\mathrm{d}\boldsymbol{\alpha}^i = \boldsymbol{\Xi}^i, \tag{8.478}$$

$$\mathrm{d}\boldsymbol{\Xi}^i = \mathbf{0}. \tag{8.479}$$

Equations (8.478) and (8.479) capture the classical static theory of defects (e.g., DeWit, 1969, 1973). As discussed in section 8.7, expressions (8.479) and (8.479) imply that $\boldsymbol{\Xi}^i$ is an *exact form*. Consequently, disclinations cannot end inside the continuum.

Example 8.29 Action in General Relativity

In Einstein's original formulation of general relativity, spacetime geometry is determined by the *Einstein–Hilbert action* (Hilbert, 1915)

$$I = \int \boldsymbol{L}, \tag{8.480}$$

where the integral is over all of spacetime, and where the *Lagrangian density*, $\boldsymbol{L}$, is defined in terms of the *Ricci scalar* (7.119), $r \equiv r_{\mu\nu}{}^{\nu\mu}$, and the volume form (8.66), $\boldsymbol{\epsilon}$, by

$$\boldsymbol{L} \equiv \frac{1}{2\kappa}\, r\, \boldsymbol{\epsilon}. \tag{8.481}$$

The constant κ is determined in terms of Newton's gravitational constant, G, and the speed of light in vacuum, c, by $\kappa = 8\pi G/c^4$. Einstein's theory of general relativity is based on *Riemannian geometry*, and the Lagrangian (8.481) is considered to be a function of the metric $g_{\mu\nu}$ only (e.g., Carroll, 2004, Section 4.1), resulting in the Einstein–Hilbert action (8.480).

More generally, in the Einstein–Cartan or Palatini formulation of general relativity based on *Riemann-Cartan geometry*, the Lagrangian density (8.481) is considered to be a function of the metric $g_{\mu\nu}$ and the affine connection one-forms $\boldsymbol{\omega}_{\sigma}^{\tau}$ (e.g., Hehl et al., 1976, Appendix), resulting in the *Palatini action*. Even more generally, in the metric-affine formulation of general relativity, the Lagrangian density is considered to be a function of the metric $g_{\mu\nu}$, the coframes $\mathbf{e}^{\mu}$, and the affine connection one-forms $\boldsymbol{\omega}_{\sigma}^{\tau}$ (e.g., Hehl et al., 1995, Section 5.1). In that case, the Lagrangian density is expressed as (e.g., Trautman, 1973; Acharya, 1999; Randono, 2010; Dudek & Garecki, 2019; Montesinos et al., 2020)

$$\boldsymbol{L} \equiv \frac{1}{2\kappa}\, \boldsymbol{\epsilon}_{\sigma\tau} \wedge \mathbf{r}^{\sigma\tau}, \tag{8.482}$$

where we have used definition (8.64), and where we have defined $\mathbf{r}^{\sigma\tau} \equiv g^{\sigma\beta}\, \mathbf{r}_{\beta}^{\tau}$.

The variation of the Lagrangian density (8.482) is

$$\delta \boldsymbol{L} = \frac{1}{2\kappa}\, (\delta \boldsymbol{\epsilon}_{\sigma\tau} \wedge \mathbf{r}^{\sigma\tau} + \boldsymbol{\epsilon}_{\sigma\tau} \wedge \delta \mathbf{r}^{\sigma\tau}). \tag{8.483}$$

Upon varying definition (5.102) of $\bar{g}$ and using the properties of the alternating symbol, we find

$$\begin{aligned} \delta\bar{g} &= \tfrac{1}{2}\, g\, \delta\bar{g}^2 \\ &= \tfrac{1}{2}\, \bar{g}\, g^{\mu\nu}\, \delta g_{\mu\nu} \\ &= -\tfrac{1}{2}\, \bar{g}\, g_{\mu\nu}\, \delta g^{\mu\nu}, \end{aligned} \tag{8.484}$$

and thus, by varying (8.64) and using (8.63),

$$\delta\epsilon_{\sigma\tau} = -\tfrac{1}{2} g_{\mu\nu}\, \delta g^{\mu\nu}\, \epsilon_{\sigma\tau} + \delta\mathbf{e}^{\mu} \wedge \epsilon_{\sigma\tau\mu}, \tag{8.485}$$

and, varying the curvature two-forms (8.292),

$$\delta\mathbf{r}^{\sigma\tau} = \delta g^{\sigma\beta}\, \mathbf{r}^{\tau}_{\beta} + g^{\sigma\beta}\, \mathrm{D}\delta\boldsymbol{\omega}^{\tau}_{\beta}. \tag{8.486}$$

Thus, the variation (8.483) becomes

$$\begin{aligned}\delta\boldsymbol{L} = \frac{1}{2\kappa}\,[(\delta^{\sigma}_{\mu}\,\delta^{\beta}_{\nu} - \tfrac{1}{2} g^{\sigma\beta}\, g_{\mu\nu})\, \delta g^{\mu\nu}\, \epsilon_{\sigma\tau} \wedge \mathbf{r}^{\tau}_{\beta} \\ + \delta\mathbf{e}^{\mu} \wedge \epsilon_{\mu\sigma\tau} \wedge \mathbf{r}^{\sigma\tau} - \delta\boldsymbol{\omega}^{\tau}_{\beta} \wedge \mathrm{D}\epsilon^{\beta}{}_{\tau} + \mathrm{d}(\epsilon^{\beta}{}_{\tau} \wedge \delta\boldsymbol{\omega}^{\tau}_{\beta})].\end{aligned} \tag{8.487}$$

The last term in this expression is an exterior derivative which integrates out in the variation of the action (8.480). Thus, we find from (8.487) the variations

$$\begin{aligned}\frac{\delta\boldsymbol{L}}{\delta g^{\mu\nu}} &= \frac{1}{2\kappa}\,(\delta^{\sigma}_{\mu}\,\delta^{\beta}_{\nu} - \tfrac{1}{2} g^{\sigma\beta}\, g_{\mu\nu})\, \epsilon_{\sigma\tau} \wedge \mathbf{r}^{\tau}_{\beta} \\ &= \frac{1}{4\kappa}\,(E_{\mu\nu} + E_{\nu\mu})\, \epsilon,\end{aligned} \tag{8.488}$$

$$\begin{aligned}\frac{\delta\boldsymbol{L}}{\delta\mathbf{e}^{\mu}} &= \frac{1}{2\kappa}\, \epsilon_{\mu\sigma\tau} \wedge \mathbf{r}^{\sigma\tau} \\ &= \frac{1}{2\kappa}\,(E_{\mu}{}^{\sigma} + \overline{E}_{\mu}{}^{\sigma})\, \epsilon_{\sigma},\end{aligned} \tag{8.489}$$

$$\begin{aligned}\frac{\delta\boldsymbol{L}}{\delta\boldsymbol{\omega}^{\mu}_{\nu}} &= -\frac{1}{2\kappa}\, \mathrm{D}\epsilon^{\nu}{}_{\mu} \\ &= \frac{1}{2\kappa}\, \Pi_{\mu}{}^{\nu\tau}\, \epsilon_{\tau}.\end{aligned} \tag{8.490}$$

Here $E_{\mu\nu}$ denotes the Einstein tensor (7.126), and we have defined the *co-Einstein tensor*

$$\overline{E}_{\mu}{}^{\nu} \equiv \overline{r}_{\mu}{}^{\nu} - \tfrac{1}{2}\, r\, \delta_{\mu}{}^{\nu}, \tag{8.491}$$

in terms of the *co-Ricci tensor*

$$\overline{r}_{\mu\nu} \equiv r_{\mu\tau}{}^{\tau\nu}. \tag{8.492}$$

The *Palatini tensor* (Appendix of Hehl et al., 1976) is defined as

$$\Pi_{\mu}{}^{\nu\tau} \equiv -\tfrac{1}{2} Q_{\mu}\, g^{\tau\nu} + Q_{\mu}{}^{\tau\nu} + \tfrac{1}{2} Q^{\nu}\, \delta_{\mu}{}^{\tau} - \overline{Q}^{\nu}\, \delta_{\mu}{}^{\tau} + g^{\beta\nu}\, C_{\mu\beta}{}^{\tau}, \tag{8.493}$$

where $C_{\mu\beta}{}^{\tau}$ is the modified torsion or Cartan tensor (7.128), and where the *Weyl vector* and *co-Weyl vector* are defined in terms of contractions of the nonmetricity tensor (7.152) as

$$Q^{\nu} \equiv Q^{\nu}{}_{\beta}{}^{\beta}, \tag{8.494}$$

and

$$\overline{Q}^{\nu} \equiv Q_{\beta}{}^{\beta\nu}. \tag{8.495}$$

The Palatini tensor (8.493) captures geometrical effects due to nonmetricity and torsion. In the absence of nonmetricity, the Palatini tensor equals the Cartan tensor, $\Pi_{\mu}{}^{\tau} = C_{\mu}{}^{\nu\tau}$.

The field equations of metric-affine relativity may be obtained by considering an action that combines the geometry action (8.480) with an action describing matter fields:

$$I = \int (\boldsymbol{L} + \boldsymbol{L}_M), \tag{8.496}$$

where $\boldsymbol{L}_M$ denotes the matter Lagrangian density. Stationarity of this action leads to the field equations of metric-affine gravity:

$$\kappa\, \sigma_{\mu\nu} = \tfrac{1}{2}\,(E_{\mu\nu} + E_{\nu\mu}), \tag{8.497}$$

$$\kappa\, \varsigma_{\mu}{}^{\nu} = \tfrac{1}{2}\,(E_{\mu}{}^{\nu} + \overline{E}_{\mu}{}^{\nu}), \tag{8.498}$$

$$\kappa\, \Delta_{\mu}{}^{\nu\tau} = \Pi_{\mu}{}^{\nu\tau}. \tag{8.499}$$

Here we have defined the symmetric *metrical energy-momentum current four-forms*

$$\sigma_{\mu\nu}\, \epsilon \equiv -2\,\frac{\delta \boldsymbol{L}_M}{\delta g^{\mu\nu}}, \tag{8.500}$$

the *canonical energy-momentum current three-forms*

$$\boldsymbol{\varsigma}_{\mu} = \varsigma_{\mu}{}^{\nu}\, \boldsymbol{\epsilon}_{\nu} \equiv -\,\frac{\delta \boldsymbol{L}_M}{\delta \boldsymbol{e}^{\mu}}, \tag{8.501}$$

and the *hypermomentum current three-forms*

$$\boldsymbol{\Delta}_{\mu}{}^{\nu} = \Delta_{\mu}{}^{\nu\tau}\, \boldsymbol{\epsilon}_{\tau} \equiv -2\,\frac{\delta \boldsymbol{L}_M}{\delta \boldsymbol{\omega}_{\nu}^{\mu}}. \tag{8.502}$$

The hypermomentum current (8.502) can be broken down in symmetric and antisymmetric parts in terms of the *spin current three-forms*

$$\mathbf{M}_{\mu\nu} = M_{\mu\nu}{}^{\sigma}\, \boldsymbol{\epsilon}_{\sigma} \equiv \tfrac{1}{2}\,(\boldsymbol{\Delta}_{\mu\nu} - \boldsymbol{\Delta}_{\nu\mu}) = \tfrac{1}{2}\,(\Pi_{\mu\nu}{}^{\sigma} - \Pi_{\nu\mu}{}^{\sigma})\, \boldsymbol{\epsilon}_{\sigma}, \tag{8.503}$$

and the *strain current three-forms*

$$\mathbf{S}_{\mu\nu} = S_{\mu\nu}{}^{\sigma}\,\epsilon_\sigma \equiv \tfrac{1}{2}\,(\boldsymbol{\Delta}_{\mu\nu} + \boldsymbol{\Delta}_{\nu\mu}) = \tfrac{1}{2}\,(\Pi_{\mu\nu}{}^{\sigma} + \Pi_{\nu\mu}{}^{\sigma})\,\epsilon_\sigma, \tag{8.504}$$

as

$$\boldsymbol{\Delta}_{\mu\nu} = \mathbf{M}_{\mu\nu} + \mathbf{S}_{\mu\nu}. \tag{8.505}$$

Using expression (8.493) for the components of the Palatini tensor, we have, specifically,

$$M_{\mu\nu}{}^{\sigma} = C_{\mu\nu}{}^{\sigma} + \tfrac{1}{2}(Q_\nu\,\delta_\mu{}^{\sigma} - Q_\mu\,\delta_\nu{}^{\sigma} + Q_\mu{}^{\sigma}{}_\nu - Q_\nu{}^{\sigma}{}_\mu - \bar{Q}_\nu\,\delta_\mu{}^{\sigma} + \bar{Q}_\mu\,\delta_\nu{}^{\sigma}), \tag{8.506}$$

$$S_{\mu\nu}{}^{\sigma} = \tfrac{1}{2}(Q_{\mu\nu}{}^{\sigma} + Q_{\nu\mu}{}^{\sigma} - \bar{Q}_\nu\,\delta_\mu{}^{\sigma} - \bar{Q}_\mu\,\delta_\nu{}^{\sigma}). \tag{8.507}$$

We conclude that the spin current (8.506) is determined by torsion and nonmetricity, whereas the strain current (8.507) is controlled by nonmetricity. The strain current may be further broken down (e.g., Hehl et al., 1995) in terms of the *dilation current*

$$S^{\sigma} \equiv \Pi_\nu{}^{\nu\sigma}, \tag{8.508}$$

and the traceless *shear current*

$$D_{\mu\nu}{}^{\sigma} \equiv S_{\mu\nu}{}^{\sigma} - \tfrac{1}{4}\,S^{\sigma}\,g_{\mu\nu}. \tag{8.509}$$

In the absence of nonmetricity, that is, in Einstein–Cartan general relativity, the strain current vanishes, $S_{\mu\nu}{}^{\sigma} = 0$, and the spin current equals the modified torsion, $M_{\mu\nu}{}^{\sigma} = C_{\mu\nu}{}^{\sigma}$. In that case, the Einstein and co-Einstein tensors are equal, $E_\mu{}^{\sigma} = \bar{E}_\mu{}^{\sigma}$, and the Palatini tensor equals the Cartan tensor, $\Pi_\mu{}^{\tau} = C_\mu{}^{\nu\tau}$. Thus, we have the Einstein–Cartan field equations

$$\kappa\,\sigma_{\mu\nu} = \tfrac{1}{2}\,(E_{\mu\nu} + E_{\nu\mu}), \tag{8.510}$$

$$\kappa\,\varsigma_\mu{}^{\nu} = E_\mu{}^{\nu}, \tag{8.511}$$

$$\kappa\,M_{\mu\nu}{}^{\tau} = C_{\mu\nu}{}^{\tau}. \tag{8.512}$$

In this case, the first field equation (8.510) is simply the symmetric part of the second field equation (8.511), which implies that the metrical energy-momentum current is the symmetric part of the canonical energy-momentum current, $\sigma_{\mu\nu} = \tfrac{1}{2}(\varsigma_{\mu\nu} + \varsigma_{\nu\mu})$. We encountered the Einstein–Cartan field equations (8.511) and (8.512) previously in example 7.7. In that example, we identified the Einstein tensor, $E_\mu{}^{\nu}$, with the energy-momentum tensor, $\sigma_\mu{}^{\nu}$, and the Cartan tensor, $C_{\mu\nu}{}^{\tau}$, with the spin tensor, $M_{\mu\nu}{}^{\tau}$. In other words, in Einstein–Cartan general relativity the canonical energy-momentum current equals the (non-symmetric) energy-momentum tensor, $\varsigma_\mu{}^\nu = \sigma_\mu{}^\nu$, and the hypermomentum current equals the spin tensor, $\Delta_{\mu\nu}{}^\tau = M_{\mu\nu}{}^\tau$.

Finally, in the absence of nonmetricity and torsion, the Einstein tensor is symmetric, $E_{\mu\nu} = E_{\nu\mu}$, the metrical energy-momentum current is identical to the canonical energy-momentum current and the energy-momentum tensor, $\sigma_{\mu\nu} = \varsigma_{\mu\nu}$, and the Cartan tensor vanishes, $C_{\mu\nu}{}^\tau = 0$. In that case, we are left with the Einstein field equations

$$\kappa\, \sigma_{\mu\nu} = E_{\mu\nu}. \tag{8.513}$$

GLOSSARY

$\mathcal{B}$	referential or material manifold
$\mathcal{M}$	point in $\mathcal{B}$
φ , Φ	maps
x^i , $x^{i'}$	spatial coordinates, $i, i' = 1, 2, 3$
$\varphi^i(x^{i'})$, $\Phi^{i'}(x^i)$	transformations
$\mathbf{e}_i = \partial_{x^i} = \partial_i$	spatial basis vector
$\lambda^i{}_{i'}$	transformation matrix
$\Lambda_i{}^{i'}$	inverse transformation matrix
$\mathbf{e}^i = \mathrm{d}x^i$	spatial basis one-forms
$[\mathbf{e}_i, \mathbf{e}_j] = \tau^k_{ij}\,\mathbf{e}_k$	Lie bracket with structure coefficients τ^k_{ij}
$\otimes$	tensor product
$\cdot$	dot product; contraction between adjacent covariant and contravariant indices
$:$	double-dot product; contraction between two adjacent covariant and contravariant indices
$\mathbf{I} = \delta^i{}_j\,\mathbf{e}_i \otimes \mathbf{e}^j$	identity tensor
$\mathbf{T}^t$	transpose of a rank-two tensor
$e_{i'}{}^i$	tetrad transformation matrix
$\widehat{\mathbf{T}}$	symmetric part of a rank-two tensor
$\widetilde{\mathbf{T}}$	antisymmetric part of a rank-two tensor
$\overline{T}^{i_1\cdots i_p}{}_{j_1\cdots j_q}$	components of a tensor density of weight one
$\underline{T}^{i_1\cdots i_p}{}_{j_1\cdots j_q}$	components of a tensor capacity of weight one
$\underline{\epsilon}_{ijk\cdots}$, $\overline{\epsilon}^{ijk\cdots}$	Levi-Civita symbols
$\mathbf{g} = g_{ij}\,\mathbf{e}^i \otimes \mathbf{e}^j$	metric tensor
$\mathbf{g}^{-1} = g^{ij}\,\mathbf{e}_i \otimes \mathbf{e}_j$	inverse of metric tensor
$\overline{g}$	square root of the determinant of the metric tensor
$\underline{g}$	square root of the determinant of the inverse metric tensor
$\lVert\mathbf{u}\rVert$	norm of vector $\mathbf{u}$
$\hat{\mathbf{u}}$	unit vector
$\mathbf{T}^\dagger$	adjoint of $(1, 1)$ tensor
$\epsilon = \overline{g}\,\underline{\epsilon}_{ijk\cdots}\,\mathbf{e}^i \otimes \mathbf{e}^j \otimes \mathbf{e}^k$	volume form

$\hat{\mathbf{n}}\,dS = \epsilon(\,\cdot\,, \mathbf{u}, \mathbf{w})$	surface element spanned by vectors **u** and **w**
φ^*	pullback
φ_*	pushforward
$\boldsymbol{\nabla}$	covariant derivative
Γ^i_{jk}	connection coefficients
$\mathrm{div}\,\mathbf{T} = \boldsymbol{\nabla}\cdot\mathbf{T}$	divergence of a tensor field
$\mathbf{u}\cdot\boldsymbol{\nabla} = \boldsymbol{\nabla}_{\mathbf{u}}$	directional derivative
$\mathbf{t}$	torsion tensor
$t_{ij}{}^k$	elements of the torsion tensor
$\mathbf{r}$	curvature tensor
$r_{ijk}{}^\ell$	elements of the curvature tensor
$\omega^{k'}_{ij'}$	frame connection coefficients
$\tilde{\nabla}$	mixed covariant derivative
$r_{ij} = r_{kij}{}^k$	Ricci tensor
$r = g^{ij}\,r_{ij}$	Ricci scalar
$E_i{}^j = r_i{}^j - \frac{1}{2}\,r\,\delta_i{}^j$	Einstein tensor
$C_{ij}{}^k = t_{ij}{}^k + t_{\ell i}{}^\ell\,\delta^k{}_j - t_{\ell j}{}^\ell\,\delta^k{}_i$	Cartan or modified torsion tensor
$\mathbf{Q} = \boldsymbol{\nabla}\mathbf{g}$	nonmetricity tensor
d_t	Euler derivative
$\mathcal{L}_{\mathbf{u}}$	Lie derivative relative to vector field **u**
$\mathscr{L}_{\mathbf{u}}$	autonomous Lie derivative relative to vector field **u**
$\wedge$	wedge product
$\mathbf{u}\,\rfloor\,\omega = \mathbf{i}_{\mathbf{u}}\omega$	interior product between a vector **u** and a *k*-form ω
$*\omega$	Hodge dual or Hodge star of a *k*-vector ω
$\boldsymbol{\Sigma} = \mathbf{i}_{\hat{\mathbf{n}}}\epsilon = \hat{\mathbf{n}}\cdot\epsilon$	surface two-form
d	exterior derivative
ω^j_i	connection one-forms
$\mathbf{t}^i$	torsion two-forms
$\mathbf{r}^i_j$	curvature two-forms
D	exterior covariant derivative
$\mathfrak{L}_{\mathbf{u}}$	covariant Lie derivative relative to vector field **u**
$\mathbf{Q}_{ij} = \mathrm{D}g_{ij}$	nonmetricity one-forms
$\boldsymbol{L}$	Lagrangian density
I	action
$\mathbf{J}$	Noether current

Continuum Mechanics Glossary

$\mathcal{S}$	spatial manifold
$\mathcal{s}$	point in $\mathcal{S}$
r^i	Eulerian spatial coordinates in $\mathcal{S}$, $i=1,2,3$
t	Eulerian time
X^I	Lagrangian spatial coordinates in $\mathcal{S}$, $I=1,2,3$
$\mathcal{B}$	referential manifold
$\mathcal{M}$	point in $\mathcal{B}$
T	Lagrangian time
X^I	referential coordinates in $\mathcal{B}$ at $T=0$
$\varphi^i(X^I, T)$	particle motion
$\Phi^I(x^i, t)$	inverse motion
$\mathbf{e}_i = \partial_{r^i} = \partial_i$	Eulerian basis vectors
$\mathbf{e}_I = \partial_{X^I} = \partial_I$	Lagrangian basis vectors
$\mathbf{v} = v^i\, \mathbf{e}_i$	material velocity
${F^i}_I$	deformation gradient
${(F^{-1})^I}_i$	inverse of deformation gradient
$\mathrm{d}r^i$	Eulerian basis one-forms
$\mathbf{e}^i$	Eulerian frame basis
$\mathrm{d}X^I$	Lagrangian basis one-forms
$\mathbf{e}^I$	Lagrangian coframe basis
σ	stress tensor
p	pressure
$\mathbf{c}$	elastic tensor
$\mathbf{c}^{-1}$	compliance tensor
ε	infinitesimal strain tensor
$\mathbf{d}$	strain deviator tensor
$\mathbf{M}$	seismic moment tensor
g_{ij}, g_{IJ}	Eulerian and Lagrangian metric components
g^{ij}, g^{IJ}	Eulerian and Lagrangian metric components
$\overline{g}, \overline{G}$	square root of the determinant of the Eulerian/Lagrangian metric components
$\underline{g}, \underline{G}$	square root of the determinant of the inverse of the Eulerian/Lagrangian metric components
$\epsilon_{ijk} = \overline{g}\, \underline{\epsilon}_{ijk}$	Eulerian components of volume form
$\epsilon_{IJK} = \overline{G}\, \underline{\epsilon}_{IJK}$	Lagrangian components of volume form
$\hat{\mathbf{n}}\, dS = \boldsymbol{\epsilon}(\,\cdot\,, \mathbf{u}, \mathbf{w})$	surface element spanned by vectors **u** and **w**
$\mathbf{v} \cdot \hat{\mathbf{n}}\, dS = \boldsymbol{\epsilon}(\mathbf{v}, \mathbf{u}, \mathbf{w})$	volume transport by vector field **v** through a surface element spanned by vectors **u** and **w**
$\mathring{\mathbf{e}}_I$	referential basis vectors in $\mathcal{B}$
$\mathring{\mathbf{e}}^I$	referential basis one-forms in $\mathcal{B}$
$\mathring{\mathbf{g}} = \mathring{g}_{IJ}\, \mathring{\mathbf{e}}^I \otimes \mathring{\mathbf{e}}^J$	referential metric in $\mathcal{B}$
$\overline{\mathring{G}}$	square root of determinant of referential metric
$\underline{\mathring{G}}$	square root of determinant of inverse referential metric
$\mathring{\boldsymbol{\epsilon}} = \mathring{\epsilon}_{IJK}\, \mathring{\mathbf{e}}^I \otimes \mathring{\mathbf{e}}^J \otimes \mathring{\mathbf{e}}^K$	referential volume form in $\mathcal{B}$
$\hat{\mathring{\mathbf{n}}}\, d\mathring{S} = \mathring{\boldsymbol{\epsilon}}(\,\mathring{\cdot}\,, \mathring{\mathbf{u}}, \mathring{\mathbf{w}})$	referential surface element spanned by vectors $\mathring{\mathbf{u}}$ and $\mathring{\mathbf{w}}$ in $\mathcal{B}$

$J = \overline{G}\,\underset{\sim}{\mathring{G}}$	Jacobian of the motion
$\mathbf{C}$	Cauchy-Green tensor
$\mathring{\mathbf{E}}$	strain tensor in $\mathcal{B}$
$\mathbf{E}^{\mathrm{L}}$	Lagrangian strain tensor
$\mathbf{E}^{\mathrm{E}}$	Eulerian or Almansi strain tensor
$\mathbf{E}^{\log}$	logarithmic or Hencky strain tensor
$\boldsymbol{\sigma}$	Cauchy stress tensor
$\mathring{\mathbf{S}}$	second Piola-Kirchhoff stress tensor in $\mathcal{B}$
$\boldsymbol{\tau}$	Kirchhoff stress tensor
$P_i{}^I$	elements of the first Piola-Kirchhoff stress
Γ^i_{JK}	Eulerian connection coefficients
Γ^I_{JK}	Lagrangian connection coefficients
$\mathbf{G} = (\boldsymbol{\nabla}\mathbf{v})^t$	material velocity gradient
$\mathbf{D}$	deformation-rate tensor
$\mathbf{W}$	vorticity tensor
D_t	material derivative
$\overset{\circ\Omega}{\mathrm{D}}_t$	corotational derivative
$\mathcal{L}_{\mathbf{v}}$	Lie derivative relative to material velocity $\mathbf{v}$
$\sharp, \natural, \flat$	accidentals to distinguish rank-2 tensors
$\boldsymbol{\Gamma}^i_j = \Gamma^i_{kj}\,\mathbf{e}^k$	Eulerian connection one-forms
$\boldsymbol{\Gamma}^I_J = \Gamma^I_{KJ}\,\mathbf{e}^K$	Lagrangian connection one-forms
$\boldsymbol{\alpha}^i$	dislocation two-forms
$\boldsymbol{\Xi}^i_j$	disclination two-forms
$\boldsymbol{\sigma}^i$	stress two-forms
$\mathring{\mathbf{S}}^I$	stress two-forms in $\mathcal{B}$
$\mathbf{E}_i$	strain two-vectors
$\mathbf{c}^{ij}$	elastic tensor
V	volume
S	surface
$\mathbf{q}$	“q-stuff” three-form
$\boldsymbol{\rho}$	mass three-form
$\mathbf{p}^i$	linear momentum three-forms
$\mathbf{f}^i$	volume force three-forms
$\mathbf{l}^{ij}$	angular momentum three-forms
$\boldsymbol{\theta}^{ij}$	spin angular momentum three-forms
$\mathbf{n}^{ij}$	volume torque three-forms
$\boldsymbol{\nu}^{ij}$	surface torque two-forms
$\boldsymbol{\mu}^{ij}$	couple-stress two-forms
$\mathbf{e}$	internal energy three-form
h	rate of internal heating
$\mathbf{h}$	rate of internal heating three-form
$\mathbf{H}$	heat flux two-form
w_{ij}	material angular velocity
ϖ_{ij}	total angular velocity

$\Phi^i{}_j$	total rotation
$\mathbf{j}^{ijk\ell}$	inertia density three-forms
θ	absolute temperature
S	specific entropy density
$\boldsymbol{\Phi}$	Helmholtz free energy three-form
ξ	internal state variables
$\boldsymbol{L}$	Lagrangian density three-form
I	action

General Relativity Glossary

$\mathcal{S}$	spacetime event
φ, Φ	maps
$x^{\mu}, x^{\mu'}$	spacetime coordinates, $\mu, \mu' = 0, 1, 2, 3$
$\varphi^{\mu}(x^{\mu'}), \Phi^{\mu'}(x^{\mu})$	transformations
$\mathbf{e}_{\mu} = \partial_{\mu}$	basis vectors
$\lambda^{\mu}{}_{\mu'}$	transformation matrix
$\Lambda^{\mu'}{}_{\mu}$	inverse transformation matrix
$\mathbf{e}^{\mu} = \mathrm{d}x^{\mu}$	basis one-forms
e_a	tetrad frame basis
e^a	tetrad coframe basis
$e_a{}^{\mu}$	tetrad transformation matrix
$g_{\mu\nu}$	components of metric tensor
$g^{\mu\nu}$	components of inverse metric tensor
$\bar{g} = \sqrt{-\det(g_{\mu\nu})}$	square root of determinant of metric
$\underline{g}$	square root of determinant of inverse of metric
g_{ab}	metric in tetrads
$\epsilon_{\mu\nu\tau\sigma} = \bar{g}\,\underline{\epsilon}_{\mu\nu\tau\sigma}$	components of volume form
$\Gamma^{\tau}_{\mu\nu}$	connection coefficients
ω^{c}_{ab}	tetrad connection coefficients
$\mathbf{t}$	torsion tensor
$t_{\mu\nu}$	elements of the torsion tensor
$\mathbf{r}$	curvature tensor
$r_{\mu\nu\tau}{}^{\sigma}$	elements of the curvature tensor
$r_{\mu\nu} = r_{\tau\mu\nu}{}^{\tau}$	Ricci tensor
$\bar{r}_{\mu\nu} = r_{\mu\tau}{}^{\tau\nu}$	co-Ricci tensor
$r = g^{\mu\nu} r_{\mu\nu}$	Ricci scalar
$\breve{\nabla}_{\mu}$	modified covariant derivative
$E_{\mu}{}^{\nu}$	Einstein tensor
$\bar{E}_{\mu}{}^{\nu}$	co-Einstein tensor
$C_{\mu\nu}{}^{\tau}$	Cartan or modified torsion tensor
$\sigma_{\mu}{}^{\nu}$	stress-energy tensor
$M_{\mu\nu}{}^{\tau}$	spin tensor
$Q_{\mu\nu\tau} = \nabla_{\mu} g_{\nu\tau}$	nonmetricity tensor
$Q^{\nu} = Q_{\beta}{}^{\nu\beta}$	Weyl vector
$\bar{Q}^{\nu} = Q_{\beta}{}^{\beta\nu}$	co-Weyl vector
$\boldsymbol{\omega}^{\nu}_{\mu}$	connection one-forms
ω^{b}_{a}	tetrad connection one-forms
$\tilde{\mathrm{D}}$	mixed exterior covariant derivative
$\mathbf{t}^{\mu}$	torsion two-forms
$\mathbf{r}^{\mu}_{\nu}$	curvature two-forms
$\mathbf{Q}_{\mu\nu} = \mathrm{D}g_{\mu\nu}$	nonmetricity one-forms
$\boldsymbol{L}$	Lagrangian density three-form

$\boldsymbol{L}_M$	matter Lagrangian density three-form
I	action
$\sigma_{\mu\nu}\,\epsilon$	metrical energy-momentum current four-forms
ς_μ	canonical energy-momentum current three-forms
$\boldsymbol{\Delta}_\mu{}^\nu$	hypermomentum current three-forms
$\mathbf{M}_{\mu\nu}$	spin current three-forms
$\mathbf{S}_{\mu\nu}$	strain current three-forms

BIBLIOGRAPHY

Acharya, A., 1999, On compatibility conditions for the left Cauchy–Green deformation field in three dimensions, *J. Elast.*, **56**(2), 95–105.

Adkins, C. J., 1983, *Equilibrium Thermodynamics*, Cambridge University Press, Cambridge.

Amari, S., 1962a, A geometrical theory of moving dislocations and anelasticity, *RAAG Research Notes Ser. 3*, **52**, 1–27.

Amari, S., 1962b, A theory of deformation and stresses of terromagnetic substances by Finsler geometry, *RAAG Memoirs 3 D-XV*, pp. 257–278.

Amari, S., Uehara, T., & Oshima, N., 1961, A theory of deformation and stresses of terromagnetic substances by Finsler geometry, *RAAG Research Notes Ser. 3*, **49**, 1–24.

Appel, W., 2005, *Mathématiques pour la physique et les physiciens*, HK Éditions, Paris.

Bilby, B. A., Bullough, B. R., & Smith, E., 1956, Continuous distributions of dislocations: a new application of the methods of non-Riemannian geometry, *Proc. Roy. Soc. London*, **231**, 263–273.

Bilby, B. A. & Smith, E., 1956, Continuous distributions of dislocations: III, *Proc. Roy. Soc. London*, **236**, 481–505.

Bishop, R. L. & Goldberg, S. I., 1980, *Tensor Analysis on Manifolds*, Dover Publications, New York.

Bóna, A., Bucataru, I., & Slawinski, M. A., 2004, Material symmetries of elasticity tensors, *The Quarterly Journal of Mechanics and Applied Mathematics*, **57**(4), 583–598.

Borgnakke, C. & Sonntag, R. E., 2012, *Fundamentals of Thermodynamics*, Wiley, New York, 8th edn.

Callen, H. B., 1985, *Thermodynamics and an Introduction to Thermostatistics*, Wiley, New York, 2nd edn.

Carroll, S. M., 2004, *Spacetime and Geometry: An Introduction to General Relativity*, Addison Wesley, San Francisco.

Cartan, E., 1923, Sur les variétés à connexion affine et la théorie de la relativité généralisée (Première partie), *Annales Scientifiques de l'Ecole Normale Supérieure*, **58**, 325–412.

Cartan, E., 1924, Sur les variétés à connexion affine et la théorie de la relativité généralisée (Première partie, suite), *Annales Scientifiques de l'Ecole Normale Supérieure*, **59**, 1–25.

Cartan, E., 1925, Sur les variétés à connexion affine et la théorie de la relativité généralisée (Deuxième partie), *Annales Scientifiques de l'Ecole Normale Supérieure*, **60**, 17–88.

Cosserat, E. & Cosserat, F., 1896, Sur la théorie de l'élasticité, *Ann. de l'Ecole Normale de Toulouse*, **10**, 1.

Cosserat, E. & Cosserat, F., 1907, Sur la mécanique générale, *C. Rend. hebd. des Séances de l'Acad. des Sci.*, **145**, 1139.

Cosserat, E. & Cosserat, F., 1909, *Théorie des corps déformables*, A. Herman et Fils, Paris.

Dahlen, F. A. & Tromp, J., 1998, *Theoretical Global Seismology*, Princeton University Press, Princeton, New Jersey.

DeWit, R., 1969, Linear theory of static disclinations, in *Fundamental Aspects of Dislocation Theory*, edited by J. A. Simmons & R. deWit, vol. 1, pp. 651–673.

DeWit, R., 1973, Theory of disclinations: II. Continuous and discrete disclinations in anisotropic elasticity, *J. Res. Natl. Bur. Stand. (U.S.)*, **77**(1), 49–100.

Do Carmo, M. P., 1976, *Differential Geometry of Curves and Surfaces*, Prentice-Hall, Englewood Cliffs, New Jersey.

Do Carmo, M. P., 1994, *Differential Forms and Applications*, Springer-Verlag, Berlin.

Dubrovin, B. A., Fomenko, A. T., & Novikov, S. P., 1985, *Modern Geometry—Methods and Applications*, Springer-Verlag, New York.

Dudek, M. & Garecki, J., 2019, General relativity with a positive cosmological constant Λ as a gauge theory, *Axioms*, **8**(1), 24.

Edelen, D.G.B. & Lagoudas, D. C., 1988, *Gauge Theory and Defects in Solids*, North Holland, Amsterdam.

Edmonds, A. R., 1960, *Angular Momentum in Quantum Mechanics*, Princeton University Press, Princeton, New Jersey.

Einstein, A., 1915, Die Feldgleichungen der Gravitation, *Königlich Preussische Akademie der Wissenschaften, Sitzungsberichte*, **3**, 844–847.

Einstein, A. 1917, Sitz. König. Preuss. Akad., Kosmologische Betrachtungen zur allgemeinen Relativitätstheorie, 142–152.

Einstein, A., 1918, Prinzipielles zur allgemeinen Relativitätstheorie, *Annalen der Physik*, **55**, 241–244.

Flanders, H., 1989, *Differential Forms with Applications to the Physical Sciences*, Dover Publications, New York.

Frankel, T., 2004, *The Geometry of Physics: An Introduction*, Cambridge University Press, Cambridge, 2nd edn.

Hehl, F. W. & McCrea, J. D., 1986, Bianchi identities and the automatic conservation of energy-momentum and angular momentum in general-relativistic field theories, *Foundations of Physics*, **16**(3), 202–226.

Hehl, F. W. & Obukhov, Y. N., 2007, Élie Cartan's torsion in geometry and in field theory, an essay, *Annales de la Fondation Louis de Broglie*, **32**, 157–193.

Hehl, F. W. & Weinberg, S., 2007, Note on the torsion tensor, *Phys. Today*, **60**, 16.

Hehl, F. W., von der Heyde, P., Kerlick, G. D., & Nester, J. M., 1976, General relativity with spin and torsion: Foundations and prospects, *Rev. Mod. Phys.*, **48**, 393–416.

Hehl, F. W., McCrea, J. D., Mielke, E. W., & Ne'eman, Y., 1995, Metric-affine gauge theory of gravity: field equations, noether identities, world spinors, and breaking of dilation invariance, *Phys. Rep.*, **258**(1–2), 1–171.

Hencky, H., 1928, Über die Form des Elastizitätsgesetzes bei ideal elastischen Stoffen, *Z. Techn. Phys.*, **9**, 214–247.

Hilbert, D., 1915, Die Grundlagen der Physik, *Nachrichten von der Gesellschaft der Wissenschaften zu Göttingen—Mathematisch-Physikalische Klasse*, **3**, 395–407.

Holländer, E. F., 1960a, The basic equations of the dynamic of the continuous distribution of the dislocations I. General theory, *Czech. J. Phys. B*, **10**, 409–418.

Holländer, E. F., 1960b, The basic equations of the dynamic of the continuous distribution of the dislocations II. Interpretation of general theory, *Czech. J. Phys. B*, **10**, 479–487.

Holländer, E. F., 1960c, The basic equations of the dynamic of the continuous distribution of the dislocations III. Special problems, *Czech. J. Phys. B*, **10**, 551–560.

Jackson, J., 1998, *Classical electrodynamics*, John Wiley & Sons, New York, 3rd edn.

Jaumann, G., 1911, Geschlossenes System physikalischer und chemischer Differentialgesetze, *Akad. Wiss. Wien, Sitzber. IIa*, pp. 385–530.

Kibble, T.W.B., 1961, Lorentz invariance and the gravitational field, *J. Math. Phys*, **2**, 212–221.

Kleinert, H., 1989, *Gauge Fields in Condensed Matter*, World Scientific, Singapore.

Kleinert, H., 2008, *Multivalued Fields in Condensed Matter, Electromagnetism, and Gravitation*, World Scientific, Singapore.

Kondo, K., 1952, On the geometrical and physical foundations of the theory of yielding, *Proc. 2nd Japan Nat. Congr. App. Mech.*, pp. 41–47, published 1953.

Kondo, K., 1955a, Geometry of elastic deformation and incompatibility, *RAAG Memoirs 1 Div. C*, pp. 361–373.

Kondo, K., 1955b, Non-Riemannian geometry of imperfect crystals from a macroscopic viewpoint, *RAAG Memoirs 1 Div. D*, pp. 458–469.

Kondo, K., 1955c, Energy at plastic deformation and criterion for yield, *RAAG Memoirs 1 Div. D*, pp. 484–494.

Kondo, K., 1963, Non-Riemannian and Finslerian approaches to the theory of yielding, *Int. J. Engng. Sci.*, **1**, 71–88.

Kondo, K., 1964a, On the analytical and physical foundations of the theory of dislocations and yielding by the differential geometry of continua, *Int. J. Engng. Sci.*, **2**, 219–251.

Kondo, K., 1964b, On the variational foundations and the polycrystalline origin of the non-Riemannian theory of yielding, *Jour. Fac. Engng., Univ. Tokyo*, **27(1)**, 183–216.

Kosevich, A. M., 1965, Dynamical theory of dislocations, *Soy. Phys. USPE-KHI*, **7**, 837–854.

Lovelock, D. & Rund, H., 1989, *Tensors, Differential Forms, and Variational Principles*, Dover Publications, New York.

Malvern, L., 1969, *Introduction to the Mechanics of a Continuous Medium*, Prentice-Hall, Englewood Cliffs, New Jersey.

Marsden, J. E. & Hughes, J. R., 1983, *Mathematical Foundations of Elasticity*, Prentice-Hall, Englewood Cliffs, New Jersey (reprinted by Dover, New York, 1994).

Martin, D., 2002, *Manifold Theory: An Introduction for Mathematical Physicists*, Horwood Publishing, Chichester, West Sussex.

Maugin, G. A., 2010, Generalized continuum mechanics: what do we mean by that?, in *Mechanics of Generalized Continua: One Hundred Years after the Cosserats*, edited by Gérard A. Maugin & Andrei V. Metrikine, pp. 3–13, Springer, New York.

Maugin, G. A., 2013, *Continuum Mechanics through the Twentieth Century: A Concise Historical Perspective*, Springer, Dordrecht.

Maugin, G. A. & Metrikine, A. V., 2010, *Mechanics of Generalized Continua*, Springer, New York.

Misner, C., Thorne, K., & Wheeler, J., 2017, *Gravitation*, Princeton University Press, Princeton, New Jersey.

Montesinos, M., Romero, R., & Gonzalez, D., 2020, The gauge symmetries of $f(R)$ gravity with torsion in the Cartan formalism, *Class. Quantum Grav.*, **37**, 045008.

Munkres, J. R., 1990, *Analysis on Manifolds*, Perseus Books, New York.

Mura, T., 1963, Continuous distribution of moving dislocation, *Phil. Mag.*, **8**, 843–857.

Mura, T., 1982, *Micromechanics of Defects in Solids*, Martinus Nijhoff Publishers, The Hague.

Nabarro, F.R.N., 1967, *Theory of Crystal Dislocation*, The Clarendon Press, Oxford.

Noether, E., 1918, Invariante Variationsprobleme, *Nachrichten von der Gesellschaft der Wissenschaften zu Göttingen. Mathematisch-Physikalische Klasse*, pp. 235–257.

Noll, W., 1974, *The Foundations of Mechanics and Theromodynamics: Selected Papers*, Springer-Verlag, Berlin.

Nowacki, W., 1986, *Theory of Asymmetric Elasticity*, Pergamon Press, Oxford.

Nye, J. F., 1953, Some geometrical relations in dislocated crystals, *Acta Metallurg*, **1**, 153–162.

Penrose, R., 2004, *The Road to Reality: A Complete Guide to the Laws of the Universe*, Jonathan Cape, London.

Randono, A., 2010, Gauge gravity: a forward-looking introduction.

Rodrigues, O., 1840, Des lois géométriques qui regissent les déplacements d'un système solide dans l'espace, et de la variation des coordonnées provenant de déplacements considérés indépendamment des causes qui peuvent les produire, *J. de Mathématiques Pures et Appliquées*, **5**, 380–440.

Roychowdhury, A. & Gupta, A., 2013, Geometry of defects in solids.

Ruggiero, M. L. & Tartaglia, A., 2003, Einstein-Cartan theory as a theory of defects in space-time, *Am. J. Phys.*, **71**(12), 1303–1313.

Sakata, S., 1971, On a representation of plastic material manifold with high order anomalies by generalized diakoptical tearing, *Memoirs Sagami Inst. Techno.*, **5(1)**, 9–25.

Schey, H. M., 2004, *Div, Grad, Curl, and All That: An Informal Text on Vector Calculus*, W. W. Norton & Company, New York, 4th edn.

Schutz, B., 1980, *Geometrical Methods of Mathematical Physics*, Cambridge University Press, Cambridge.

Sciama, D. W., 1962, On the analogy between charge and spin in general relativity, in *Recent Developments in General Relativity*, pp. 415–439, Pergamon Press, Oxford.

Sciama, D. W., 1964, The physical structure of general relativity, *Rev. Mod. Phys.*, **36**, 463–469.

Sedov, L. I., 1966, *Foundations of the Non-Linear Mechanics of Continua*, Pergamon Press, Oxford.

Shimbo, M., 1971, On the effect of dilatancy with reference to the extended Einsteinian coordinates, *RAAG Research Notes Set. 3*, **167**, 1–14.

Shimbo, M., 1981, Continuum theory of defects, in *Physics of Defects*, edited by R. Balian, M. Kléman, & J.-P. Poirier, vol. 26, pp. 214–315, North-Holland, Amsterdam, Session XXXV, Les Houches, France.

Spivak, M., 1965, *Calculus on Manifolds: A Modern Approach to Classical Theorems of Advanced Calculus*, Harper Collins, New York.

Stone, M. & Goldbart, P., 2009, *Mathematics for Physics: A Guided Tour for Graduate Students*, Cambridge University Press, Cambridge.

Takeo, M. & Ito, H., 1997, What can be learned from rotational motions excited by earthquakes?, *Geophys. J. Int.*, **129**, 319–329.

Tarantola, A., 2005, *Inverse Problem Theory and Methods for Model Parameter Estimation*, SIAM, Philadelphia, Pennsylvania.

Torres del Castillo, G. F., 2012, *Differentiable Manifolds: A Theoretical Physics Approach*, Birkhauser, Basel.

Trautman, A., 1973, Einstein-Cartan theory, *Symposia Mathematica*, **12**, 139.

Trautman, A., 2006, Einstein-Cartan theory, in *Encyclopedia of Mathematical Physics*, edited by J.-P. Francoise, G. Naber, & S. T. Tsou, pp. 189–195, Elsevier.

Tromp, J., 2025, *Theoretical and Computational Seismology*, Princeton University Press, Princeton, New Jersey.

Truesdell, C. A. & Toupin, R. A., 1960, The classical field theories, in *Encyclopedia of Physics*, edited by S. Flügge, Springer-Verlag, Berlin.

Vitagliano, V., Sotiriou, T., & Liberati, S., 2011, The dynamics of metric-affine gravity, *Ann. Phys. (N.Y.)*, **326**, 1259–1273.

Wald, R. M., 1984, *General Relativity*, University of Chicago Press, Chicago.

Weile, D., Hopkins, D., Gazonas, G., & Powers, B., 2013, A convective coordinate approach to continuum mechanics with application to electrodynamics, Tech. rep., U.S. Army Research Laboratory, Report number: ARL-TR-6298.

Zaremba, S., 1903, Sur une forme perfectionée de la théorie de la relaxation, *Bull. Intern. Acad. Sci. Cracovie*, pp. 594–614.

AUTHOR INDEX

Acharya, A. 206
Adkins, C. J. 195
Amari, S. 145
Appel, W. 134

Bilby, B. A. 145
Bishop, R. L. 2
Borgnakke, C. 195
Bullough, B. R. 145

Callen, H. B. 195
Carroll, S. M. 2, 17, 101, 102, 134, 167, 206
Cartan, E. 104
Cosserat, E. 188
Cosserat, F. 188

Dahlen, F. A. 65, 81
deWit, R. 205
do Carmo, M. P. 7, 11, 127
Dubrovin, B. A. 2
Dudek, M. 206

Edelen, D. G. B. 146, 205
Edmonds, A. R. 64
Einstein, A. 104

Flanders, H. 127
Fomenko, A. T. 2
Frankel, T. 59, 133, 135, 137, 148, 178

Garecki, J. 206
Gazonas, G. 9
Goldbart, P. 2
Goldberg, S. I. 2
Gonzalez, D. 206
Gupta, A. 108, 174

Hehl, F. W. 104, 105, 107, 147, 151, 173, 174, 206, 207, 209
Hencky, H. 82
Hilbert, D. 206
Holländer, E. F. 145
Hopkins, D. 9
Hughes, J. R. 2, 79, 123, 148

Ito, H. 204

Jackson, J.D. 108
Jaumann, G. 113

Kerlick, G. D. 104, 174, 206, 207
Kibble, T. W. B. 104
Kleinert, H. 146
Kondo, K. 145, 204
Kosevich, A. M. 145

Lagoudas, D. C. 146, 205
Liberati, S. 107, 174
Lovelock, D. 2, 42, 71, 97

Malvern, L. 38, 53, 188, 194
Marsden, J. E. 2, 79, 123, 148
Martin, D. 2, 7
Maugin, G. A. 1, 188
McCrea, J. D. 107, 151, 173, 174, 206, 209
Metrikine, A. V. 188
Mielke, E. W. 107, 173, 174, 206, 209
Misner, C. 2, 23, 28, 29, 34, 184
Montesinos, M. 206
Munkres, J. R. 7, 16
Mura, T. 145

Nabarro, F. R. N. 145
Ne'eman, Y. 107, 173, 174, 206, 209
Nester, J. M. 104, 174, 206, 207
Noether, E. 181
Noll, W. 145
Novikov, S. P. 2
Nowacki, W. 188
Nye, J. F. 145

Obukhov, Y. N. 147
Oshima, N. 145

Penrose, R. 96
Powers, B. 9

Randono, A. 206
Rodrigues, O. 67
Romero, R. 206
Roychowdhury, A. 108, 174
Ruggiero, M. L. 147
Rund, H. 2, 42, 71, 97

Sakata, S. 145
Schey, H. M. 142
Schutz, B. 2, 16, 19, 27, 42, 114, 178
Sciama, D. W. 104
Sedov, L. I. 9

Shimbo, M. 145
Smith, E. 145
Sonntag, R. E. 195
Sotiriou, T. 107, 174
Spivak, M. 7, 114, 178
Stone, M. 2

Takeo, M. 204
Tarantola, A. 58
Tartaglia, A. 147
Thorne, K. 2, 23, 28, 29, 34, 184
Torres del Castillo, G. F. 7
Toupin, R. A. 82, 188
Trautman, A. 104, 105, 136, 137, 174, 206
Tromp, J. 2, 65, 81, 147, 174, 188, 205
Truesdell, C. A. 82, 188

Uehara, T. 145

Vitagliano, V. 107, 174
von der Heyde, P. 104, 174, 206, 207

Wald, R. M. 85, 95, 103
Weile, D. 9
Weinberg, S. 104, 105
Wheeler, J. 2, 23, 28, 29, 34, 184

Zaremba, S. 113

GENERAL INDEX

absolute temperature, 194
accidentals, 123
action, 180, 201, 202, 206; Hilbert, 206; metric-affine, 206; Palatini, 206
adiabatic process, 195, 198
adjoint, 57
adjoint operator, 75
affine connection one-forms, 206
Almansi strain, 82
alternating symbol, 48
Ampère's law, 109
angular momentum three-form, 188
angular velocity; material, 192
anholonomic, 31
anholonomicity, 30
antisymmetric, 39
antisymmetric tensor, 39
associative, 3
atlas, 7
autonomous Lie derivative, 118

basis, 17; anholonomic, 31; holonomic, 30, 94, 95; linearly independent, 17
basis vector, 17
beach ball, 40
beach ball representation, 39
Bianchi identities, 97, 147, 172–174; contracted, 102–105; post-Riemannian, 174
bijective, 11, 73
bulk modulus, 69
Burgers vector, 147, 204, 205

caloric equation of state, 196
canonical energy-momentum current, 208
capacity; Levi-Civita, 48; tensor, 47, 58
Cartan homotopy formula, 149, 167
Cartan magic formula, 149, 167
Cartan tensor, 104, 105
Cauchy stress; weighted, 83
Cauchy stress tensor, 38
Cauchy stress two-forms, 152, 159
Cauchy-Green tensor, 81
chart, 7, 77, 78
Christoffel symbol; first kind, 100; Lagrangian, 101; second kind, 100
circulation tube, 132, 158
classical vector calculus, 142
Clausius–Duhem inequality, 195
closed form, 139, 182
co-Einstein tensor, 207
co-Ricci tensor, 207
co-Weyl vector, 208
coframe, 43, 44, 130; orthogonal, 57
commutative, 3
commutator, 29
comoving coordinates, 9, 78
comoving volume, 183
compact, 179
compatibility, 32
compatibility conditions, 32
compliance tensor, 69
congruence, 19, 114
conjugate pair, 156
conjugate thermodynamic parameters, 196
conjugate variables, 196
connection; spin, 102, 167; torsion-free, 97
connection coefficients, 88; frames, 101; Lagrangian, 100; nonmetricity, 107; torsion-free, 100; transformation, 89
connection one-forms, 161; coordinate, 166; Eulerian, 166; Lagrangian, 145, 166; tetrad, 166
conservation law, 182
conservation of angular momentum, 188
conservation of energy, 190; with spin, 192
conservation of linear momentum, 186
conservation of mass, 186
continuity equation, 186; Maxwell's equations, 110
continuous symmetry, 181
continuum mechanics, 18, 21, 22, 27, 38, 42, 43, 45, 53, 59, 68, 77, 81–83, 90, 95, 100, 108, 112, 119, 121, 124, 145, 151, 156, 159, 166, 168, 183, 186, 188, 190, 192, 194, 195, 201, 202, 204
contortion tensor, 99, 108
contracted Bianchi identities, 102, 103
contraction, 36, 159; tensor-valued form, 159
contravariant, 34, 36
convected coordinates, 78, 115
coordinate connection one-forms, 166
coordinate system, 7; convected, 115
coordinates, 7; comoving, 9, 78; convected, 78; Eulerian, 9; Lagrangian, 9; referential, 77; spatial, 9

corotation, 112
corotational material derivative, 112; Zaremba-Jaumann, 113
cosmological constant, 105
Cosserat micropolar medium, 188
cotangent bundle, 23
cotangent space, 23
couple stress, 189; three-forms, 203
couple-stress two-forms, 189
covariant, 34, 36
covariant derivative, 85; mixed, 102; modified, 105; tensor capacity, 107; tensor density, 107; tetrad, 101; torsion-free, 97
covariant Lie derivative, 167; Eulerian, 168; Lagrangian, 168
covariant vector, 27
covector, 27
Cramer's rule, 62
cross product, 49, 60, 134
cubic symmetry, 68
curl, 141
current; dilation, 209; shear, 209; spin, 208; strain, 209
curvature, 34, 36, 108; Gaussian curvature, 96; intrinsic, 205; Ricci, 103
curvature tensor, 34, 36, 37, 41, 94–96
curvature two-forms, 169, 170
curve, 16

dark energy and cosmological constant, 105
defects, 32, 107, 145, 205; interstitial, 109; point, 109; substitutional, 109; vacancy, 109
deformation; elastic, 196, 198; plastic, 196, 198
deformation gradient, 22
deformation rate, 90
deformation tensor; Green, 53; right Cauchy–Green, 53
deformation-rate tensor, 91, 121, 123
density; Levi-Civita, 48; tensor, 47, 58
derivative; corotational material, 112; covariant, 85; Euler, 112; exterior, 24, 139; exterior covariant, 164; Lie, 114; material, 112; substantial, 112
determinant, 46, 47, 50
deviatoric strain, 69
diffeomorphism, 8, 72, 182
differentiable manifold, 7, 8, 15
differential, 24
differential forms, 127
differential geometry, 15
dilation current, 209
directional derivative, 92
disclination, 32, 96, 145–147, 205; twist, 147; wedge, 147
disclination two-forms, 147, 204, 205
disformation tensor, 108
dislocation, 32, 96, 145–147, 204; edge, 147; screw, 147
dislocation line, 147
dislocation two-forms, 147, 204, 205
distance, 51
distributive, 4
divergence, 91, 141
dot product, 37, 55
double-dot product, 38, 55, 66, 159
Dreibein, 43
dual, 23, 29
dual space, 23
dual vector, 27
duality, 23, 29
duality condition, 24
duality product, 24, 41
dummy index, 35
dynamic equations, 105; Einstein general relativity, 106; Einstein-Cartan general relativity, 106

earthquake, 39
edge dislocation, 147
Einstein general relativity, 104, 174; dynamic equations, 106; field equations, 104, 210
Einstein summation convention, 17
Einstein tensor, 103, 207
Einstein-Cartan general relativity, 104, 147, 174; dynamic equations, 106; field equations, 105, 209
Einstein-Hilbert action, 206
elastic deformation, 196, 198
elastic tensor, 67; isotropic, 68
electric charge density, 109
electric displacement, 111
electric field, 109
endomorphism, 58
entropy, 194
equation of state, 195; caloric, 196
Euclidean space, 7
Euler angles, 64
Euler derivative, 112, 113, 149
Euler-Lagrange equations, 181
Eulerian coordinates, 9, 77
Eulerian strain, 82
event, 10, 29
exact form, 139, 142, 205
explosion, 40
exterior covariant derivative, 164; Eulerian, 166; Lagrangian, 166; mixed, 166
exterior derivative, 24, 139; covariant, 164
exterior differential forms, 127
exterior product, 130, 159; Grassmann product, 131; wedge product, 131

Faraday tensor, 143
Faraday's law, 109
field equations, 104, 206; Einstein general relativity, 104, 210; Einstein-Cartan general relativity, 105, 209; metric-affine gravity, 208
first law of thermodynamics, 190; with spin, 192
first Piola-Kirchhoff stress, 83; two-forms, 154
first postulate, 55
flat $\flat$, 123
flow of matter, 21
fluid, 69
flux, 133, 150
focal hemisphere, 39
form; closed, 139, 182; connection, 161; exact, 139, 142, 205; integration, 174; Lie derivative, 148; operations, 130; stress, 151; tensor-valued, 147, 150; vector-valued, 147, 150
form-versions of Maxwell's equations, 143
four-velocity, 29, 45, 54
frame, 43; inertial, 106; nonspinning, 106

frames; inertial; nonspinning, 106; orthogonal, 57
Frank vector, 147, 204, 205
Frobenius inner product, 38
Frobenius norm, 38, 47
fundamental theorem of calculus, 178
fundamental theorem of exterior calculus, 179

Galilean reference frame, 9
Galilean spacetime, 95
gas, 69
Gauss's theorem, 180
Gauss's law; electric field, 109; magnetic field, 109
general relativity, 7, 10, 36, 37, 41, 43, 44, 52, 54, 63, 94, 104, 105, 108, 137, 173, 206; Einstein, 174; Einstein field equations, 104; Einstein-Cartan, 174; Einstein-Cartan field equations, 105; instein-Cartan, 147; metric-affine, 174; metric-affine field equations, 208
generalized Kronecker delta symbol, 60
generalized Stokes's theorem, 178
geodesic, 93
geodesic equation, 93
geometry; metric-affine, 206; Riemann-Cartan, 206; Riemannian, 206
Gibbs relation, 196
gradient, 141; deformation, 22
gravitation; Einstein, 174; Einstein-Cartan, 107, 174; metric-affine, 107, 174
Green deformation tensor, 53
Green's Theorem, 179

Hamilton's principle, 201
Hausdorff manifold, 11
heat, 190
heat flux, 191
heat flux two-form, 194
helicity, 96
Helmholtz free energy, 195
Hencky strain, 47, 82
Hilbert action, 206
Hodge dual, 134
Hodge star, 134
holonomic, 31
holonomic basis, 94, 95
holonomicity, 30
holonomy, 96
homeomorphism, 72
Hooke's law, 38, 67, 159; form version, 159
hypermomentum current, 208
hypocenter, 39

identity tensor, 46
implosion, 40
incompatibility, 32
incompatibility tensor, 32
incompatible motion, 145
incompressibility, 69
induction equation, 110, 111
inertia density tensor, 193
inertial frame, 106
inertial frames; nonspinning, 106
inertial reference frame, 9
infinitesimal strain tensor, 38, 68, 82
injective, 8, 73
inner product, 37, 52
integral; line, 174, 175; surface, 175, 176; volume, 177
integral curve, 19
interior product, 133
internal energy density, 190
internal heating, 191
internal state variables, 196
interstitial defect, 109
intrinsic curvature, 205
intrinsic spin, 188, 192
intrinsic torsion, 204, 205
inverse of rank-2 tensors, 50, 63
inverse problems, 58
irreversible process, 195, 196
isotropic symmetry, 68

Jacobian, 13
Jacobian of the motion, 79, 124

k-form, 127, 129, 130, 133
k-vector, 133
kinetic energy; rotational, 192
kinetic energy density, 190
Kirchhoff stress, 83
Kronecker delta symbol, 20; generalized, 60
Kronecker determinants, 60
Kronecker tensor, 46

Lagrangian connection coefficients, 100
Lagrangian coordinates, 9, 77
Lagrangian density, 180, 206; four-form, 202; matter, 208; three-form, 201
Lagrangian strain, 82
Legendre transformation, 196
Leibniz's rule, 87, 183
Levi-Civita capacity, 48, 58
Levi-Civita density, 48, 58
Levi-Civita pseudotensor, 48, 58, 59
Levi-Civita symbol, 48
Lie bracket, 29
Lie derivative, 112, 114, 115, 149; autonomous, 118; covariant, 167; covariant, 167; Euler derivative, 113; form, 148; function, 120; general tensor, 119; geometrical interpretation, 117; Levi-Civita tensor, 123; metric tensor, 120; one-form, 118; vectors, 114
Lie dragging, 117
line integral, 175
line of nodes, 64
linear momentum three-form, 186
linear momentum two-form, 186
linear space, 3; external operation, 4; distributivity, 4; internal operation, 3; associativity, 3; commutativity, 3; linear transformation, 3; null element, 4; opposite element, 4; properties, 3; tensor, 3; vector, 7; vector space, 4; zero element, 4
linear transformation, 3–5; linear space, 3
linear vector dipole; compensated, 40
linearity, 86
logarithmic strain, 82
Lorentz factor, 55
Lorentz force, 111

Maclaurin series, 47
macroscopic Maxwell's equations, 111
magnetic diffusivity, 111
magnetic field, 109

magnetization field, 111
magnetizing field, 111
manifold, 7; chart, 7; coordinates, 7; differentiable, 7, 15; Hausdorff, 11; orientable, 13; atlas, 13; chart, 13; circle, 13; Jacobian, 13; Klein bottle, 13; Möbius strip, 13; tangent space, 13; referential, 77; spacetime, 104; spatial, 77, 78; surface; orientable, 13
map, 8, 71; between manifolds, 72; bijection, 71; bijective, 11; injective, 8, 71; one-to-one, 8; surjective, 71
maps between manifolds, 76
mass flux two-form, 186
mass three-form, 186
material angular velocity, 192
material defects, 32
material derivative, 112, 120
material point, 77
material velocity, 18, 21, 90, 95, 112, 121, 124
matter Lagrangian density, 208
Maxwell's equations, 108; form version, 144; in matter, 111; in vacuum, 110; macroscopic, 111
Mercator series, 47
metric signature, 54
metric tensor, 51, 63; covariant derivative, 98; inverse, 53, 63; Riemannian, 52
metric-affine action, 206
metric-affine geometry, 206
metric-affine gravitation theory, 107, 108, 174; field equations, 208
metrical energy-momentum current, 208
micropolar medium, 188
Minkowski metric, 54
Minkowski spacetime, 54, 167
mixed covariant derivative, 102
mixed exterior covariant derivative, 166
modified covariant derivative, 105
modified torsion tensor, 104, 105
modulus; bulk, 69; shear, 69
moment tensor, 39, 40
monifold; referential, 154
monoclinic symmetry, 68
motion, 12; incompatible, 145; Jacobian, 79

Nanson's relation, 81
natural ♮, 123
no-more continuum, 145
nodal-plane ambiguity, 40
Noether current, 182, 183
nondegenerate, 52
nonholonomic, 31
nonmetricity, 107, 108, 173, 174
nonmetricity one-forms, 173
nonmetricity tensor, 107
nonspinning frame, 106
norm, 56; Frobenius, 47
normal, 137
normal strain two-vector, 158
normal traction two-forms, 153
normal vector, 137
notation, 15, 37, 47

objective rate, 114
Ohm's law, 110
one-form, 23; basis, 24; connection, 145, 161; Eulerian, 166; Lagrangian, 166; field, 23; nonmetricity, 173; norm, 56; spin connection, 167; strain, 158
one-forms; affine connection, 206; spin connection, 167
one-to-one, 8, 73
open set, 8
orientability, 13
orientable manifold, 13
oriented, 179
oriented surface, 59
orthogonal group, 64; special, 64
orthogonal transformations, 64
orthogonality, 56
orthotropic symmetry, 68
outer product, 35; tensor product, 35

Palatini action, 206
Palatini tensor, 207
Palatini torsion tensor, 104
parallel transport, 91, 92, 114; analytic formulation, 92; geometrical interpretation, 92
particle rotation, 199
permeability, 110
permittivity, 109

Piola transformation, 82
Piola-Kirchhoff stress; first, 83; second, 83; two-forms, 154
plastic deformation, 196, 198
plasticity, 198
Poincaré's lemma, 139, 182
point defects, 107–109
polarization field, 111
post-Riemannian Bianchi identities, 174
product; cross, 134; dot, 37; double-dot, 38, 159; duality, 24; exterior, 130, 131, 159; inner, 37; interior, 133; outer, 35; wedge, 128, 130, 131
product space, 16
proper time, 29
pseudoform, 133, 135
pseudoscalar, 60
pseudotensor, 46; Levi-Civita, 59
pseudotensor capacity, 48
pseudotensor density, 48
pull down operator, 133
pullback, 72, 144, 159, 178
pushforward, 74, 75, 144
Pythagorean theorem, 56

radiation pattern, 39
reference frame; Galilean, 9; inertial, 9
referential coordinates, 77
referential manifold, 77, 154
referential state, 77
relative tensors, 47
relativity, 54
reversible process, 195, 196
Reynolds transport theorem, 183
Ricci curvature, 103
Ricci identity, 95
Ricci scalar, 103, 206
Ricci tensor, 37, 103; symmetry, 106
Riemann tensor, 35, 95, 103
Riemann-Cartan geometry, 206
Riemannian geometry, 206
Riemannian manifold; differentiation, 99; Levi-Civita pseudotensor, 60; metric tensor, 51, 55, 60
Riemannian metric; definition, 52
right Cauchy-Green tensor, 53

rigid rotation, 114
rigid translation, 114
rigidity, 69
Rodrigues formula, 66, 67
rotation, 46, 63, 202; angle, 65; axis, 65; particle, 199
rotation tensor, 63
rotational kinetic energy, 192

scalar, 34, 35, 38; field, 35
screw dislocation, 147
second law of thermodynamics, 194, 195
second Piola-Kirchhoff stress, 82, 83; two-forms, 155
second Piola-Kirchhoff stress two-forms, 156
second postulate, 54
seismology, 38–40, 68
semicolon notation, 86
sharp ♯, 123
shear current, 209
shear modulus, 69
smooth, 11
source; dip-slip, 40; explosion, 40; eyeball, 40; fried-egg, 40; implosion, 40; strike-slip, 40; thrust, 40
source mechanism, 39; double-couple, 39; non-double-couple, 39
spacetime, 7; Galilean, 95; general relativity, 10; Minkowski, 167
spacetime manifold, 10, 29, 104
spatial coordinates, 9
spatial manifold, 77, 78
spatial point, 77
special orthogonal group, 64
special relativity, 54; first postulate, 55; second postulate, 54
specific entropy density, 194
speed of light, 110
spin, 192
spin angular momentum, 188
spin angular momentum three-forms, 192
spin connection, 102, 167
spin connection one-forms, 167
spin current, 208
spin tensor, 105
Stokes's theorem, 178, 180; classical, 180
strain, 156, 158; Almansi, 82; deviator, 69; deviatoric, 69; Eulerian, 82; form version, 156; Hencky, 47, 82; Lagrangian, 82; logarithmic, 82; two-vectors, 158
strain current, 209
strain deviator, 69
strain energy density, 68
strain one-forms, 158
strain tensor, 158; infinitesimal, 38, 68
strain two-vector, 157, 159; normal, 158
stress, 38; couple, 189; first Piola-Kirchhoff, 83; Kirchhoff, 83; second Piola-Kirchhoff, 82; three-form, 203; vector-valued two-form, 151
stress tensor; Cauchy, 38; weighted Cauchy, 83
stress-energy tensor, 104
structure coefficients, 30, 31, 43
substitutional defect, 109
summation convention, 17
surface, 137
surface integral, 175, 176
surface one-form, 59
surface torque, 188
surface two-form, 137
surjective, 71, 73
symmetric tensor, 39
symmetry; cubic, 68; isotropic, 68; monoclinic, 68; orthotropic, 68; transversely isotropic, 68; triclinic, 68
symmetry classes, 68

tangent bundle, 16
tangent space, 15, 77
Taylor series, 47
tensor, 33; addition, 35; adjoint, 57; alternating, 58; antisymmetric, 39; capacity, 47, 58; Cartan, 104; co-Einstein, 207; contortion, 99, 108; contraction, 36; curvature, 94–96; deformation-rate, 91; density, 47, 58; determinant, 50; disformation, 108; Einstein, 103, 207; elastic, 67; exponential, 46; field, 35; identity, 46; incompatibility, 32; inertia density, 193; inverse, 50, 63; Kronecker, 46; Levi-Civita, 58; logarithm, 46; metric, 51; modified torsion, 104; moment, 39; nonmetricity, 107; operations, 35; Palatini, 207; Palatini torsion, 104; Ricci, 103; Riemann, 95, 103; rotation, 63; skewsymmetric, 133; strain, 158; symmetric, 38; torsion, 94, 96; trace, 37; traction, 153; transformation, 41; transpose, 38, 40; two-point, 83; valence, 34; vorticity, 91
tensor capacity, 47, 58; covariant derivative, 107; weight, 47
tensor density, 47, 58; covariant derivative, 107; weight, 47
tensor field, 35; divergence of, 91
tensor product, 35; outer product, 35
tensor-valued form, 96, 147, 150
tensors; relative, 47
tetrad, 43, 101, 102, 106, 130; covariant derivative, 101
tetrad connection one-forms, 166
thermodynamic mechanical parameter, 196
thermodynamics, 190
three-form, 129, 132; angular momentum, 188; canonical energy-momentum current, 208; disclination, 205; hypermomentum current, 208; internal energy, 190; internal heating, 191; kinetic energy density, 190; kinetic energy density energy, 192; linear momentum, 186; mass, 186; spin angular momentum, 192; spin current, 208; strain current, 209; volume, 135; volume force, 186; volume torque, 188
time derivative; Euler, 112
time dilation, 55
torque; surface, 188; volume, 188
torque two-forms, 192
torsion, 108; intrinsic, 204, 205
torsion form, 104
torsion tensor, 94, 96; modified, 105
torsion two-forms, 162, 163
torsion-free connection, 97
total angular velocity, 192
total rotation, 202
trace, 37

traction, 153; two-forms, 153
traction tensor, 153
traction two-form, 159; normal, 153, 154; shear, 153, 154
transpose, 38, 40; (1,1) tensor, 40
transversely isotropic symmetry, 68
triclinic symmetry, 68
true strain, 82
twist disclination, 147
two-form, 128; Cauchy stress, 151, 152, 154, 157, 159, 187; couple-stress, 189; curvature, 169, 170; disclination, 147, 204, 205; dislocation, 147, 204, 205; first Piola-Kirchhoff stress, 154; heat flux, 191, 194; linear momentum, 186; mass flux, 186; Piola-Kirchhoff stress, 154; second Piola-Kirchhoff stress, 155, 156; surface, 137; tensor-valued, 96; torque, 188, 192; torsion, 162, 163; traction, 153, 154, 159; vector-valued, 96, 150
two-point tensor, 83, 154
two-vector; strain, 157, 159

unit vector, 56
universal electric constant, 109
universal magnetic constant, 110

vacancy defect, 109
valence, 34
variational principle, 180
vector, 15, 16; basis, 17; field, 19; length, 56; norm, 56; normal, 137; unit, 56
vector calculus, 142
vector field, 19; constancy, 92
vector space, 4, 17; dimension, 5; basis, 5; linear space, 4
vector-valued form, 96, 147, 150
vector-valued two-form, 150
velocity, 183
Vielbein, 43
Vierbein, 43, 45
Volterra cut-and-weld protocols, 147
volume, 135
volume element, 135
volume force three-forms, 186
volume form, 59, 124, 135; properties, 137
volume integral, 177
volume torque, 188
volume transport, 60
vorticity, 90, 192
vorticity tensor, 91

wedge disclination, 147
wedge product, 128, 130; exterior product, 131; Grassmann product, 131
weighted Cauchy stress tensor, 83
Weyl vector, 208
worldline, 29

Zaremba-Jaumann rate, 113